MW00914174

CONFIDENTIALITY, DISCLOSURE AND DATA ACCESS:
Theory and Practical Applications for Statistical Agencies

CONFIDENTIALITY, DISCLOSURE, AND DATA ACCESS:

Theory and Practical Applications for Statistical Agencies

PAT DOYLE
U.S. Bureau of the Census, Washington, DC.

JULIA I. LANE
The Urban Institute, Washington DC.

JULES J. M. THEEUWES
University of Amsterdam, The Netherlands.

LAURA V. ZAYATZ
U.S. Bureau of the Census, Washington DC.

2001

ELSEVIER

Amsterdam - London – New York – Oxford – Paris – Shannon - Tokyo

ELSEVIER SCIENCE B.V.
Sara Burgerhartstraat 25
P.O. Box 211, 1000 AE Amsterdam, The Netherlands

First edition 2001

Library of Congress Cataloging in Publication Data
A catalog record from the Library of Congress has been applied for.

ISBN: 0-444-50761-2

∞ The paper used in this publication meets the requirements of ANSI/NISO Z39.48-1992 (Permanence of Paper).
Printed in The Netherlands.

To:

Megan and Jonathan Cohen-Doyle,

Dennis, Britta, Ian, and Annika Glennon,

Frans Theeuwes,

and

Tim, Nicholas, and Joseph Zayatz

Contents

Foreword

Confidentiality, Disclosure, and Data Access: Theory and Practical Applications for Statistical Agencies

Technology is rapidly changing our world of data in three major ways. First, we are able to capture in digital form many new types of measurements, which in time will be amenable to more sophisticated analysis. These measurements range from digital audio and video tapes, to functional magnetic resonance images of the brain, to three-dimensional laser images of the body, to information on one's DNA.

Second, information technology has made collection, storage, transfer, analysis, and retrieval of large amounts of information relatively quick and inexpensive—and therefore ubiquitous. Today we can link very large datasets by matching information common to individuals, and we are discovering more efficient ways of analyzing even separate terabyte datasets through distributed computer systems.

Third, we are able to disseminate a profusion of information to multitudes virtually at once and to particularize the information according to the needs of each individual who might use it. And we are on the verge of new information technology to achieve feats heretofore inconceivable.

But this very same technology brings increased risks—risks that disclosures from a data file will fail to protect the confidentiality of some individual or business in that file and thereby make that individual or business identifiable. In response to these risks, statistical agencies may feel forced to become even more restrictive in allowing their data to be used for research and public policy purposes.

The balance between the risk of disclosure and the benefits of research is a precarious one. Small changes in the risk of disclosure translate into large changes in research and public policy benefits. If statistical agencies become more risk-averse, much valuable research will suffer. Some data may not be made available for research at all.

As professionals who are knowledgeable about data, we must be concerned with how to protect confidentiality in order to reap the full benefit of data collected. This volume tells us how. It brings together what we know from research and practice. It builds on the work of many researchers, as well as the panel study and workshops of the Committee on National Statistics at the National Academy of Sciences—National Research Council.

The volume describes and compares the methods that have been developed to protect confidentiality while at the same time providing access to data, through various means that either alter the data or restrict access to them. It also discusses, in the case of education statistics, legislative remedies that serve as the underpinnings for data access at remote sites. The volume is a consummate exposition of just how far we are able to go to achieve the win-win situation of confidentiality *and* data access with different approaches.

If we are struggling with confidentiality of today's data, imagine our challenges with the data and information technology of the future. Yet this technological revolution also holds the solution to these challenges. It enables the research and public policy community and the government statistics community to work toward a common vision: a comprehensive, integrated network of knowledge and information systems that will promote the conduct of research on social and organizational change and its determinants and also foster synergistic decision making by individuals, organizations, and public policy leaders at all levels—local and regional as well as national.

We have the technological wherewithal today to take steps in the direction of this vision. Examples range from geographic information systems, to computerized emergency telephone numbers (911 in the United States) and property tax systems, to global positioning systems, to digitized topographic maps. What is stopping us from bringing together the data we have already, and planning for the data of the future, is the fear of violating confidentiality. This volume is the base from which we can look to the future and address the predicaments of responsible data access. It is an invaluable resource.

Miron Straf
National Research Council—National Academy of Sciences
National Science Foundation
President-Elect (2001), American Statistical Association

Confidentiality, Disclosure, and Data Access: Theory and Practical Application for Statistical Agencies
Pat Doyle, Julia I. Lane, Jules J.M. Theeuwes and Laura M. Zayatz (Eds)

Chapter 1
Introduction*

1. The Trade-off Dilemma

There is a fundamental tension at the heart of every statistical agency's mission. Each is charged with collecting high quality data to inform national policy and enable statistical research. This goal necessitates dissemination of both summary data and microdata. Each is also charged with protecting the confidentiality of survey respondents—not only because of legal and ethical mandates, but because public trust and perceptions of that trust are important contributors to data quality and response rates.

Protecting confidentiality necessitates perturbing or summarizing the data in some fashion so that the individual respondent cannot be identified. Greater protection of confidentiality means that the data, which cost so much to collect and produce, are likely to become less valuable. The resulting trade-off dilemma, which could well be stated as protecting confidentiality (avoiding disclosure) but maximizing data quality and data access, has become more complex as both technological advances and public perceptions have changed in this Information Age. In sum, while statistical agencies go to great lengths to collect high quality data, the necessity of protecting confidentiality results in some data quality compromises. This book describes new theoretical, practical, and technological responses to the challenges that statistical agencies face.

What are these challenges, and how have they changed the world within which agencies collect and protect data? A partial list would include an increasing demand by policy makers for timely, relevant information; an increasing demand by academic researchers for microdata; the public dissemination of state- and local-level administrative records; and possibly most important, the increasing data collection by the private sector.

* The editors would like to thank the following reviewers: Nancy Bates, Larry Cahoon, Cynthia Clark, James Fagan, Gerald Gates, Nancy Gordon, Jennifer Guarino, Joan Marie Hill, Frederick Knickerbocker, Larry Long, Paul Massell, Randall Neugebauer, Carole Popoff, Juanita Rasmann, Arnie Reznek, Stephanie Shipp, Phil Steel, Sam Hawala, Judith Waldrop, Diane Willimack, and Tommy Wright. We also thank Felicity Skidmore, for excellent editing assistance, and Jayne Sutton of EEI Communications for her technical expertise in preparing the final manuscript.

Information is critically important to policy makers—it permits the management of the economy and it informs public debate. Data ranging from inflation rates to unemployment rates, crime statistics, and healthcare statistics are all important components of our information infrastructure. As the economy becomes increasingly more complex, however, and the interactions among households, businesses, and governments more entangled, the demand for data describing these interactions has increased. At the same time, the cost and burden on respondents has increased, so statistical agencies have turned more often to administrative data to respond to user needs. Some state and local governments provide drivers' license records and property ownership information on-line. The Netherlands no longer conducts traditional population censuses; instead, it relies on administrative records to count the size of the population. Although these uses provide high value added from existing data without imposing additional burden on respondents, they do pose special confidentiality challenges. This book addresses some of the solutions to these challenges.

External researchers increasingly demand access to detailed data at the micro level for two reasons. First, no microdata on business issues are available to the public from statistical agencies, because these data are protected from disclosure, yet important new statistics on job creation and job destruction can only be derived from such data. Second, increasingly stringent disclosure protection has meant that publicly available data on households and individuals are increasingly too geographically aggregated, or income levels too strictly top-coded, to allow full policy analysis.

Statistical agencies have an incentive to meet the needs of researchers not only because it fulfills their core mission of data dissemination, but also because research access can often lead to the core improvement of statistical data products. Analytically sophisticated projects, mounted by external researchers, can be used to assess whether the data produced by statistical agencies are of sufficient quality for policy makers to act on the results. In this sense, giving external researchers access to data can act to complement the standard data quality control checks used by statistical agency staff.

In short, the process of creating analytical results (and hence, if the research is successful, informing policy makers) is integral to evaluating and improving the data that statistical agencies produce—quite apart from the scientific value contributed by the analysis of the data themselves. As the economy becomes more complex and the questions posed by policy makers more detailed, there is an increasing need for external researchers to have access to micro-level data without violating disclosure guidelines. Again, some chapters in this book explain how this can be done.

Data collection by the private sector has soared. This phenomenon is likely to interact in complex ways with both the ability of statistical agencies to protect the data they collect and the public perception of that ability. On the one hand, the vast expansion of private data collection creates a much more sophisticated master file for potential intruders. Latanya Sweeney (Chapter 3) gives a sense of just how vast private collections are in her example of Catalina Marketing. This company began in March 1996 to store the shopping patterns of 143 million shoppers each week from 11,000 supermarkets nationwide. By July 1998, it had amassed a 2-terabyte database with 18 billion rows of data. Widespread awareness of these types of data collections may make the public more concerned by the increased potential for re-identification. On the other hand, because statistical agencies are much more heavily regulated than the private sector and heavy punishments apply for disclosure, the public may be reassured. Chapters in this book inform us of both the ways in which statistical agencies have responded to the technological challenges and the effect their response has had on public perceptions.

How have statistical agencies adapted to these challenges? The historical approach has been to protect data in three ways: statistical analysis, data protection, and different access modalities. Statistical analysis protects confidentiality because information about each respondent is used only in the context of information collected from other respondents. That is, information is not obtained based on the experience of just one or two individuals, but rather on trends observed in groups of individuals, and the safety in numbers approach is at work to protect individual respondents.

Data protection occurs when data producers take steps to modify or suppress information that might identify an individual either directly or indirectly before they make it available to the public for analysis. This process, which is also called disclosure protection, has evolved from the simple suppression of information on unique observations to complex statistical methods affecting a large amount of the information collected. The growth in the level of sophistication of data protection mirrors the expansion of technology that could potentially permit people to circumvent the protection process. Some of the most technical chapters in the book deal directly with new advances in this area.

New access modalities have also evolved—from licensing, to remote access, to secure remote sites. The book has chapters on each of these topics.

How does the public perceive the way in which statistical agencies respond to these challenges? Perceptions matter, both because statistical agencies need to maintain response rates and data quality and because they need to reassure the public that they are fulfilling their legal and ethical requirements. If the public has become more sensitive to data privacy concerns in response to private sector actions, then statistical agencies need to respond accordingly.

The assessment of public perceptions can also be an important tool in guiding data dissemination decisions. For example, new access modalities such as restricted access sites are potentially important new dissemination mechanisms. Their establishment should be evaluated, however, not only for the technical protection and the resulting data quality issues but also in the context of public perception of these approaches. It may well be that the public perceives them as being much safer than the release of public use files, particularly given the wealth of private data available, and statistical institutes can tailor their response appropriately. Three of the chapters in this book explicitly address the perception issues, for businesses and for households.

This book provides an overview of how statistical agencies have risen to the new challenges confronting them and could rise to future challenges. It begins with two chapters that delineate the new technological challenges faced and summarize the types of approaches that different statistical agencies use to protect data. The subsequent set of chapters review and develop new state-of-the-art techniques that directly address these statistical disclosure techniques from both theoretical and practical perspectives. The next section's chapters describe alternative access modalities. The book concludes with chapters that update our knowledge on perception issues.

2. Fundamental Concepts

In any technical monograph like this one, the material is most efficiently presented if basic technical terms are used. So that all readers are able to appreciate the material in this book, we provide a discussion of underlying concepts and some technical definitions. The brochure *Confidentiality and Data Access Issues Among Federal Agencies* is the source of much of the material that follows. The authors thank the publisher of the brochure, the Confidentiality and Data Access Committee, sponsored by the Federal Committee on Statistical Methodology.

Data Structures

Data and thus disclosure methods come in two basic forms: microdata that refer to individual units and aggregate estimates from survey or census responses. Two data formats commonly used to present aggregate estimates are tables of frequency counts and tables of aggregate magnitude data.

Microdata files consist of individual records that contain values of variables for a single person, a business establishment, or another individual unit. Public use microdata files are released to the public for research and analytical purposes after being subjected to procedures that limit the risk of disclosure.

Frequency count tables count the number of respondents with specified characteristics. For example, a two-dimensional frequency count table may have rows corresponding to race categories and columns corresponding to age groupings. An individual table cell at the intersection of a given row and a given column would indicate the number of residents of a certain geographic area with that race and age combination.

Tables of *aggregate magnitude data* are analogous to frequency count tables in that they are defined by cross-classification of categorical variables. However, the cells contain aggregate values, over the corresponding respondents, of some quantity of interest. For example, a two-dimensional table on income defined by race and age would contain total incomes in each race-by-age cell.

A special case of tabular data are systems that permit generation of tables on demand through tabulations from an on-line query to a statistical database using the Internet or other forms of remote access. Data users may create their own tabulations by customized queries to the database.

Disclosure Limitation Methods

Different approaches to disclosure protection apply depending on the type of underlying data to be protected.

In the case of microdata, three common types of procedures are applied to prevent disclosure of confidential information. First, information that directly reveals the identity of the respondents is suppressed. Second, information that may indirectly reveal the identity of a respondent is suppressed. This can be accomplished by reducing the variation within the data through rounding, top- and bottom-coding, collapsing response categories, and suppressing information such as detailed geography. Third, some uncertainty can be introduced into the reported data. This can be accomplished by altering the underlying data through swapping of reported values among similar respondents, adding predetermined random noise to the data, and performing other more structured randomization of the data.

In the case of aggregate magnitude data and some frequency count data, some information is suppressed. *Primary cell suppression* is withholding information in a cell because its publication would explicitly or implicitly reveal confidential or sensitive information. When a table contains cells that represent sums of either rows or columns, primary cell suppression alone does not always protect the confidential data. Original values of primary cells may be determined exactly or within a narrow range through subtraction. When this occurs, it is necessary to perform *complementary cell suppression* to protect the primary suppressions from disclosure. In addition to cell suppression, agencies may use rounding, data swapping, or some other type of noise addition to protect frequency count data.

Systems that support custom tabulations of sensitive data also provide disclosure protection so that all data extracted directly from these systems are adequate for public dissemination. An issue unique to this form of data access is that of complementary disclosure. If two or more separate requests are combined outside the system, it might create a disclosure problem that did not exist for the individual requests.

Aside from table-on-demand systems, data producers can provide the opportunity for outside researchers to create custom tables and models from non-public data through *licensing, secure sites,* and *remote access.* Licensing is an arrangement whereby an institution and its researchers sign formal agreements to protect the data according to the laws and policies governing the data collection agencies. In return they can receive the data for a limited period of time. Secure sites represent offices under the data production agencies' control that are made available to researchers who formally agree to follow the laws and policies of the data collection agencies. They also agree to subject their results to disclosure review before the results are removed from the offices. Remote access is an arrangement whereby researchers provide computer programs to the data collection agencies, which execute them and review the results for disclosure violations before sending them to the researcher.

3. The Contribution of the Book

Overview

The book starts with two overview chapters that summarize current practical approaches in key North American and European countries to protect confidentiality while maximizing access to economic and demographic data. They also present the current challenges in protecting the confidentiality of data on individuals that arises from the ever-increasing volume of information available from public and private sources and the ever-increasing sophistication of technology in accessing and linking this information.

Flóra Felsö, Jules Theeuwes, and Gert. G. Wagner (Chapter 2) conducted a survey of several national statistical offices on disclosure control methods in use, receiving responses from Canada, the Czech Republic, Denmark, Estonia, France, Germany, Hungary, Italy, Lithuania, the Netherlands, New Zealand, Norway, Sweden, and four agencies in the United States. The authors begin their chapter with a literature review focusing mainly on *Statistical Policy Working Papers #2* (Federal Committee on Statistical Methodology 1978) and *#22* (Federal Committee on Statistical Methodology 1994). These papers discuss disclosure limitation procedures for various types of data, including microdata, frequency count data, and aggregate magnitude data from both demographic and establishment surveys and censuses. The survey questions were based on *Statistical Policy Working Paper #22* and again covered all types of data.

In summarizing the survey results, the authors found that most agencies use a threshold rule (quite often the threshold is 3) for determining which cells in a frequency count table are sensitive (potential disclosure problems), and these agencies typically use cell suppression to protect such cells. For aggregate magnitude data, most agencies use a technique called the N-K rule to determine which cells are sensitive and use cell suppression and/or table reconstruction to protect them. Many agencies do not release microdata to the public. Those that do eliminate obvious identifiers, limit geographic detail, and limit the number and detail of variables. There was no consensus on what software packages should be used for purposes of limiting disclosure.

The explosion of private data collection is documented by Latanya Sweeney (Chapter 3). She describes three approaches to data collection in the private data sector today: collect more, collect specifically, and collect it if you can. The implications of this behavior in terms of the potential to identify respondents are troubling because they potentially lead to many more match keys that could be used to link publicly funded masked microdata to privately funded or administrative databases with identifiers. For example, historically, birth certificate information had only 7 to 15 fields of information, but today more than 100 fields of information are collected about each child's birth. This situation clearly poses a threat to statistical agencies because they could potentially be used in conjunction with statistical data products in attempts to uncover a survey or census respondent's information. Sweeney speculates that past practices may no longer be applicable guides because of the amount of data now being collected and current technology. Based on an examination of some information released by the federal government that was not subject to strict disclosure controls, she also speculates that 'current policies and practices support crude decisions'. Fortunately, Felsö, Theeuwes, and Wagner (Chapter 2) demonstrate that most efforts to protect confidentiality are more stringent than the ones Sweeney has studied, so that her conjecture does not apply across the board.

Public Access Through Data Manipulation

The second group of chapters focuses on the problem of providing public access to high quality information to inform public debate and research without disclosing information on individuals or businesses. This approach is one of data manipulation to mask the underlying information in the data to avoid disclosure while minimizing the impact of that manipulation on estimates derived from the public data. These chapters

- Describe different approaches to measuring the disclosure risk associated with uses of data on individuals and households and test the success of those methods in preventing disclosure.
- Discuss disclosure methods used to protect information included in microdata and the impact of these disclosure methods on the estimates produced from the microdata (known as information loss).
- Summarize disclosure methods for frequency count data, identify methods for measuring information loss, and quantify the loss under various disclosure limitation techniques.
- Summarize exploratory research to develop connections between current and proposed selected disclosure rules and disclosure limitation methods for economic tabular data and the secondary cell suppression problem as a mathematical statement of the problem of avoiding disclosure but maximizing access for tabular data.

- Present methods used to assess disclosure risk for aggregate data. The chapters address issues such as the appropriate primary suppression rules for tabular data to avoid disclosing a particular institution's response and approaches to reduce disclosure risk for economic tabular data, information loss resulting from cell suppression and/or recoding, and secondary cell suppression software.
- Discuss the use of linked data; that is, datasets created through matching of information from two or more sources such as data on employers from establishment sources and data on employees from surveys of individuals. The authors pay particular attention to the public release of parameters of models estimated using data that cannot be released for disclosure reasons.

In understanding disclosure risk, it is important to decompose the different factors that contribute to that risk. Mark Elliot (Chapter 4) does this, focusing on disclosure risk for public use microdata files, which are typically produced from demographic surveys or censuses. He discusses an 11-point disclosure attempt scenario in which he decomposes the risk into, for example, the motivation for an intruder to make such an attempt, the means necessary to carry out the intrusion, and the effect of data divergence between the intruder's information and the target file. In so doing, he identifies the critical elements that contribute to the risk of disclosure for each given file: the sampling fraction, level of detail on variables, level of geographic detail, number of key variables that might be used for linking purposes, and data divergence. He describes two reidentification studies that indicate the difficulties faced by an intruder identifying an individual in demographic microdata—in particular, the large number of false positive matches. He also develops a data intrusion simulation that could be used in an attempt to measure the disclosure risk of a microdata file, noting that disclosure risk assessment for microdata is still a young and complex research area.

We noted earlier that statistical agencies face a fundamental trade-off between data quality and data protection. If decisions must be made about data protection, it would be enormously helpful to have some measure of this trade-off. Josep Domingo-Ferrer and Vicenç Torra (Chapters 5 and 6) provide an overview of the disclosure limitation methods for both continuous and categorical variables in microdata. Several types of available methods are described, and the authors measure the effect of each method on the disclosure risk of the data and the resulting information loss. They measure disclosure risk using distance-based record linkage (linking the original data to the masked data), probabilistic record linkage, and interval disclosure. They measure information loss for continuous data in terms of mean square error, mean absolute error, and mean variation between covariance, correlation, and other matrices used for data analysis. They measure information loss for categorical data in terms of a comparison of contingency tables and entropy-based measures. The methods tested for continuous data were additive noise, data distortion by probability distribution, resampling, micro aggregation, lossy compression, and rank swapping. Of those methods, rank swapping seemed to perform best. The methods tested for categorical data were top- and bottom-coding, global recoding, and post-randomization (PRAM). Of these methods, top- and bottom-coding and global recoding all performed well, while PRAM did not.

George T. Duncan and colleagues (Chapter 7) provide an equally interesting framework for disclosure risk and information loss for frequency count data via a graph they call an R-U confidentiality map. They present methods of auditing tables and sets of related tables[1] to look for potential disclosure problems. These methods include linear and integer programming, generalizations of Frechet and Bonferroni bounds, and a generalization of Buzzigoli and Giusti's shuttle algorithm. The authors describe methods for protecting frequency count data, including sampling, cell suppression, local suppression, rounding, data swapping, simulated tables, and Markov perturbation, and they note advantages and disadvantages of each. They compare cell suppression, rounding, and Markov perturbation with different parameters via the confidentiality map and show that determining the best procedure may depend on the level of risk an agency finds acceptable.

While the previous set of chapters looked at the disclosure issues associated with demographic data, Lawrence H. Cox (Chapter 8) examines disclosure risk in aggregate magnitude data in the context of economic surveys and censuses. This disclosure poses special problems because large firms are very easily identified if industry or geographic detail is provided (the classic example is General Motors in Detroit, but one could just as easily think of Microsoft in Washington State). Cell suppression is the method typically used to protect this type of data. The author discusses the typical structure of the data and how sensitive cells (primary suppressions) are identified via different ways of quantifying risk. He then describes complementary suppression and various ways to calculate information loss, including number of cells suppressed and total value suppressed. Until recently, complementary cell suppression was done for one primary suppression at a time. Fischetti and Salazar have developed and tested a method for performing complementary cell suppression for all primary suppressions simultaneously. This method can reduce oversuppression and allow for more data to be published. Their algorithm has some drawbacks, the largest of which is that it protects data at the establishment level while most agencies must protect data at the company level. Cox has developed a similar algorithm designed to protect data at the company level. This algorithm needs to be examined and improved upon based on computational considerations.

While Cox provides a very theoretical approach to the core disclosure problems posed by economic data, Sarah Giessing (Chapter 9) provides a description from a practitioner's point of view. Her chapter provides numerous examples with different ways to quantify risks and differences between various approaches—particularly discussing minimum size requirements for complementary suppressions, one cell's capacity for protecting another, cell suppression patterns, and protection intervals around suppressed cells. The author discusses

[1] This is particularly important because often individual tables do not reveal individual identities, but can present problems when examined in conjunction with one another.

heuristic approaches to complementary cell suppression such as the hypercube approach, the network flow approach, the linear programming approach, and the integer linear programming approach. She then mentions the currently available cell suppression software systems and highlights some key attributes of that software, such as computing time and resource requirements, data structure and software implementation, the ability to process linked tables, and the ability to assign preferences to choose or not choose certain cells as complementary suppressions.

John Abowd and Simon Woodcock (Chapter 10) present methods for disclosure limitation of longitudinal linked data. Longitudinal linked microdata contain observations from two or more related sampling frames with measurements for multiple time periods from all units of observation. The prototypical longitudinal linked data set that they consider contains observations about individuals, work histories, and employers. They present methods for disclosure limitation of parameter estimates obtained from analyses of such data, as well as conditional expectations such as contingency tables and summary statistics. They also present a method for disclosure limitation of the microdata itself that is based on multiple imputation techniques developed for missing data. Disclosure limitation of longitudinal linked data is complex because of the requirement that new data be disclosure-proofed in a manner consistent with both the underlying microdata and previous disclosure-proofed releases. In the particular application they consider, this complexity is intensified by the fundamental differences in the statistical properties of data on individuals and data on businesses.

Remote Access to Non-Public Data

All the techniques described thus far necessarily involve data manipulation or suppression and are likely to reduce the quality of estimates to be produced from data sources. As a result, statistical agencies have begun to investigate other methods that allow use of data while protecting confidentiality of the respondents. These methods allow the data to be used in an environment controlled by the data-producing agency and require that its use be subject to the same legal and ethical protections placed on the agency itself. This group of chapters

- Introduces the process of licensing whereby institutions and researchers outside the data-producing agencies temporarily gain access to data at their site by agreeing to conform to the legal protections surrounding those data that are imposed on the data-producing agency.
- Describes secure sites, where the data remain under the control of the data-producing agencies and researchers come to an agency office to access it. Such sites are an increasingly popular means of providing researchers access to respondent-level microdata while protecting the confidentiality of the data.

- Demonstrates an approach to house the data at the data-producing agency and allow remote access by researchers through an intermediary controlled by the agency that guarantees all use conforms to the law.

Data licensing is a way to provide access to data when they cannot be released to the public because of confidentiality concerns. It is described in some detail by Marilyn M. Seastrom (Chapter 11). A number of U.S. statistical agencies currently use licensing, but she focuses on the licensing system at the National Center for Education Statistics. She describes in detail the license application, required security procedures, who can access the licensed data, publishing requirements, security inspections, and the termination of licenses. The author discusses various U.S. laws and regulations that agencies use in licensing and then compares how various agencies implement and enforce licensing agreements. She gives examples of both major and minor violations of licensing agreements. She concludes by recommending that all agencies that license data perform periodic inspections of the licensed sites and develop and maintain a database application that allows the agency to readily access records of licensed files and authorized users for each agreement.

Probably the most important access modality developed in the past decade is that of restricted access sites. These sites permit statistical agencies to respond to the microdata needs of researchers, avoid the linkage problems posed by the Internet, and address potential perception problems that might be associated with other access modalities. Timothy Dunne (Chapter 12) discusses the establishment and management of such secure research sites to provide access to data when they cannot be released to the public because of confidentiality concerns. He notes that an agency must first decide if it legally has the authority to establish such a site and under what conditions the data may be accessed. The next steps, described in detail, are to choose the physical location and establish security and personnel there. The agency must then focus on what data will be available at the secure site and how they will be managed. Dunne then addresses project management issues, including project selection, formal agreements that must be established between the researcher and the agency, researcher training, and reviewing results for potential disclosure problems. If the secure site is at a non-agency location, the agency must consider how a site is awarded and how it will be managed. The author highlights many benefits of establishing secure sites, including the development of a community of skilled data analysts, the development of new data and statistical products, analysis of longitudinal and/or linked data, and feedback to the agency on methodological aspects and data quality issues of the data being analyzed.

Michael Blakemore (Chapter 13) presents the potentials and the perils of remote access. He stresses how rapidly developing information technologies offer increasing potential for unrestricted information flow. At the same time, he notes, the development of information technology increases the costs to data custodians of making mistakes, one cost being a potential loss of trust on the part of data providers. Remote access has three key components: the network, along with the physical information technology infrastructure; the software, ranging from security systems to encryption; and finally, the organization context. These key elements are highlighted. The chapter concludes with a set of case studies on ways in which remote access has been implemented.

Perceptions

Regardless of the extent and success of the measures used to protect individual information, some people still believe their data cannot be protected, and this perception could have a detrimental impact on their participation in surveys. This series of chapters

- Summarizes the research on the public's attitudes and perceptions toward privacy and confidentiality.
- Addresses how individuals' beliefs about disclosure of personal information are influenced by historic mistrust of government in groups considered hard to enumerate.
- Discusses perceptions among businesses, organizations, and institutions, contributing to the growing body of literature in the demographic sector on the effect of the perceptions of protection and perceptions of harm on respondents' willingness to participate in surveys.

The chapter by Eleanor Singer (Chapter 14) examines in some detail what the public believes about the confidentiality of data collected by statistical agencies and how it regards the prospect of data sharing among federal agencies. Singer also examines changes in the public's beliefs and attitudes over time and how these beliefs may affect response rates to demographic surveys and censuses. The chapter is based primarily on four surveys undertaken (in 1995, 1996, 1999, and 2000) by the U.S. Census Bureau that tracked attitudes about privacy and confidentiality, primarily in relation to the census of population and housing. The author describes in detail the methods used, trends in beliefs about confidentiality and attitudes toward privacy, trends in attitudes toward sharing of data among federal statistical agencies, predictors of privacy-related attitudes, and the relationship between attitudes and behavior. Looking across the four surveys and from the perspective of five years, one can see distinct patterns of change with respect to knowledge and awareness of the census itself and with respect to knowledge about Census Bureau confidentiality practices. There is a secular increase in knowledge about confidentiality, which is paralleled by a significant increase over time in the percentage of respondents who would be bothered if their census data were provided to anyone outside the Census Bureau. Interestingly enough, these changes are *not* paralleled by increasing distrust of data usage or increasing concerns about privacy or by declining trust in the government.

What affects public response to requests for information in government surveys? Eleanor R. Gerber summarizes research on this topic (Chapter 15). The core focus is on modeling respondents' decisions to provide (or refuse to provide) information: how this decision is made, what factors are taken into account, and what other concerns are evoked in considering this decision. Exploratory qualitative techniques (ethnography) rather than numerical assessments were used for this work. The author finds that people like to feel that they are in control of information about them. This feeling affects their attitudes toward data sharing and toward different modes of questionnaire administration. Perception of a legitimate need for the information (including benefiting society as a whole) is a critical factor in whether a respondent decides to provide that information. These data strongly suggest that trust (or lack thereof) in assurances of confidentiality is only one element in a complex set of attitudes toward privacy in general. The public perception is that data are widely exchanged among government agencies, and many people view this exchange as a loss of control of their information. The author recommends stressing the legitimate need for the information being collected and addressing data use concerns at the time of data collection.

Although a great deal of research has been targeted at understanding people's views of confidentiality, very little has addressed business perceptions of confidentiality protections. Nick Greenia and colleagues (Chapter 16) describe the results of one of the first surveys conducted in this area. The survey focused on the sensitivity of the individual data items, perceived benefits of the data collection, cost of the data collection, and the protection provided to the respondents. The results of this survey show that a wide variety of businesses distrust the government, but a large number of businesses would actually be amenable to some sharing of data among agencies and the release of older and less sensitive business data to the general public.

4. Implications: 'There's No Data Like No Data'

The trade-off dilemma mentioned in the opening paragraphs of this introductory chapter is central to this book. Statistical agencies want to protect the confidentiality of survey respondents and avoid disclosure while at the same time maximizing data quality and data access. While the fundamental, hard-to-solve tension between these two objectives has always existed, the tension is exacerbated by the increased ability of modern society to generate more information and the expanded desire for fast and accurate information on complex societal problems. Finding solutions to the confidentiality problems thus posed should be paramount because the natural tendency of statistical agencies, when faced with uncertainty about the impact of the release of information, is to err on the side of protecting confidentiality. The concomitant risk, which is the reduced quality or quantity of publicly available data, is a very real one without alternative legal approaches to permit access to data for research directed toward public policy issues.

This book addresses some issues associated with the core trade-off dilemma. In particular, it provides an up-to-date overview of tools that are available for extending access to data users. It discusses the wide array of instruments that are available to statistical agencies as they seek resolution of their trade-off dilemma. These instruments range from use of disclosure limitation procedures when releasing data to providing remote access through data licensing and restricted access sites. It is clear that some tools are more disclosure-proof and less rich in information than others. We could draw a line with perfect disclosure protection at one end and complete and full disclosure at the other, with the different tools located somewhere on this line. This book presents evidence that for different data, and in different countries, different choices will be made on this line. The book also shows that these same tools could be ranked in terms of how well they can solve the trade-off dilemma. This situation raises two key questions: Are there tools that are superior to other instruments in that they provide a higher level of confidentiality for the same amount of information richness? Or is it possible to find tools that provide a higher informational content for the same amount of confidentiality?

While we hope this book has gone some of the way toward furthering understanding of the issues involved, more research is needed. Though the research in the book examines information loss in general, it does not address more complex disclosure analysis—for example, disclosure-proofing the results of a set of program benefit simulations when each component of the derivation is slightly adjusted. In addition, different utility metrics, or loss metrics, could be used to quantify disclosure risk and data quality loss. New access modalities, such as simulated access sites, could usefully be explored. Other research, such as research into the area of business perceptions, could be extended and potentially codified by statistical agencies.

The bottom line is that confidentiality is not an arcane topic of little policy interest. Governments and taxpayers pay billions of dollars to statistical agencies to provide decision makers with high quality data. Although there have been no documented cases of disclosing a respondent's identity in the nearly 40 years in which the U.S. Census Bureau, in particular, has released anonymous microdata files, researchers specializing in this area continue to pursue more and more sophisticated anonymization techniques. In addition, disclosure review boards are making increasingly conservative decisions about data release. Users should pay attention to the challenges faced by statistical agencies and work constructively with the agencies to find workable solutions to the core trade-off dilemma. Agencies and users should work together to promote legislative, regulatory, and dissemination policies and practices that facilitate timely and cost-effective access to data for statistical research and policy analysis but do not permit full and open access by all of the public for any use. If confidentiality issues are not fully addressed in constructive and proactive ways, users face the very real risk of losing access to high quality data.

References

Federal Committee on Statistical Methodology (1978) *Report on Statistical Disclosure and Disclosure-Avoidance Techniques (Statistical Policy Working Paper #2)*, Washington, D.C.: U.S. Department of Commerce, Office of Federal Statistical Policy and Standards.

—— (1994) *Report on Statistical Disclosure Limitation Methodology (Statistical Policy Working Paper #22)*, Washington, D.C.: U.S. Office of Management and Budget, Statistical Policy Office.

Confidentiality, Disclosure, and Data Access: Theory and Practical Application for Statistical Agencies
Pat Doyle, Julia I. Lane, Jules J.M. Theeuwes and Laura M. Zayatz (Eds)

Chapter 2

Disclosure Limitation Methods in Use: Results of a Survey*

Flóra Felsö
SEO Amsterdam/Economics–University of Amsterdam

Jules Theeuwes
SEO Amsterdam/Economics–University of Amsterdam

Gert. G. Wagner
German Institute for Economic Research (DIW) in Berlin and European University in Viadrina

1. Introduction

An important objective of this book is to review new state-of-the-art techniques that address disclosure limitation or control. This chapter summarizes the extent to which these methods have been implemented and are currently in use by statistical agencies. We review how well statistical agencies have kept pace with new advances and whether disclosure control—combined with user-friendly access to the data—is recognized as key to the heart of the mission of national statistical agencies.

Our review is based on the responses to a survey sent to national statistical agencies in North America, New Zealand, the European Union (EU) countries, and countries aspiring to be EU members. We also discussed our survey results with two experts in the field, one in Europe and one in the United States. We were interested mainly in the methods, how agency policy has evolved, the recent changes in approaches, and expected future changes.

We start with a review of the literature on previous surveys of disclosure control methods in use. We then explain the rationale for our choice of survey questions and describe our sample 'design' and respondents. In the rest of the chapter, we

* We would like to thank Eric Shulte Nordholt (CBS, Netherlands) and Alvan O. Zarate (NCHS) for their comments and their time and for sharing their expertise in the area of disclosure limitation methods.

summarize the results of the survey for tabular data and microdata. For tabular data, we distinguish between counts or frequencies and magnitude data. For microdata, we distinguish between demographic data and business (establishment) data. For demographic data, we further distinguish between census data (*i.e.,* total counts) and non-census or sample databases.

2. Previous Surveys of Disclosure Control Methods in Use

The literature reviewed here is a useful starting point for our discussion, although previous studies were more limited—either in their focus on agency practice only as a side issue or in their country coverage.

The first comprehensive work on disclosure control (called 'avoidance' at that time) was the *Statistical Policy Working Paper #2* published in 1978 by the Subcommittee on Disclosure-Avoidance of the U.S. Federal Committee on Statistical Methodology (FCSM 1978). The intent of this paper, which included an appendix containing descriptions of their disclosure avoidance practices prepared by seven federal statistical agencies, was to help managerial and technical staff of U.S. federal agencies to achieve 'appropriate disclosure avoidance'.

Because comparatively little was known about disclosure and there was no widely accepted definition of 'disclosure' (p.1), the authors of the working paper had to begin by developing a framework in which disclosure practices could be reviewed and evaluated. In spite of this lack of formal definition, they found that several major federal statistical agencies had, in fact, developed a variety of disclosure avoidance techniques for tabulations and for microdata, although little attention had been given to developing explicit policies for what disclosures were or were not acceptable (p.32).

The authors of *Working Paper #2* recommended that all statistical agencies formulate and apply policies and procedures designed specifically to avoid unacceptable disclosures. Rather than develop a uniform set of rules, given the wide variations across agencies in the information released, they advised agencies to apply a test of reasonableness: that no information about a specific individual or other entity should be disclosed in a manner that could harm a respondent (whether an individual or an organization) (pp.39–42). The authors also argued that special care should be taken for releases based on a complete file—as opposed to a sample—and for small area data (p.42). Further, they recommended clear assignment of individual responsibilities for compliance with the disclosure avoidance policies chosen (p.42). More than two decades later, several (non-U.S.) statistical agencies have not yet followed this advice.

This initial work in 1978 was followed up in early 1992, when Hermann Habermann of the Statistical Policy Office of the U.S. Office of Management and Budget organized an ad hoc interagency committee on disclosure risks. A new subcommittee was formed to look at methodological issues, to analyze results of an informal survey of agency practices, and to develop recommendations for improvement. This subcommittee was the predecessor of the FCSM's Subcommittee on Disclosure Limitation Methodology, established in 1993. The name change from disclosure 'avoidance' to disclosure 'limitation' reflects the realization that the zero-risk condition for disclosure is an impossibly high standard (FCSM website, www.fcsm.gov).

The goal of the new subcommittee was to update *Statistical Policy Working Paper #2,* by describing and evaluating existing disclosure limitation methods, providing recommendations and guidelines for the selection and use of disclosure limitation techniques, and encouraging the development, sharing, and use of new methods and specialized software. The result was *Statistical Policy Working Paper #22* (FCSM 1994).

Chapter II of that paper gave a simple description of disclosure limitation methods for tabulations and microdata and provided a 'guideline for good practice for all agencies' (p.3). We used this guideline as the basis for *our* survey, as described below. Chapter III of *Statistical Policy Working Paper #22* summarized the disclosure control methods used by 12 major federal agencies and programs, based on information from a number of sources. The main source consisted of the 12 responses to Herman Habermann's 1992 request that each statistical agency provide an up-to-date description of its current practices, standards, and research plans for tabular data and microdata. Supplementary sources included statistical agency responses to a request of the Panel on Confidentiality and Data Access, Committee on National Statistics (results published in Jabine 1993) and the seven federal statistical agency reports in the appendix to *Statistical Policy Working Paper #2.*

The main conclusion of *Working Paper #22* was that most of the agencies did have standards, guidelines, or formal review mechanisms to ensure that adequate disclosure analyses were performed and appropriate statistical disclosure methods applied before the release of tabular data or microdata (p.37). Agencies varied widely in the specificity of their rules, however, with some applying a couple of simple rules and others specifying a much more detailed set of rules.

For frequency and aggregate magnitude data, most agencies were found to apply a minimum cell size and some type of concentration rule. Minimum cell sizes of three were almost invariably used, because each member of a cell of size two can derive the value of the other member (p.37).

Only half of the responding agencies had established disclosure limitation procedures for microdata. The authors of *Working Paper #22* cited Jabine (1993) as evidence that procedures for microdata were not parameter-driven like those for tabulations. Rather, they required judgments that take into account the following: whether resources are available 'that might be needed by an "attacker" to identify individual units', the expected number of unique records in file, geography, and the number of variables in cases, for example, where characteristics of the local area could identify the location, and thereby an individual. They found that top-coding was a commonly used method to prevent disclosure of individuals (or other entities) with extreme values, and that blurring, noise, and rounding were all applied to prevent disclosure (p.38).

Like the FCSM, Eurostat also released a guide on disclosure limitation (Eurostat 1996). Its purpose was not to recommend the choice of a particular method, but rather to list alternatives for use depending on the type of data release and the intended level of data protection.

This Eurostat guide was followed in 1997 by a survey of disclosure practices among the member states of the EU carried out by Holvast & Partner (Holvast 1999). The main finding of the EU survey was that all national statistical institutes were aware of the importance of confidentiality and had implemented the necessary safeguards and that 10 out of 14 considered mathematical and computing aspects important (p.201). The survey found little difference in the official treatment of demographic versus business data. The major differences turned out to depend on the nature of the data (*e.g.,* frequencies versus aggregate magnitude data). The report on the EU survey highlighted the specific disclosure problems faced by small countries, where even simple sector statistics can be plagued by the limited numbers of firms.

The next follow-up was a 1998 survey done by the United Nations Economic Commission for Europe (Luige and Meliskova 1999), among the transition countries—14 (Central) Eastern European countries (EEC) and 5 countries from the Commonwealth of Independent States (CIS). Most EEC statistical agencies responded that they considered the protection of data to be very important; none, however, reported a systematic approach to statistical data protection. The problem for the EEC group was not in the legal framework; the required general legal basis for statistical disclosure control already existed in most of those countries. The main difficulties were in the organizational and technical implementation of the legally defined confidentiality principles. In CIS countries, even the legal base did not exist. Little attention had been given in either group of countries to mathematical and computing aspects of data confidentiality.

About half the 16 countries in the 1998 survey reported that they disseminated public use files of their labor force survey. Other forms of demographic data release, such as detailed tabular data (12 of the 16 countries), and microdata for research (9 of the 16), were used more frequently. The most frequently used disclosure limitation technique for demographic data was categorization of variables. EEC respondents reported also using rounding, subsampling, microaggregation, imputation of missing values, and top-coding, often using several techniques in parallel. CIS respondents, in contrast, reported virtually no use of these techniques.

Access to business data was much more restricted than access to demographic data. Only three countries in the 1998 survey released public use files on business statistics. Other countries used forms of release similar to demographic data, such as tabular data and—less frequently—microdata for research and synthetic data files. On-site access was provided for business statistics sometimes, but less frequently than for demographic data. The disclosure limitation methods reported for business data were somewhat different from those for labor force data. Categorization of variables, imputation of missing values, and microaggregation were employed in about half the countries. Subsampling, top-coding, rounding, and the dominance rule were also reported. Data swapping and the use of special software were applied in one country only. Adding noise was not used at all.

In a 2000/2001 update (UNECE 2001) to the 1998 survey, all statistical offices said they now recognized the importance of protecting statistical data. This increased attention is attributable in part to the Population Census in 2000 and agricultural censuses conducted in some transition countries, and in part to increasing gaps between rich and poor and growing criminality, which has made the public in those countries very sensitive to confidentiality issues, especially concerning individual privacy.

According to the 2000 survey, the major agency focus was still on the legislative and administrative aspects of disclosure control, with mathematical aspects receiving less attention, as in the previous survey, and data protection still done on an *ad hoc* basis. The disclosure control methods reported by the transition countries in 2000 are summarized in Tables 1 and 2.

The most popular methods in 2000 for tabular data were minimum cell count rules (usually three), geographical or population thresholds (releasing data only for areas above a particular spatial or population threshold), recoding data into broad categories, and dominance rules (if fewer than a certain number of units—usually two or three—account for at least a minimum of the cell total).

With respect to microdata, the 2000 survey of transition countries indicated that statistical agencies are no more willing to allow other institutions access to original data than they were in 1998. This is at least in part because pressure on statistical agencies to release microdata in EEC and CIS countries is not as intense as in Western countries. The most frequently used forms of microdata release were 'microdata for research' and synthetic files. Those that did release microdata often assumed that omitting explicit identification variables (namely addresses) was sufficient to avoid disclosure. Some respondents said they released public use files, one on demographic data and three on business data.

Table 1. Disclosure Limitation Techniques Used for the Release of Demographic Data

Geographic	Tabular Data: Geographic or population thresholds	Tabular Data: Minimum cell count rules	Tabular Data: Dominance rules	Tabular Data: Rounding	Tabular Data: Adding noise, blurring	Tabular Data: Re-coding variables	Microdata: Geographic or population thresholds	Microdata: Sampling	Microdata: Top- and bottom-coding	Microdata: Recoding data into broad categories	Microdata: Data swapping	Microdata: Deletion of sensitive records	Microdata: Deletion of data items	Microdata: Micro-aggregation
Bulgaria		+	+			+			+	+			+	
Czech Republic		+	+									+		
Estonia		+	+	+										+
Hungary	+	+		+		+							+	
Latvia	+												+	
Lithuania	+	+		+				+	+			+		+
Poland	+					+	+			+				
Romania	+	+				+	+	+		+			+	+
Slovakia	+	+	+			+	+	+	+	+			+	+
Slovenia	+	+	+	+				+		+			+	
The Former Yugoslav Republic of Macedonia														
Yugoslavia	+													
Azerbaijan														
Belarus														
Kyrgyzstan										+				+
Russia														
Turkmenistan														
Total														

Source: UNECE Secretariat (2001), *Statistical Data Confidentiality in the Transition Countries: 2000/2001 Winter Survey*

Table 2. The Disclosure Limitation Techniques Used for the Release of Business Data

	Tabular Data						Microdata							
Geographic	**Geographic or population thresholds**	**Minimum cell count rules**	**Dominance rules**	**Rounding**	**Adding noise, blurring**	**Re-coding variables**	**Geographic or population thresholds**	**Sampling**	**Top- and bottom-coding**	**Recoding data into broad categories**	**Data swapping**	**Deletion of sensitive records**	**Deletion of data items**	**Micro-aggregation**
Bulgaria		+	+			+			+	+			+	
Czech Republic		+	+											
Estonia														
Hungary		+											+	
Latvia	+	+	+			+	+						+	
Lithuania	+	+	+	+	+	+	+	+	+	+	+	+	+	+
Poland	+	+					+			+				+
Romania								+					+	+
Slovakia	+	+				+	+	+	+	+				+
Slovenia	+	+	+											
The Former Yugoslav Republic of Macedonia														
Yugoslavia														+
Azerbaijan														
Belarus														
Kyrgyzstan										+	+			+
Russia														
Turkmenistan														
Total														

Source: UNECE Secretariat (2001), *Statistical Data Confidentiality in the Transition Countries: 2000/2001 Winter Survey*

The general conclusion of the UNECE 2000 survey was that the transition countries need technical assistance, software, and training in particular.

At about the same time the most recent UNECE survey was done, a survey of international users of official microdata (KVI 2001a) was undertaken by an expert committee (Kommission zur Verbesserung der Statistischen Infrastruktur in Zusammenarbeit zwischen Statistik und Wissenschaft [KVI])[1], convened by the German Federal Ministry of Education and Research (BMBF) to analyze Germany's statistical infrastructure. This survey asked users of microdata their opinion on the user-friendliness of different dissemination and disclosure limitation approaches and their general judgment on the possibilities of microdata analysis in their home country.

The KVI survey was sent to 16 experts in eight countries (one sociologist and one economist in each country). Fifteen of the 16 experts responded. The countries selected were similar to those included in our survey—among them Canada, France, Germany, the Netherlands, Norway, Sweden, the United Kingdom, and the United States. This overlap is hardly surprising given that these are the most active countries with respect to microdata release.

According to the KVI survey, the major complaint of users of official microdata was not about limitations due to disclosure control but high user costs, bad documentation, and nontransparent access policies.

Whereas statistical microdata are treated as a 'public good' in the United States and Canada, in some European countries the (high) costs of disclosure control are charged to individual users. In the United States, Canada, and the Netherlands, minimal user charges cover just the marginal costs of dissemination, and the costs of disclosure control are funded through different channels. In the United States, for example, the statistical agencies cover the cost of producing public use or scientific use files out of their own budgets. In Canada, the universities pay Statistics Canada a flat rate that covers the cost of producing public use and scientific use files, in return for which university researchers can obtain the data without fee. In the Netherlands, the National Science Foundation (WSA) pays a flat rate to Statistics Netherlands (CBS) for data production and researchers pay a marginal fee per data source.

Even if user costs are low, as emphasized in the KVI survey report, unclear rules and weak (or nonexistent) documentation can still limit access. All respondents in all countries judged the microdata documentation provided by the official statistical agencies themselves as inferior (in need of improvement) or nearly nonexistent. Special service agencies with independent funding, such as those established in Norway and the Netherlands, were quoted as models for transparent service and reasonable documentation.

---

1 For a written summary in English of the work of KVI, see KVI (2001b).

3. The Motivation for a New Survey

Our survey was designed to identify the disclosure control methods currently used by national statistical agencies in a wide range of countries.

The survey questions[2] were based mainly on the 'Guidelines for Good Practice' given in *Statistical Policy Working Paper #22* (FCSM 1994). We used that framework to ask simple questions (mostly yes/no) about specific methods. Respondents were also invited to comment wherever they wished to share personal thoughts or comments. Many of them enclosed papers that they or colleagues had written, which gave us valuable extra information.

In the survey, we ask about tabular data and microdata separately. For tabular data, we make a further distinction between disclosure methods for counts or frequencies and those for aggregate magnitude data. For microdata, we distinguish between demographic microdata and business microdata. Within demographic data, we differentiate between census and non-census (or sample) data.

At the end of the questionnaire, we ask about the evolution of disclosure limitation methods in the recent past, whether the agency has ever experienced disclosure-related problems, and whether it has publications on this topic. We also ask about the use of specific software packages. The intent of these final questions was to trace how disclosure control practices have evolved and to identify any implicitly or explicitly formulated disclosure control methods. We also wanted to explore whether disclosure control is a matter of concern within the statistical agencies. As noted, we discussed the implications of our findings with two disclosure limitation experts, one from Europe and one from the United States.

4. Sample 'Design' and Distribution

We chose to distribute the survey by e-mail, for several reasons. A major factor was speed. This was important because of the laborious process we had to go through to find our respondents. Because we did not have an address list of staff members within the statistical agencies currently dealing with these issues, we began with participant lists from conferences dealing with confidentiality issues. Obviously, many of the addresses were no longer valid. In addition, some persons we reached were no longer in charge of disclosure control, in which cases we asked them to forward the questionnaire to colleagues who could help us. The first mailing was sent to 25 national statistical agencies.

The general objective was to get as many responses as possible, with particular emphasis on covering the following countries: Canada, France, Germany, the Netherlands, Norway, Sweden, the United Kingdom, and the United States. This turned

[2] See Appendix.

out to be quite an ambitious aim, but thanks to the keen interest of statistical agencies in disclosure issues and the kind cooperation of many staff members, we largely succeeded in our objective[3].

For reasons mentioned before, as well as simple nonresponse, we did not always find the best point of entry to an agency. In these cases, we had to ask help from the contributors to this book and their many connections to bring us in contact with the right people at the right places. All these efforts have resulted in responses from the national statistical agencies of Canada, the Czech Republic, Denmark, Estonia, Germany, Hungary, Italy, Lithuania, the Netherlands, New Zealand, Norway, Sweden, and the United States (in the latter from four different sources).

Our impression was that most of the time, (chief) statisticians in charge of many statistical procedures, including but not restricted to policies related to disclosure, completed the questionnaire. Only in a minority of cases had someone been appointed specifically to coordinate disclosure-related problems. Some respondents noted that these responsibilities are rather new and often not yet fully centralized.

We cannot claim that our results are representative of national statistical agencies in general because our survey suffers from severe selectivity in response. Those agencies that have an interest in the topic and have already implemented specific disclosure control measures have probably responded in greater number than those less interested or less far along. In addition, the 16 questionnaires that were returned are not equally complete. Some questionnaires were the collective response of different departments of an agency, with each section answered by the staff member in charge of the specific issue. But some left out certain sections, in particular on microdata for the census or microdata on establishments, simply because they do not release that type of data or it was not the responsibility of the person who filled out the questionnaire.

5. The Results

Tabular Data: Tables of Counts or Frequencies

Tables of counts or frequencies usually refer to demographic data (on individuals or households). Those tables display numbers or fraction of individuals by category.

The first survey question on tables of counts and frequencies was whether sampling and weighting, without publishing the weight, was used as a statistical disclosure limitation method. Only four agencies reported using this technique (see Figure 1).

---

[3] We wish to thank the many people involved in the survey for their kindness and their help, sometimes above and beyond the call of duty; in particular, Juergen Chlumsky, Mark Elliot, Virág Erdei, Francis Kramarz, Julia Lane, Mark Schipper, Rainer Winkelmann, and Virginia de Wolf.

Most agencies reported that they apply a threshold rule where a cell in a table of frequencies is defined to be sensitive to the number of respondents if less than some specified number. The most frequently reported threshold was 3 (nine agencies). Others said that their threshold varies by application and gave 4, 5, 10, 30, and 75 as thresholds. Some commented that they choose minimum cell sizes largely for quality purposes (*e.g.*, to avoid big sampling errors) rather than disclosure avoidance.

All agencies that apply the threshold rule (and answered the question) apply cell suppression. Other techniques are much less widespread, and many respondents noted that they use different techniques for different applications. Interestingly, random rounding is used occasionally but never as a sole protection method.

Figure 1. Disclosure Methods for Tables of Counts or Frequencies

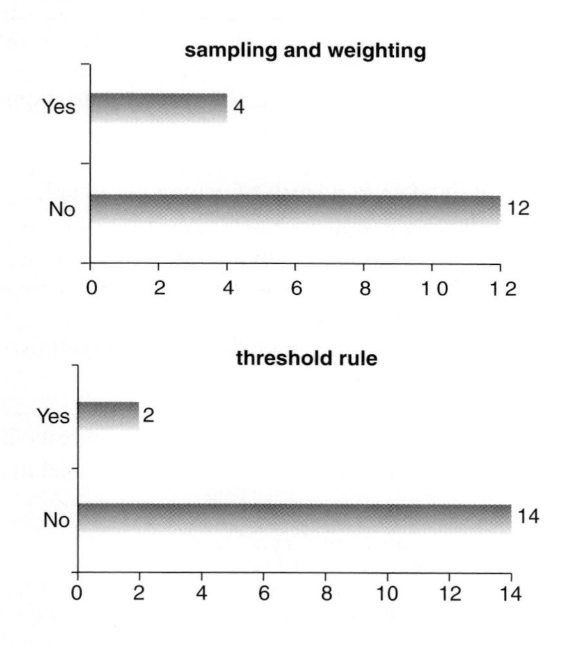

Figure 1. Disclosure Methods for Tables of Counts or Frequencies (Continued)

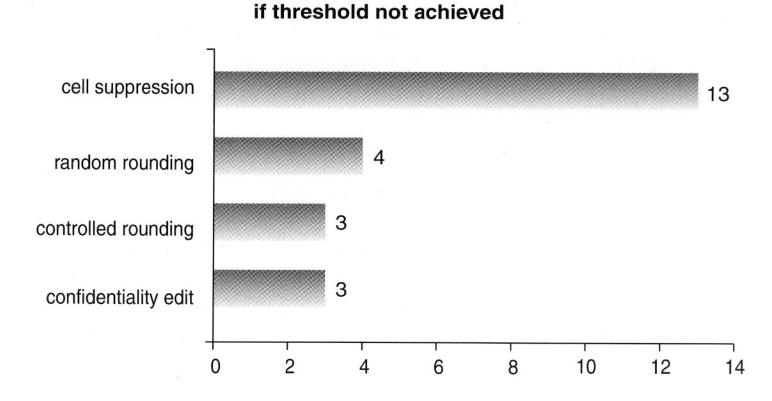

Special rules reported by respondents include special restrictions on the level of detail that can be provided in a table, such as

- other population thresholds;
- rules against publication of certain levels of geographic detail;
- collapsing of categories;
- release only of total frequency counts and percentages;
- prohibition of cross-tabulations with levels of detail that would allow spontaneous recognition of a population unique[4];
- combining of categories, which may include top- and bottom-coding as well as collapsing intermediate categories;
- application of the dominance rule[5]; and
- (for highly confidential frequency data) application of a threshold rule to the difference between number of respondents to the (sub-) marginal in each dimension of the table.

Tabular Data: Aggregate Magnitude Data

Responses from a total of 13 agencies are used in this section. One of the agencies reported that it does not release aggregate magnitude data, one did not comment, and one reported that all items are still being studied.

Magnitude data are business data reporting non-negative quantities about certain establishments or similar entities. The distribution of these values is likely to be skewed, with a few entities having very large values. Disclosure limitation for

[4] For example, the 'occupation' of someone being 'member of parliament' and 'location' 'small town' would enable recognition of a unique population, even if the weighted up frequency was around 100.

[5] For example, if two units account for more than 80 percent in terms of turnover.

these types of data concentrates on making sure that the values reported by the largest, most visible respondents cannot be estimated too precisely. To this end, primary suppression rules (see box) have been developed to determine whether a cell could reveal sensitive information, and to make it difficult for one of the respondents in a survey to estimate the value reported by another respondent with precision. We asked the statistical agencies how they identify these sensitive cells.

Primary suppression rules

(n,k) rule: "Regardless of the number of respondents in a cell, if a small number (n or fewer) of these respondents contribute a large percentage (k percent or more) of the total cell value, then the so-called n respondent, k percent rule of cell dominance defines this cell as sensitive." (page 48 in the *Statistical Policy Working Paper #22*).

p-percent rule: "Approximate disclosure of magnitude data occurs if the user can estimate the reported value of some respondent too accurately. Such disclosure occurs, and the table cell is declared sensitive, if upper and lower estimates for the respondent's value are closer to the reported value than a prespecified percentage, p." (page 46)

pq rule: "In the derivation for the p-percent rule, one assumes that there was a limited prior knowledge about respondent's values. Some believe that agencies should not make this assumption. In the pq rule, agencies can specify how much prior knowledge there is by assigning a value q, which represents how accurately respondents can estimate another respondent's value before any data are published ($p<q<100$)." (page 47)

The most widespread technique used by respondents to identify sensitive cells is the (n,k) rule (eight agencies; see Figure 2). One respondent said that in the future that agency would shift to the p-percent rule (currently applied by three agencies). Another agency noted that it had already made this change. Only one respondent reported that the agency applies the pq rule. Two agencies reported that they use another method. One of these said it uses the threshold rule. The other reported using a comprehensive set of special guidelines, as follows. (Respondents could report more than one method.)

- In no table should all cases of any line or column be found in a single cell.
- In no case should the total figure for a line or column of a cross-tabulation be less than three.
- In no case should the quantity figure be based on fewer than three cases.
- In no case should a quantity figure be published if one case contributes more than 60 percent of the amount.

- In no case should the data on an identifiable case, or any of the kinds of data listed in the preceding items, be derivable through subtraction or other calculation from the combination of tables published on a given study.
- Data published by the agency should never permit disclosure when used in combination with other known data.

After the sensitive cells have been identified, an agency can choose either to restructure the table by collapsing cells until no sensitive cell remains or to use cell suppression. Ten agencies reported that they restructure the table, some routinely and some occasionally (*e.g.,* when the number of primary cell suppressions would be large). All 13 respondents said they practice cell suppression, although rare cases were reported in which no suppression is applied, even though individuals or other entities may potentially be identified. These are cases that have a tradition of full disclosure (NCHS 1980, p.19).

There is also an administrative way to avoid cell suppression, used occasionally by 6 of the 13 agencies: obtaining permission from a respondent in the sensitive cell to publish the cell. One agency commented that in rare cases it may request such a 'waiver', for instance for a large public company where similar information is already in the public domain.

Figure 2. Tables of Magnitude Data

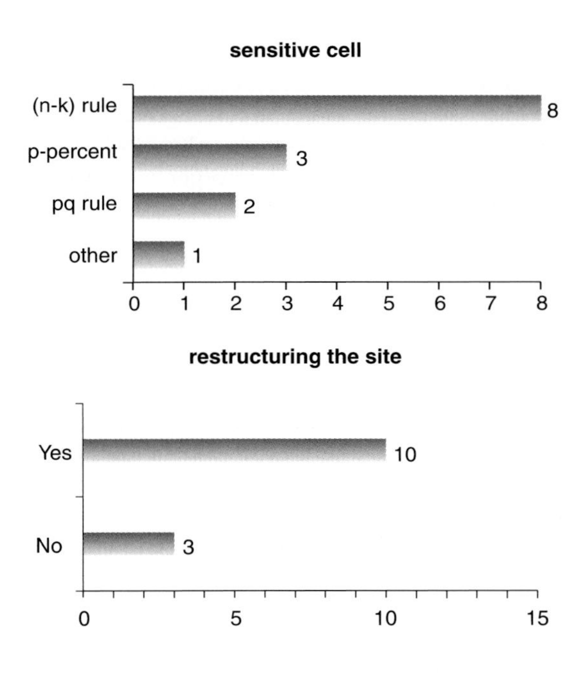

Figure 2. Tables of Magnitude Data (Continued)

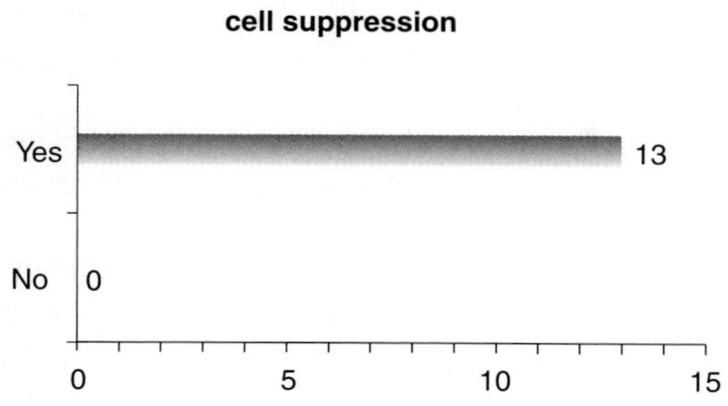

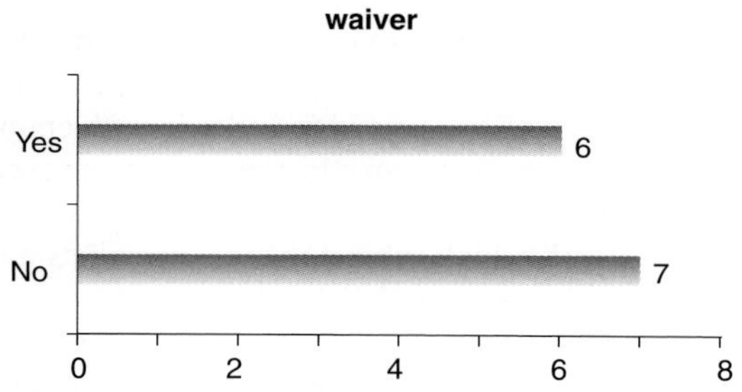

cell suppression

Yes 13
No 0
0 5 10 15

waiver

Yes 6
No 7
0 2 4 6 8

Demographic Microdata

Most agencies reported that they do not release microdata to the public but do provide access to researchers under certain circumstances, mostly in controlled environments—in exceptional cases even outside the premises of the agency.

Nine agencies reported that they release microdata from a census either for public use or for restricted research purposes. One respondent wrote that it is not yet releasing microdata from its census, but it is planning to do so in the future with all identifiers removed. Another respondent is in the process of creating an on-site working facility for researchers.

Respondents also reported that their main tools for protecting census microdata are to exclude obvious identifiers, limit geographic detail, and limit the number of variables on the file (see Figure 3). Other methods that were reported (but are not included in Figure 3) are

- Issuing multiple files, one with more detailed geography and less detailed characteristics and the other with less detailed geography and more detailed characteristics.
- Grouping, by splitting continuous variables into ranges to reduce detail, as opposed to top- or bottom-coding.
- Deleting sensitive records.

Thirteen agencies released non-census demographic microdata. They protect all files by excluding obvious identifiers and limiting geographic detail. Limiting the number of variables (often customizing the file to provide only the variables needed for the research project) and top- and bottom-coding are also applied quite frequently.

Other methods that were reported (also not listed in Figure 3) include

- Deleting sensitive records, deleting sensitive items, recoding into broad categories, sampling, microaggregation.
- Imputing blanks/missing values in unsafe records when other methods fail to protect these records.
- Local suppression.
- Grouping.
- Eliminating any variables that can be used to link to external sources containing individual identifiers or more geographic detail than can be released on the microdata files.

One agency reported that any file destined for release to the public is systematically reviewed by a disclosure review team composed of subject matter specialists, sampling experts, and experts in disclosure analyses. Procedures are codified in a disclosure potential 'checklist', which subject matter specialists are required to fill out.

Another agency reported an interesting practice with respect to data. For any given year of data released, it provides data on all respondents, but it limits the data it releases to selected years.

One respondent reported that it is sometimes possible to release a very limited number of variables. A single request, or perhaps even several, can be considered, as long as disclosure risk analysis does not indicate confidentiality problems. However, the agency assumes that any release to any member of the public is a release to the entire public, so that successive releases of information on a few variables are regarded as cumulative, even though the data were released to different parties. Hence, at some point the marginal release is seen as an unacceptable disclosure risk.

One agency reported that it offers analytical programming services, meaning that users submit computer programs electronically and receive output that is subject to disclosure review before delivery to the user. There is also an automated version of this service, called remote access, provided through a modem.

Several statistical agencies also reported that they sometimes give short-term appointments to outside researchers, which require them to sign confidentiality agreements.

Figure 3. Microdata of Demographic Data

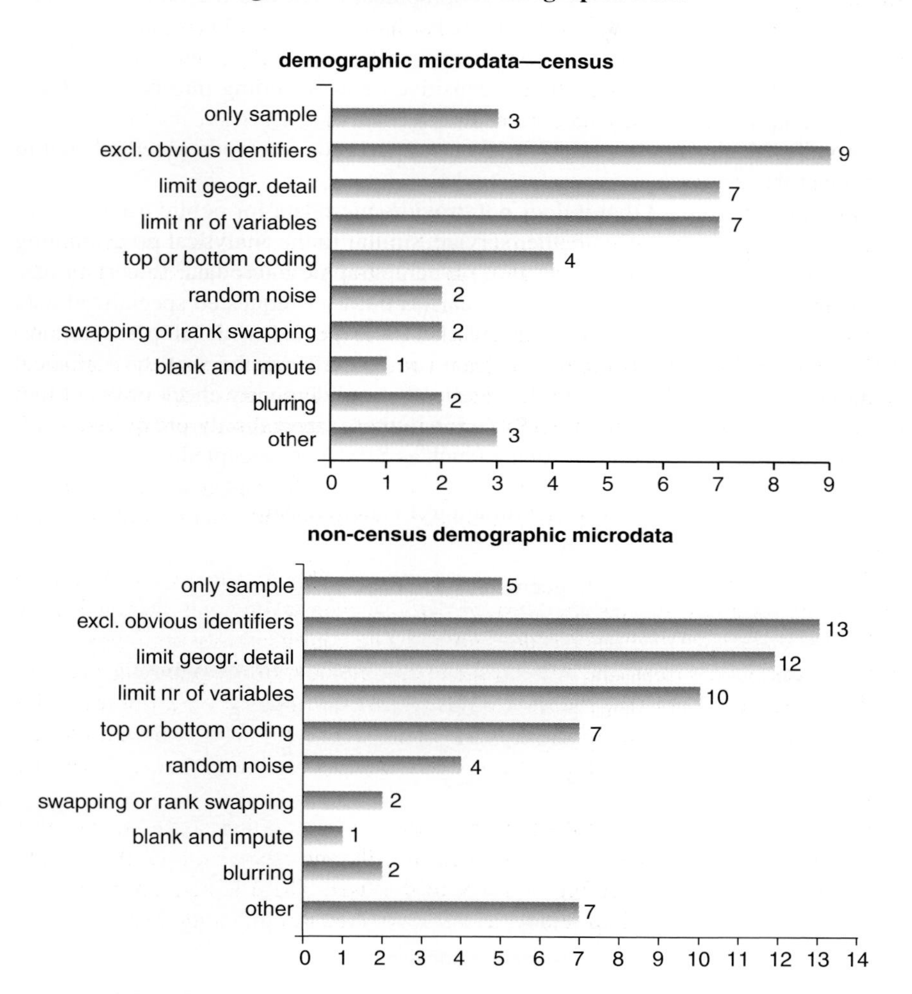

Economic Microdata

There are virtually no public use microdata files released for business data. Statistical agencies do provide access to business microdata for research purposes, however. The problem with this type of data is that the information collected about business establishments is primarily magnitude data. Their distribution is skewed and large establishments would be easily identifiable with the use of other publicly available information.

Eight agencies reported that they provide researchers access to establishment microdata (see Figure 4). All of them exclude obvious identifiers when providing such access. Most of them also limit geographical detail and the number of variables in the file. Techniques on our survey list include

- grouping;
- deleting sensitive records, deleting sensitive items, recoding into broad categories, sampling, microaggregation; and
- imputing blanks/missing values in the unsafe records when other methods fail to protect these records.

One agency reported that it does not provide microdata for public use or scientific use, but it does provide another service, similar to the analytical programming service described in the earlier section on demographic microdata. In certain (exceptional) cases, the statistical institute carries out user-requested specialized data analysis. In these cases, the user (researcher) is expected to develop a computer program to do the analysis, which is then run by staff members of the statistical agency on the original microdata file. Staff do a confidentiality check of the output before it is sent to the researcher. To facilitate these checks, only programs developed on the basis of standard software (such as SPSS) are accepted.

Figure 4. Microdata of Economic Data

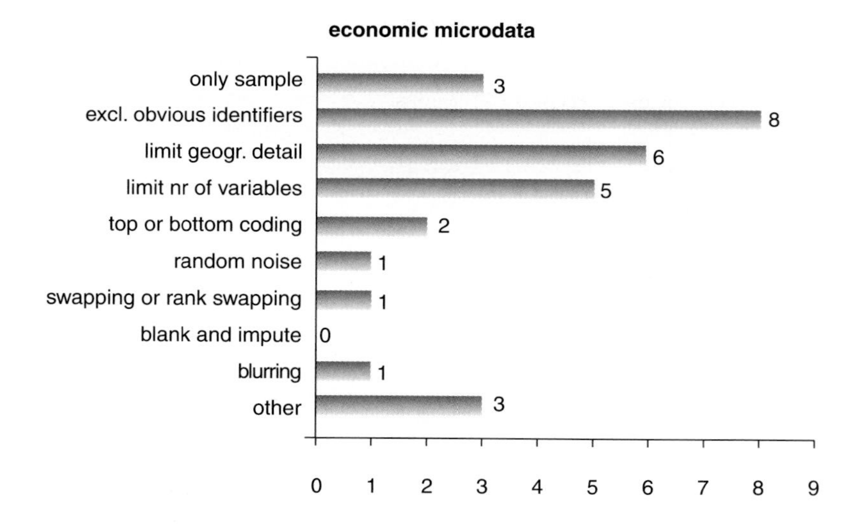

Software Packages in Use

No clear consensus emerged from our respondents on which software packages should be used (see Figure 5). In fact, there are no software packages in widespread use. The only program used by more than one agency in our sample is μ-ARGUS. Two statistical agencies reported that they use 'homemade' programs.

Some agencies commented on the software packages on our survey list. One reported the following: 'We evaluated CONFID in 1996 and decided that it would be difficult to integrate into the agency's system. We evaluated ACSSuprs in 1997–98. It is promising, but some questions remain as to availability of ongoing support. We also evaluated τ-ARGUS in 1997–98. It is promising as well, but the cell suppression methodology is not complete. We will keep up-to-date on developments. μ-ARGUS is not a major issue for the agency, but we would be interested if we released confidential microdata'.

Another agency reported that it does not use any specific software packages now, but it is intending to implement τ-ARGUS and μ-ARGUS in the future. τ-ARGUS and μ-ARGUS are also being tested for potential use by three other respondents, although one of them remarked that packages that are more easily integrated into the agency's general production system (based on SAS) would be more convenient.

Figure 5. Software Packages in Use

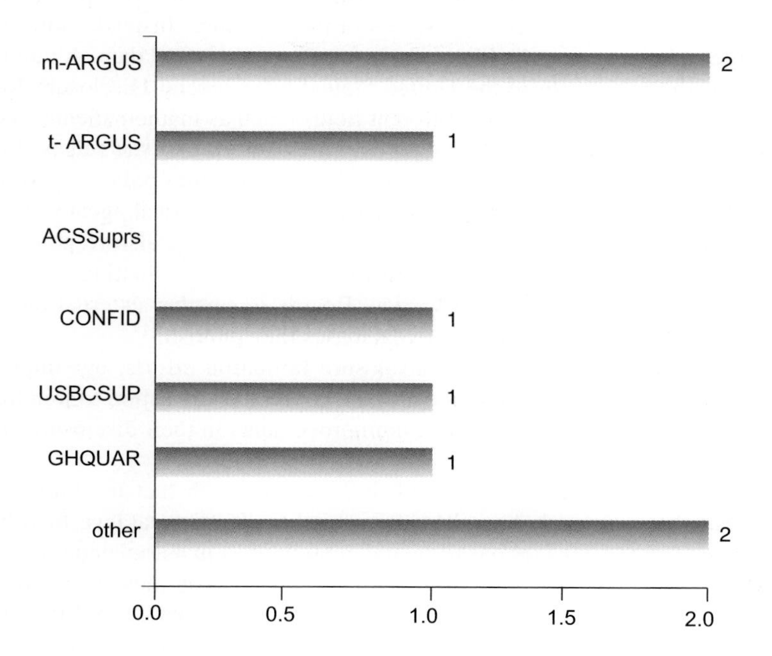

The software packages τ-ARGUS and μ-ARGUS have emerged from the Statistical Disclosure Control (SDC) project of the Fourth Framework Programme of the European Union. New versions of the ARGUS packages are being developed by the methods and informatics department of Statistics Netherlands, as part of the Computational Aspects of Statistical Confidentiality project in the Fifth Framework Programme of the European Union. For τ-ARGUS, for example, research is being done in the field of disclosure limitation for linked tables, and more options for secondary suppressions and consecutive years of the same survey. In the case of μ-ARGUS, the difference between protecting microdata for research and protecting public use files is of interest (Schulte Nordholt 2001, pp.11–12).

We have also been notified of the collective effort of the Confidentiality and Data Access Committee (CDAC) and the FCSM to develop an audit software for large tabular presentations. To this end, the Subcommittee on Disclosure Auditing has been formed to develop a framework for which the Energy Information Administration (EIA) has taken the lead (EIA 2001). The software, currently referred to as Disclosure Audit System (DAS), is designed to assist in releasing the most detailed data possible with acceptably small risk of confidential data identification.

6. Evolution of Disclosure Limitation Methods Used

With some exceptions, statistical agencies generally seem not to apply the rather complicated disclosure probability models in the literature. Instead, more or less *ad hoc* rules are set and applied to a whole family of statistical data. A number of statistical agencies (mostly in the United States) have special Disclosure Review Boards consisting of experts from different fields (such as mathematicians, economists, sampling experts) which examine data files before they are released to the public. These boards are responsible for disclosure limitation policy regarding all publicly available data products of the relevant federal statistical agency. They review cases of concern (sometimes all publications). They are also responsible for communication inside and outside the institute and for coordinating research. A major concern of these Disclosure Review Boards is whether external databases exist that can potentially be matched to databases they publish.

Concerning the technical side of disclosure limitation efforts, our impression from this and other surveys is that most of the countries who aspire to join the European Union have experienced dramatic improvements in their disclosure control practices over the past 10 years.

For the other countries we surveyed, it is not so much that the "rules" have changed but that new tools have become available. The appearance of different specialized software packages has triggered great interest in automated procedures. Most agencies are experimenting or gathering knowledge on these software packages but so far are cautious about installing them, mainly because of difficulties in integrating them into their current systems. The most frequently mentioned software packages are the ARGUS packages, but even these are typically still being studied and tested. The UNECE studies give the same impression. The CDAC/FCSM audit software is not yet available and may be subject to restrictive access in the future. Some agencies are attempting to develop "homemade" programs. We also find suggestive evidence of a shift from interest in blanking (cell suppression) and imputation to data swapping, and from the (n,k) rule to the p-percent rule.

Apart from the challenge of automated systems, a strong external influence will surely have an effect on future research and development: the great increase in the demand for access to microdata by researchers and others. This in itself is the consequence of the development of statistical software packages. Many agencies provide access to microdata now. Some have established data laboratories and access sites on the premises of the statistical agency, where researchers are provided access in a controlled environment. These developments have challenged and will continue to challenge traditional confidentiality issues. In this context, it is interesting to report that three of our respondent agencies have experienced disclosure-related problems in the recent past.

References

Energy Information Administration [EIA] (2001) *Example of a Table Suppression Audit. Confidentiality and Data Access Committee (CDAC): Demonstration of Disclosure Limiting Software (DLS).* Handout at a demonstration to CDAC.

Erdei, Virág (2001) "A statisztikai adatok vedelme a magyar Központi Statisztikai Hivatalban," Paper Presented at the Joint Eurostat/UNECE Work Session, Skopje, March 2001.

Eurostat (1996) *Manual on Disclosure Control Methods,* Luxemburg: Office for Publications of the European Communities.

Federal Committee on Statistical Methodology [FCSM] (1978) *Report on Statistical Disclosure and Disclosure-Avoidance Techniques (Statistical Policy Working Paper #2),* Washington, D.C.: U.S. Department of Commerce, Office of Federal Statistical Policy and Standards.

—— (1994) *Report on Statistical Disclosure Limitation (Statistical Policy Working Paper #22).* Washington, D.C.: U.S. Office of Management and Budget, Statistical Policy Office.

Holvast, Jan (1999) 'Statistical Dissemination, Confidentiality and Disclosure', in *Eurostat: Statistical Data Confidentiality—Proceedings of the Joint Eurostat/UNECE Work Session on Statistical Data Confidentiality Held in Thessaloniki in March 1999.*

Hoy, E., M.M. McMillen, F. Scheuren, G.D. Stamas, G.T. Therriault, and A.O. Zarate (2000) 'Panel on Disclosure Review Boards of Federal Agencies: Characteristics, Defining Qualities and Generalizability', paper presented at the Joint Statistical Meetings, August 17, Indianapolis, Ind.

Jabine, T.B. (1993) 'Procedures for Restricted Data Access', *Journal of Official Statistics,* 9(2), pp.537-89.

Kommission zur Verbesserung der Statistischen Infrastruktur [KVI] (2001a) *Wege zur Verbesserung der Statistischen Infrastruktur (Gutachten der Kommission),* Baden-Baden, mimeo.

— (2001b; English version of 2001a) 'Ways Towards an Improved Informational Infrastructure', *Schmollers Jahrbuch—Journal of Applied Social Science Studies*, 121(2) (in press).

Luige, Tiina, and Jana Meliskova (1999) 'Confidentiality Practices in the Transition Countries', in *Eurostat: Statistical Data Confidentiality—Proceedings of the Joint Eurostat/ UNECE Work Session on Statistical Data Confidentiality Held in Thessaloniki in March 1999.*

National Center for Health Statistics [NCHS] (1980) *Staff Manual on Confidentiality.* Under revision.

Schulte Nordholt, E. (2001) 'Progress in the Implementation of SDC Methods and Techniques in Central and Eastern Europe—List of Key Issues for Discussion. Paper presented at the Joint ECE/Eurostat Work Session on Statistical Data Confidentiality, Skopje, 14–16 March 2001.

United Nations Economic Commission for Europe [UNECE] Secretariat (2001) *Statistical Data Confidentiality in the Transition Countries: 2000/2001 Winter Survey,* paper presented on the Joint ECE/Eurostat Work Session on Statistical Data Confidentiality, Skopje, 14–16 March 2001.

Zarate A.O. (1998) 'Legal, Administrative and Statistical Aspects of Confidentiality Procedures at the National Center for Health Statistics Presentation', paper presented as expert testimony on issues of 'privacy and confidentiality' for the public Meeting on the President's Initiative on Immunization Registries, Atlanta, 16 July 1990.

Appendix A: Survey of Methods Used by Statistical Agencies to Protect Confidentiality of Data and Maximize Access

Explanatory Notes

There is a fundamental tension at the heart of every statistical agency's mission. Each is charged with collecting high quality data to inform national policy and enable statistical research. This necessitates dissemination of both summary and microdata. Each statistical agency is also charged with protecting the confidentiality of survey respondents. This often necessities blurring the data to reduce the probability of reidentification of individuals. The trade-off dilemma, which could well be stated as protecting confidentiality but maximizing access, has become more complex as both technological advances and public perceptions have altered in the information age.

We intend to make a review of how statistical disclosure techniques have kept pace with these changes.

The **objective** of this **questionnaire** is to gather information on practical approaches protecting confidentiality of respondents but maximizing access to demographic and economic data. We are interested in the methods in use and also in the evolution of the applied policy, the changes in approaches in the recent past, and relevant discussion of expected future changes.

On the contrary to the topic, we will not treat your answers as confidential. We do want to make an overview of the applied techniques. The results of our survey will be published in a book, *Confidentiality, Disclosure, and Data Access: Theory and Practical Approaches for Statistical Agencies,* to be edited by Pat Doyle, Julia Lane, Jules Theeuwes, and Laura Zayatz and to be published by Elsevier North Holland in the beginning of 2002. The book will contain technical chapters on the techniques of disclosure limitation methods as well as two chapters on the practical approaches protecting confidentiality used by national statistical agencies. We will have one chapter on demographic and one on economic data.

Of course, if you desire to share personal thoughts or critiques with us that are not part of the policy of your agency we will keep that information confidential if requested.

Answering this questionnaire will take approximately 20 minutes of your time. Your contribution will be highly appreciated.

Please fill in into this file.
If necessary please duplicate specific sections.
You are kindly requested to fill in this questionnaire and return it to floraf@seo.fee.uva.nl.
In case you have specific difficulties answering the questions, you may contact Prof. Gert G. Wagner: (00)-49-30-89789-290 (gwagner@diw.de) concerning demographic data or Prof. Jules Theeuwes or Flóra Felsö: (00)-31-20-6242412 (theeuwes@seo.fee.uva.nl or floraf@seo.fee.uva.nl) concerning economic data.

Questionnaire

0. General

Can we have your name?:

And your e-mail address, telephone and fax number?:

Which agency do you work for?:

Which department?:

We are interested in the methods in use and also in the evolution of the applied techniques, changes in the recent past, and relevant discussion of future changes. Mostly, you can simply answer the questions with yes or no, but you are also welcome to comment wherever you feel you want to share something with us.

I. Disclosure methods for table of counts or frequencies

1) Do you use 'sampling and weighting' (not publishing the weight) as a statistical disclosure limitation method?

2) Do you apply a threshold rule where a cell in a table of frequencies is defined to be sensitive if the number of respondents is less than some specified number?
   If yes, how many respondents are required?

3) What method do you apply if the threshold is not achieved? Do you use
   - cell suppression?
   - random rounding?
   - controlled rounding?
   - confidentiality edit?

4) Do you use other special rules that impose restrictions on the level of detail that can be provided in a table?
   - If yes, what special rules?

II. Tables of magnitude data

1) How do you identify sensitive cells? Do you use the
   - (n,k) rule?
   - p-percent rule?
   - pq rule?
   - other…

2) If a sensitive cell is identified do you restructure the table?

3) If a sensitive cell is identified do you apply suppression?

4) If a sensitive cell is identified do you try to get a 'waiver' of the promise to protect sensitive cells?

III. Microdata

Please distinguish between demographic data (surveys/register data of persons and households) and economic data (surveys/register data of firms).

III.1 Demographic Data

Do you release microdata of the census?

If you release both microdata of the census and non-census data, please duplicate the section hereunder and answer the questions separately for census data and non-census data.

For public use of microdata or for microdata for research do you:
- include data from only a sample of the population?
- exclude obvious identifiers?
- limit geographic detail?
- limit the number of variables on the file?
- apply top- or bottom-coding?
- add random noise (adding or multiplying by random numbers)?
- apply swapping or rank swapping (also called switching)?
- apply blank and impute (electing records at random, blanking out selected variables and imputing for them)?

- blur the data (aggregating across small groups of respondents and replacing one individual's reported value with the average)?
- other...

III.2 Economic Data

For public use of microdata or for microdata for research do you:
- include data from only a sample of the universe?
- exclude obvious identifiers?
- limit geographic detail?
- limit the number of variables on the file?
- apply top- or bottom-coding?
- add random noise (adding or multiplying by random numbers)?
- apply swapping or rank swapping (also called switching)?
- apply blank and impute (electing records at random, blanking out selected variables and imputing for them)?
- blur the data (aggregating across small groups of respondents and replacing one individual's reported value with the average)?
- other...

IV. The evolution of disclosure limitation methods

1) What where the major changes in your policy in the past ten years, and why? Since when do you apply the combination of methods described above?

2) Have you ever experienced disclosure-related problems?

3) Do you use specific software packages, such as:
   - GHQUAR?
   - USBCSUP?
   - CONFID?
   - ACSSuprs?
   - τ-ARGUS?
   - μ-ARGUS?
   - other...

4) Do you or your institute have publications on this topic? Please note the title and authors.

5) Finally, can we contact you for further information?

Thank you for your contribution!

Confidentiality, Disclosure, and Data Access: Theory and Practical Application for Statistical Agencies
Pat Doyle, Julia I. Lane, Jules J.M. Theeuwes and Laura M. Zayatz (Eds)

Chapter 3
Information Explosion

Latanya Sweeney
Carnegie Mellon University

1. Introduction

Society is experiencing unprecedented growth in the number and variety of data collections as computer technology, network connectivity, and disk storage space become increasingly affordable. Data holders operating autonomously and with limited knowledge are left with the difficulty of releasing information that does not compromise privacy, confidentiality, or national interests. In many cases the survival of the database itself depends on the data holder's ability to produce anonymous data because not releasing such information at all may obstruct the goals for which the data were collected. On the other hand, failing to provide proper protection within a release may create circumstances that harm the public or others. Ironically, the broad availability of public information makes it increasingly difficult to provide data that are effectively anonymous.

Table 1. Estimated Growth in Data Collections in Illinois
(per encounter, in bytes)

Examples	1983	1996
Each birth	280	1,864
Each hospital visit	0	663
Each grocery visit	32	1,272

Sources: Washington State Department of Public Health, Massachusetts Department of Public Health, National Association of Health Data Organizations, and Giant Eagle grocery database.

Table 1 shows examples of the expansion of data collected from 1983 to 1996 for some person-specific encounters in Illinois. The values are the number of bytes (letters, digits, and other printable characters) that were stored for each person per encounter in the collection shown. Table 2 describes how the estimates used in Table 1 were computed. The values shown in Tables 1 and 2 are the number of printable characters that are reserved for providing the information.

The following sections further explain these figures and show that they are representative of many experiences in most states.

Table 2. Estimations of the Sizes of Some Person-Specific Data Collections
(per encounter)

1983	1996
1983 Birth certificate (280) based on field sizes: 40 Name 80 Address 80 Parent's names 80 Hospital name & address	1996 Birth certificate (1,864) based on field sizes: 40 Name 80 Address 80 Parent's names 80 Hospital name & address 1,584 Birth characteristics
1983 Health care cost data (0) based on field sizes: *no such collection existed*	1996 Health care cost data (663) based on field sizes with noted codes expanded to include textual description: 263 Primary fields (see Figure 2-6) 80 Expand hospital name, patient name, location 160 Diagnosis codes described (8 x 20) 160 Procedure codes described (8 x 20)
1983 Grocery purchases (32) based on field sizes: 18 Subtotal, tax, total 14 Date and time	1996 Grocery purchases (1,272) based on field sizes and average number of items expanded to include textual notation: 240 Amount (6) x number of items (40) 18 Subtotal, tax, total 14 Date and time 120 Name, address 80 Payment information 800 Item description (20) x number of items (40)

Sources: Calculated from Washington State Department of Public Health, Massachusetts Department of Public Health, National Association of Health Data Organizations, and Giant Eagle grocery database.

2. Growth in Birth Certificate Information

As shown in Table 3, birth certificate information historically had only 7 to 15 fields of information, but today in Illinois (as well as in most states) more than 100 fields of information are collected about each child's birth, even though the parents may receive only the traditional few fields.

Table 3. Minimal Set of Birth Certificate Fields

Field name
Child's first name
Child's middle name (sometimes or initial)
Child's last name
Day, month and year of birth
City and/or county of birth (sometimes hospital)
Father's name
Mother's name (including maiden name)

Source: Massachusetts Department of Public Health, Registry of Vital Records and Statistics.

Table 3 contains the minimal list of fields available on a birth certificate from almost any state or county in the United States, post-1906 (Massachusetts Department of Public Health, Registry of Vital Records and Statistics 1999).

Table 4. Typical Set of Birth Certificate Fields, Post-1925

Field name
Child's first name
Child's middle name (sometimes or initial)
Child's last name
Day, month, and year of birth
City and/or county of birth (sometimes hospital)
Father's name
Mother's name (including maiden name)
Place of birth (address and town/city)
Mother's age and address
Mother's birthplace (town/city, state, county)
Mother's occupation
Mother, number of previous children
Father's age and address
Father's birthplace (town/city, state, county)
Father's occupation

Source: Massachusetts Department of Public Health, Registry of Vital Records and Statistics.

Table 4 contains the typical list of fields available on a birth certificate from most states or counties in the United States, post-1925. This list was provided from the Commonwealth of Massachusetts (Massachusetts Department of Public Health, Registry of Vital Records and Statistics 1999).

Table 5. Typical Set of Electronic Birth Certificate Fields in 1999
Starting, Fields 1–60

Field#	Size	Field name	Field#	Size	Field name
1	1	File Status	31	3	Mother's State of Birth
2	50	Baby's First Name	32	7	Mother's Residence Address
3	50	Baby's Middle Name	33	2	Mother's Residence Direction
4	50	Baby's Last Name	34	20	Residence Street Address
5	1	Baby's Suffix Code	35	10	Residence Type
6	3	Baby's Suffix Text	36	2	Residence Extension
7	8	Baby's Date of Birth	37	10	Residence Apartment #
8	5	Baby's Time of Birth	38	20	Mother's Town of Residence
9	1	1AM/PM Indicator	39	1	Mother's Residence in City Limits
10	1	Baby's Sex			
11	3	Blood Type	40	14	Mother's County of Residence
12	1	Born Here?	41	3	Mother's State of Residence
13	40	Place of Birth	42	10	Mother's Residence Zip Code
14	1	Facility Type	43	38	Mother's Mailing Address
15	20	City of Birth	44	19	Mother's Mailing City
16	20	County of Birth	45	2	Mother's Mailing State
17	6	Certifier's Code	46	10	Mother's Mailing Zip Code
18	30	Certifier's Name	47	1	Mother Married?
19	1	Certifier's Title	48	50	Father's First Name
20	30	Attendant's Name	49	50	Father's Middle Name
21	1	Attendant's Title	50	50	Father's Last Name
22	23	Attendant's Address	51	1	Father's Suffix Code
23	19	Attendant's City	52	9	Father's Suffix Text
24	2	Attendant's State	53	9	Father's Social Security Number
25	10	Attendant's Zip Code			
26	50	Mother's First Name	54	8	Father's Date of Birth
27	50	Mother's Middle Name	55	3	Father's State of Birth
28	50	Mother's Last Name	56	14	Mother's Origin
29	9	Mother's Social Security Number	57	14	Mother's Race
			58	2	Mother's Elementary Education
30	8	Mother's Date of Birth			
			59	2	Mother's College Education
			60	11	Mother's Occupation

Source: Genesis Systems (1999).

Table 6. Typical Set of Electronic Birth Certificate Fields in 1999

Continued, Fields 61–120

Field#	Size	Field name
61	11	Mother's Industry
62	14	Father's Origin
63	14	Father's Race
64	2	Father's Elementary Education
65	2	Father's College Education
66	11	Father's Occupation
67	11	Father's Industry
68	1	Plurality
69	1	Birth Order
70	2	Live Births Still Living
71	2	Live Births Now Dead
72	4	Month/Year Last Live Birth
73	2	Number of Terminations
74	4	Month/Year Last Termination
75	1	Baby's Weight Unit
76	5	Baby's Weight
77	6	Date of Last Normal Menses
78	1	Month Prenatal Care Began
79	2	Total Number of Visits
80	2	Apgar Score – 1 Minute
81	2	Apgar Score – 5 Minute
82	2	Estimate of Gestation
83	6	Date of Blood Test
84	22	Laboratory
85	1	Mother Transferred In
86	30	Facility Mother Transferred From
87	1	Baby Transferred Out
88	30	Facility Baby Transferred To
89	1	Tobacco Use During Pregnancy
90	3	Number of Cigarettes/Day
91	1	Alcohol Use During Pregnancy
92	3	Number of Drinks/Week
93	3	Mother's Weight Gain
94	1	Release Info For SSN
95	6	Operator Code
96	12	Hospital ID
97	1	Sent to Romans
98	1	Sent to APORS
99	16	Other Certifier Specify
100	12	Temporary Audit Number
101	16	Other Facility Specify
102	16	Other Attendant Specify
103	1	Mother's Race
104	1	Father's Race
105	2	Mother's Origin
106	2	Father's Origin
107	1	Attendant Same YN
108	1	Mailing Address Same YN
109	1	Capture Father's Info YN
110	2	Mother's Age
111	2	Father's Age
112	12	Baby's Hospital Med. Rec.
113	1	High Risk Pregnancy YN
114	1	Care Giver (For Chicago)
115	1	Record Selected For Download
116	1	Downloaded
117	1	Printed
118	12	Form Number
MEDICAL RISK FACTORS		
119	1	Anemia
120	1	Cardiac Disease

Source: Genesis Systems (1999).

Table 7. Typical Set of Electronic Birth Certificate Fields in 1999

Field#	Size	Field name
121	1	Acute/Chronic Lung Disease
122	1	Diabetes
123	1	Genital Herpes
124	1	Hydramnios/Oligohydramnios
125	1	Hemoglobinopathy
126	1	Hypertension, Chronic
127	1	Hypertension, Preg. Assoc.
128	1	Eclampsia
129	1	Incompetent Cervix
130	1	Previous Infant 4000+ Grams
131	1	Previous Preterm or SGA Infant
132	1	Renal Disease
133	1	Rh Sensitization
134	1	Uterine Bleeding
135	1	No Medical Risk Factors
136	40	Other Medical Risk Factors
OBSTETRIC PROCEDURES		
137	1	Amniocentesis
138	1	Electronic Fetal Monitoring
139	1	Induction of Labor
140	1	Stimulation of Labor
141	1	Tocolysis
142	1	Ultrasound
143	1	No Obstetric Procedures
144	40	Other Obstetric Procedures
COMPLICATIONS OF LABOR & DELIVERY		
145	1	Febrile (>100 or 38C)
146	1	Meconium Moderate, Heavy
147	1	Premature Rupture (>12 Hrs)
148	1	Abruptio Placenta
149	1	Placenta Previa
150	1	Other Excessive Bleeding
151	1	Seizures During Labor
152	1	Precipitous Labor (<3 Hrs)
153	1	Prolonged Labor (>20 Hrs)
154	1	Dysfunctional Labor
155	1	Breech/Malpresentation
156	1	Cephalopelvic Disproportion
157	1	Cord Prolapse
158	1	Anesthetic Complications
159	1	Fetal Distress
160	1	No Complications of L&D
161	40	Other Complications of L&D
METHOD OF DELIVERY		
162	1	Vaginal
163	1	Vaginal After Previous C-Section
164	1	Primary C-Section
165	1	Repeat C-Section
166	1	Forceps
167	1	Vacuum
ABNORMAL CONDITIONS OF NEWBORN		
168	1	Anemia
169	1	Birth Injury
170	1	Fetal Alcohol Syndrome
171	1	Hyaline Membrane Disease/ RDS
172	1	Meconium Aspiration Syndrome
173	1	Assisted Ventilation <30
174	1	Assisted Ventilation >30
175	1	Seizures
176	1	No Abnormal Conditions of Newborn
177	40	Other Abnormal Condition of Newborn
CONGENITAL ANOMALIES OF CHILD		
178	1	Anencephalus
179	1	Spina Bifida/Meningocele
180	1	Hydrocephalus

Source: Genesis Systems (1999).

Table 8. Typical Set of Electronic Birth Certificate Fields in 1999

Field#	Size	Field name
181	1	Microcephalus
182	40	Other CNS Anomalies
183	1	Heart Malformations
184	40	Other Circ./Resp. Anomalies
185	1	Rectal Atresia/Stenosis
186	1	Tracheo-Esophageal Fistula/Esophageal Atresia
187	1	Omphalocele/Gastroschisis
188	40	Other Gastrointestinal Ano.
189	1	Malformed Genitalia
190	1	Renal Agenesis
191	40	Other Urogenital Anomalies
192	1	Cleft Lip/Palate
193	1	Polydactyly/Syndactyly/ Adactyly
194	1	Club Foot
195	1	Diaphragmatic Hernia
196	40	Other Musculoskeletal/ Integumental Anomalies
197	1	Down's Syndrome
198	40	Other Chromosomal Anomalies
199	1	No Congenital Anomalies
200	40	Other Congenital Anomalies
CODE STRIP		
201	1	Record Complete YN
202	1	Record Type
203	4	Facility ID
204	4	City of Birth
205	3	County of Birth
206	2	Mother's State of Birth
207	2	Mother's State of Residence
208	4	Mother's Town of Residence
209	3	Mother's County of Residence
210	2	Father's State of Birth
211	14	Certifier's License Number
212	6	Laboratory ID Number
213	4	Mother Xfer Code
214	3	Mother Xfer County Code
215	4	Baby Xfer Code
216	3	Baby Xfer County Code
217	4	Year of Birth
218	7	Certificate #
219	1	Unique Code
220	8	File Date
221	2	Community Area
222	4	Census Tract
223	2	Century of Last Live Birth
224	2	Century of Last Termination
225	2	Century of Last Menses
226	2	Century of Blood Test

Source: Genesis Systems (1999).

Tables 5, 6, 7, and 8 show the dramatic increase in fields being collected today in reporting a live birth. This particular schema is from the state of Illinois, but it is typical of most states, even though exact specifications differ from state to state. In the state of Washington, as in most states, the filings of live births and fetal deaths are completed by the hospital or birth attendant and then forwarded to the state department of public health (Washington State Department of Public Health 1999). Electronic birth certificate systems, which are computer programs that facilitate the filing, have enabled this dramatic increase in data collection. These systems, which were implemented around 1984, are currently used by more than 32 state departments of public health and 2,000 hospital and birthing centers. They account for more than 50 percent of all births in the United States (Genesis Systems, Inc. 1999). The information in database systems is not entered by typing the values in from scratch. Instead, much of the information is entered by selecting a value from a limited list of options, and other values may be directly transferred from the hospital's information system, reducing both data entry time and errors. An electronic birth certificate system usually creates a 'hard copy' birth certificate as well as a computer file of live births. The hard copy serves as an original birth certificate. Until recently, the computer file was usually sent to the state's central repository by diskette or by direct modem transmission. Now it is done via the Internet.

The additional fields of information included in the reporting of live births have contributed to new reports on birth characteristics, infant and maternal mortality, birth weight and gestational age, and adequacy of prenatal care (Massachusetts Department of Public Health, Bureau of Health Statistics, Research and Evaluation 1999). Without discounting the usefulness of such information, it is important to note that the growth in the volume of information being collected has been tremendous: from about 7 fields of information in 1906, to around 15 fields from 1925 to 1980, to more than 200 by 1999. Clearly, technology has provided the means to make these collections practical.

3. Growth in Healthcare Cost Data

The Illinois Health Care Cost Containment Council (IHCCCC) did not collect healthcare cost data in 1983, but today a record of each person's hospital visit is recorded. IHCCCC reports more than 97 percent compliance by Illinois hospitals in providing the information (State of Illinois Health Care Cost Containment Council 1998). Table 9 contains a sample of the kinds of fields of information that are not only collected but also disseminated. The fields shown in Table 9 are provided to researchers needing detailed patient-specific data to further knowledge; the data are considered useful for measuring access, quality, and outcomes.

Table 9. Illinois Health Care Cost Containment Council Research Health Data

Field#	Field description	Size	Field#	Field description	Size
1	HOSPITAL ID NUMBER	12	26	MDC CODE	2
2	PATIENT DATE OF BIRTH (MMDDYYYY)	8	27	TOTAL CHARGES	9
3	SEX	1	28	ROOM AND BOARD CHARGES	9
4	ADMIT DATE (MMDYYYY)	8	29	ANCILLARY CHARGES	9
5	DISCHARGE DATE (MMDDYYYY)	8	30	ANESTHESIOLOGY CHARGES	9
6	ADMIT SOURCE	1	31	PHARMACY CHARGES	9
7	ADMIT TYPE	1	32	RADIOLOGY CHARGES	9
8	LENGTH OF STAY (DAYS)	4	33	CLINICAL LAB CHARGES	9
9	PATIENT STATUS	2	34	LABOR-DELIVERY CHARGES	9
10	PRINCIPAL DIAGNOSIS CODE	6	35	OPERATING ROOM CHARGES	9
11	SECONDARY DIAGNOSIS CODE	16	36	ONCOLOGY CHARGES	9
12	SECONDARY DIAGNOSIS CODE	26	37	OTHER CHARGES	9
13	SECONDARY DIAGNOSIS CODE	36	38	NEWBORN INDICATOR	1
14	SECONDARY DIAGNOSIS CODE	46	39	PAYER ID	19
15	SECONDARY DIAGNOSIS CODE	56	40	TYPE CODE	11
16	SECONDARY DIAGNOSIS CODE	66	41	PAYER ID	29
17	SECONDARY DIAGNOSIS CODE	76	42	TYPE CODE	21
18	SECONDARY DIAGNOSIS CODE	86	43	PAYER ID	39
19	PRINCIPAL PROCEDURE CODE	7	44	TYPE CODE	31
20	SECONDARY PROCEDURE CODE	17	45	PATIENT ZIP CODE	5
21	SECONDARY PROCEDURE CODE	27	46	Patient Origin COUNTY	3
22	SECONDARY PROCEDURE CODE	37	47	Patient Origin PLANNING AREA	3
23	SECONDARY PROCEDURE CODE	47	48	Patient Origin HSA	2
24	SECONDARY PROCEDURE CODE	57	49	PATIENT CONTROL NUMBER	
25	DRG CODE	3	50	HOSPITAL HSA	2

The National Association of Health Data Organizations reported that 37 of the 50 states (or 74 percent) had legislative mandates to gather information from patient records (National Association of Health Data Organizations 1996) similar to that described in Table 9. In addition, by 1996, 17 states reported they had started collecting ambulatory care (outpatient) data from hospitals, physician offices, clinics, and so on. These data collection programs began in an effort to help reduce healthcare costs. Many states have subsequently distributed copies to researchers, sold copies to industry, and made versions available to the public. While there are many possible sources of patient-specific data, these represent a class of data collections that were not available before 1983.

4. Growth in Supermarket Transaction Data

Private sector information about individuals has expanded also. For example, in 1983 supermarket transactions consisted only of summary price information and were not linked to individuals. Today in many supermarkets in Illinois, the complete list of purchased items is often stored along with the identity of the consumer. This increase in the volume of data collected about individuals from supermarket purchases is the topic of discussion in this section.

As shown in Table 2, when a consumer purchased items from a supermarket in 1983, the only recorded information left behind could be roughly described as an inventory debit and a record of the total amount purchased and the amount of tax paid. There was no record of the identity of the consumer or of the consumer's personal habits and behaviors such as goods typically purchased and the times and days the consumer had shopped. Analyses of consumer behaviors were based on aggregated sales, so it was nearly impossible to identify multiple purchases from the same customer. Inferred person-level shopping patterns drawn from the data therefore included uncertainty.

This has all changed with today's computer technology, which makes it possible to reduce the uncertainty dramatically. Consumer transactions can be stored and analyzed, so that information about each consumer's lifestyle, behavior, beliefs, and habits can usually be revealed. Watching an individual consumer's purchases week after week provides clues about demands on the consumer's time, economic status, life experiences, and lifestyle. The rest of this section discusses the evolution of this transformation.

Catalina Marketing's Data Collection

In March 1996, Catalina Marketing, Inc., began a data collection enterprise that now stores the shopping patterns of an estimated 143 million shoppers each week from more than 11,000 supermarkets nationwide ('Get to Know Your Customer' 1998). By July 1998, the 2-terabyte database had reportedly 18 billion rows of data. Catalina's objective was to use technology to improve its ability to measure consumer behavior.

Retailers obtained chain-specific information from Catalina over the World Wide Web. After they entered a username and password, retailers queried the data from a browser and the results were displayed in HTML or e-mailed back in a comma-delimited format for easy use in spreadsheet programs. Such reports were useful in developing time-of-day specials, designing direct mail campaigns, and assessing traffic flow through checkout lanes ('Get to Know Your Customer' 1998). However, in these early years, results were typically compiled from transactions in which the consumer was not only anonymous but also independent across multiple transactions. That is, multiple purchases attributable to the same consumer were not recognizable as such in the data.

Supermarket Loyalty Cards

As Catalina Marketing began its data collection, retailers became aware that with the help of personally identifying cards and state-of-the-art database technology, retailers such as supermarkets could analyze millions of transactions quickly to identify their best customers and build loyalty through special rewards such as discounted prices. These are called loyalty programs, and the accompanying card is termed a loyalty card.

This is how a loyalty program usually works. When a consumer applies for a loyalty card, information is collected. Table 10 shows examples of the kinds of fields of information requested on applications for loyalty cards from some major supermarket chain stores in the Washington, D.C., California, and Massachusetts areas in 1998. All consumers are normally eligible (with no restrictions on teenagers, for example, who may purchase condoms and other sensitive items), and almost always they are guaranteed acceptance by merely completing the application. The information provided on the application is not customarily reviewed for accuracy. Some programs use only one loyalty card per household, but other programs seek to have each person in each household have his or her own card. When the consumer reaches the checkout counter, the cashier asks for a loyalty card. If the consumer has a loyalty card, the card is scanned or the identifying number found on the card is manually entered into the computerized register. Purchased items are then scanned as usual, with savings deducted automatically. In most loyalty card programs, the final receipt includes an itemized list of the savings that resulted from using the card. Consumers who have no loyalty card are charged the higher, nondiscounted prices.

Table 10. Typical Set of Fields on Loyalty Card Applications

Field name	Food Lion	Fresh Fields	Safeway	Star Market
Name	yes	yes	yes	yes
Home street address	yes	yes	yes	yes
Home city	yes	yes	yes	yes
Home state	yes	yes	yes	yes
Home ZIP	yes	yes	yes	yes
Home phone number	yes	yes	yes	yes
Social Security number				yes
Additional data sometimes requested				
Birth date			yes	yes
ZIP code of work place		yes		
Other stores where you shop	yes	yes		
Number of people in household	yes	yes		
Age of each person in household	yes	yes		
How much do you spend each week	yes	yes		
Additional data for accepting checks				
Bank			yes	yes
Bank account number			yes	yes

Source: Loyalty card applications of the grocery stores listed.

Here are examples of price differences based on the use of a loyalty cards in the Georgetown Safeway in Washington, D.C., on or about December 1998 ('Bargains at a Price' 1998). A bag of bagels was $1.59 with no loyalty card and $0.99 with a loyalty card. A frozen gourmet dinner was $4.19 with no loyalty card and $2.99

with a loyalty card. While these savings are dramatic, overall reported savings tend to be 20 to 40 percent.

Idaho-based Albertson's, which operates in 37 states, did not have a loyalty program in 1997. In a press release, company officials implied that discounted prices based on loyalty cards at other supermarkets might be comparable to its everyday low prices or promotional prices that did not require a card. Albertson's implication was that supermarkets that offer loyalty cards impose an inflated price (or penalty) on consumers who do not use their loyalty card.

Ideally, with a loyalty program, retailers could become better at serving individual consumers because discounts and product selections could become specific to each consumer. But this new approach tugs at two historical traditions. The first is the issue of privacy with respect to collecting so much consumer-specific data. The second concerns moving away from charging everyone the same price for the same product to charging a price based on the retailer's perceived value of the consumer. This section is limited to a discussion of the data collection itself.

When supermarket chains combine the kind of detailed transactions data kept by Catalina Marketing with the demographic information gathered from their own loyalty-card programs, they can create highly targeted marketing campaigns and cultivate better relationships with customers. In fact, targeted, scanner-based campaigns seem to have a redemption rate double that of direct mail ('Get to Know Your Customer' 1998).

As of December 1998, the Food Marketing Institute reported that 6 of 10 supermarket companies either already collected customer transaction data electronically or planned to do so soon, compared with 3 out of 10 in 1993 ('Bargains at a Price' 1998). The implication of this trend is that more than half the supermarkets in the United States could soon require consumers to be cardholders in order to receive discounted prices.

5. Opting Out

So far in this chapter, I have provided examples of data collection activities that have expanded as the supporting technology became readily available. The data collections mentioned are representative of many other data collections under way, as discussed in the next section. In addition, most privacy discussions consider whether the individual has the ability to decide not to participate in the data collection—to 'opt out' of it.

The collections of birth and health data described earlier provide no such possibility, being dictated by regulation and law and intended to capture the entire population. In contrast, a consumer can decide to pay a higher price and not use a loyalty card at the supermarket; or, alternatively, a consumer can provide false information when applying for such a card. But these actions are not wholly satisfactory. The first option requires consumers who want to opt out to pay (by means of higher prices) for privacy that they historically enjoyed for free; the second option encourages deliberately deceptive action by otherwise responsible consumers. Individuals may risk unknown future uses of their purchasing patterns that could become personally damaging. The primary beneficiaries of loyalty programs appear to be manufacturers and retailers, who can better avoid waste in promoting products. Newer technology for privacy protection integrated with effective policy can offer better solutions in these kinds of situations, giving individuals who want privacy protection an alternative to opting out.

6. Behaviors in Data Collecting Today

The previous sections give examples of recent behavioral tendencies in the collection of person-specific data. These informally observed 'trends' are enumerated below and then discussed further.

- **Behavior 1.** Given an existing person-specific data collection, expand the number of fields being collected. I casually refer to this as the 'collect more' trend.
- **Behavior 2.** Replace an existing aggregate data collection with a person-specific one. I casually refer to this as the 'collect specifically' trend.
- **Behavior 3.** Given a question or problem to solve, or merely provided the opportunity, gather information by starting a new person-specific data collection related to the question, problem, or opportunity. I casually refer to this as the 'collect it if you can' trend.

All three tendencies result in more and more information being collected on individuals. Discussions of each behavior follow.

Behavior 1: Collect More

The increase in the number of fields collected as part of the birth certificate is an example of this behavior. In fact, I found increases in many other older established person-specific collections. Table 11 contains an overview in which 13 of 21 (or 62 percent) data collections that have historically collected person-specific data expanded the number of fields being collected between 1983 and 1996.

Table 11. Expansion in Some Historic Person-Specific Collections 1983–96

Old Collections	1983	1996
bank account	•	•
birth certificate	•	À
census survey	•	À
credit card	•	À
credit history	•	À
driver license	•	À
legal actions	•	À
medical record	•	À
marriage license	•	À
military service	•	•
motor vehicle registration	•	•
phone calls	•	•
professional license	•	À
property (and tax) records	•	•
public assistance	•	À
real estate	•	•
recreational license	•	À
selective service	•	•
tax filings	•	À
voting list	•	•
worker's compensation	•	À
Percentage that increased	•	62%

Note: Census survey refers to local census information.
Source: Survey by author.

Behavior 2: Collect Specifically

In places where tabular statistics were once the form of reporting or sampling the method of collection, person-specific data collection is becoming the new standard. While use of survey data remains common, there are strong incentives to use data based on the full population. The discussion on supermarket transactions illustrates how decisions that once relied on aggregate statistical data now use highly detailed, person-specific data.

Another example is student-specific educational data. By 1996 in some states, recordings of student-specific data for kindergarten through 12th grade had begun. These collections typically include days absent, number of school lunches consumed, immunizations, allergies, and so forth for each student. In 1983, this information was provided in aggregate student-body summaries at the school level. Under the new collection practice, this information is specific to the student and shared in that form. These examples demonstrate the growing number of new entity-specific data collections.

Behavior 3: Collect It if You Can

The discussion on healthcare cost data collected by states was representative of new person-specific data collections that have recently begun or are being initiated, but that did not exist in 1983. There are many other examples as well, including the National Directory of New Hires and Immunization registries.

National Directory of New Hires. The General Accounting Office (GAO) found in a 1994 report that nonpayment of child support contributed to childhood poverty as well as to increases in the numbers of families receiving welfare (General Accounting Office 1994). The GAO said more than one-fifth of America's children lived in poverty in 1994, and it estimated that half of those children would live in single-parent families at some point in their lives. To help obtain the financial support that parents owe their children and to reduce welfare costs, Congress passed the Personal Responsibility and Work Opportunity Reconciliation Act of 1996. This act mandated the establishment of new resources at the federal and state levels to assist state child support enforcement agencies and included provisions for the establishment of a National Directory of New Hires as well as state directories of new hires.

The goal of these newly created worker-specific data collections, as at their inception in 1996, is to monitor all individuals with jobs in order to track down parents who owe child support. That is, these collections do not merely contain information about Americans found to be delinquent parents, they include information on virtually all working Americans, the vast majority of whom have not been accused of being delinquent parents.

Employers must file timely reports on every person they hire and, quarterly, the wages of every worker (U.S. Department of Health and Human Services 1997). Table 12 contains a list of the fields employers are required to report. In addition, states must regularly report all people seeking unemployment benefits and all child-support cases, even if the parents and children involved do not receive public assistance or ask for help in collecting support.

Table 12. Fields Collected in National Directory of New Hires

Field name	Reported when newly hired	Updated quarterly on all employees
Employee name	yes	yes
Employee SSN	yes	yes
Employee address: street	yes	
Employee address: city	yes	
Employee address: state	yes	
Employee address: ZIP	yes	
Employer name	yes	yes
Employer address: street	yes	yes
Employer address: city	yes	yes
Employer address: state	yes	yes
Employer address: ZIP	yes	yes
Federal employer identification number (FEIN)	yes	yes
Employee wage amount	yes	yes
Reporting period	yes	yes
Additional Fields States Can Require Be Reported		
Employee date of birth	may be required	
Employee date of hire	may be required	
Employee state of hire	may be required	

Source: U.S. Department of Health and Human Services, Office of Child Support Enforcement.

As of 1997, more than 7.4 million delinquent parents owed more than $43 billion in past child support. These registries are credited with increasing payments from $12 billion to $14.4 billion in 1998 and helping to locate more than 1.2 million delinquent parents ('Uncle Sam Has All Your Numbers' 1999). As of June 1999, the registry had information on individuals in almost 12 million families involved in child support cases.

Immunization Registries. Children can be immunized against most serious childhood diseases at little or no cost to parents. The problem is that many of these diseases appear so rarely nowadays that parents conclude the immunizations are not necessary; instead they are concerned about possible side effects and the increased number of immunizations required. Absence of immunizations can make society vulnerable to disease outbreaks that can present serious health risks.

For example, a measles epidemic broke out among Rutgers University students in the spring of 1994. Measles is a deadly viral disease, highly contagious and airborne. Students entering a university should have had two measles inoculations, but many have had only one ('Students Need Two Measles Shots' 1997). According to the Centers for Disease Control and Prevention (CDC), 18,000 cases were reported in 1989 and 27,000 in 1990.

To encourage parents to have their children immunized at a proper age, state and national registries were created. The objective is to maintain a record of each child's immunization history so that when a child appears at a clinic or a physician's office, the registry can be consulted and the proper immunizations can be administered. Table 13 shows the fields that, by legislation, must be collected in Texas and serves as an example of the kinds of fields maintained in immunization registries.

Table 13. Fields Collected in Immunization Registry in Texas

Field name
CHILD INFORMATION
Child's name (first, middle, last)
Child's address: street
Child's address: city
Child's address: state
Child's address: ZIP
Child's Social Security number (if available)
Child's gender
Child's date of birth
Mother's maiden name
HEALTHCARE PROVIDER'S INFORMATION
Healthcare provider's name (first, middle, last)
Healthcare provider's business address: street
Healthcare provider's business address: city
Healthcare provider's business address: state
Healthcare provider's telephone
Healthcare provider's business address: ZIP
VACCINE INFORMATION
Date vaccine was administered
Vaccine lot number (if known)
Dose or series number (if known)
Name of vaccine manufacturer (if known)

Source: Texas Immunization Registry.

Child-specific immunization registries began around 1997, and as of March 2000, 21 of the 50 states (or 42 percent) had a law authorizing the creation of an immunization registry. The specifics vary from state to state, but records from the electronic birth certificate database often seed immunization registries with new records. Copies of the information are forwarded from the state to the national database maintained by CDC, and in some cases—Texas is one—copies are made available to the local public health department, the child's physician, the school in which the child is enrolled, and the childcare facility in which the child is enrolled ('Chapter 100 Immunization Registry' 1998).

Other New Collections. Not all new collections result from legislative mandate. There are many in the private sector as well, and of course, technology makes more and more collections possible.

Using the World Wide Web opens many new sources of person-specific collections. On-line companies can track what customers look at, where they go, and how long they linger on a particular page. One of the leaders is Amazon.com, the on-line bookseller. It tracks the purchases of more than 4.5 million customers and then offers them suggestions about what they might enjoy, based on personal reading patterns correlated by the computer. Another company, DoubleClick, Inc., links together traces of locations where customers and potential customers have been to construct their browsing profiles.

New technology keeps emerging that fuels demand for more collections. For example, Visionics, Inc., has a product that, given a collection of identified faces, such as those available from driver's license photos or earlier surveillance photos, can automatically locate faces in complex scenes, to track and identify individuals by matching faces to those stored in the collection. At casinos in Las Vegas, gaming investigators have used the technology to identify, in real time, known casino cheaters, card counters, and others. Similar technology is deployed on some streets in England.

Another example is global positioning systems installed in vehicles to record where the vehicle has been, and when and how long it was there. A version of the system is being tested by an insurance company in Texas to help set car insurance premiums based on actual travel patterns.

There is no doubt that a tremendous amount of information is already being collected on individuals, collections are expanding, and new ones are being created at an accelerating rate. Technology is the catalyst; thus, the behavior is expected to continue.

7. Disk Storage per Person

In an attempt to characterize the growth in person-specific information, I introduce a new metric termed disk storage per person (DSP), which is measured in megabytes per person (where megabytes is 10^6 bytes, abbreviated MB). Global DSP (GDSP) is the total hard disk drive space (in MB) of new units sold in a year divided by the world population in that year. I do not consider removable or temporary storage mediums. I do not include used and refurbished drives if they were first purchased in previous years even though they may be operational during the evaluation year. And I recognize that disk storage is used for more than person-specific data. So the GDSP metric is a crude measure of how much disk storage could possibly be used to collect person-specific data on the world population. This value appears to be increasing dramatically with time.

In 1983, one million fixed drives with an estimated total of 90 terabytes (10^{12} bytes written TB) were sold worldwide; 30MB drives had the largest market segment ('Disk/Trend Report 1983' 1983). In 1996, 105 million drives, totaling 160,623 terabytes, were sold, with 1 and 2 gigabyte (10^9 bytes, abbreviated GB) drives leading the industry ('Rigid Disk Drive Sales' 1997). By 2000, with 20GB drives leading the industry, rigid drives sold for the year were projected to total 2,829,288 terabytes ('Rigid Disk Drive Sales' 1997). A summary of these storage values appears in the top row of the table in Figure 1.

The world population was 4.5 billion (10^9) in 1983, and roughly 5.8 billion in 1996, and it was expected to be 6 billion by 2000[1]. Therefore, in 1985, there were 20,000 bytes per person ($GDSP_{1985} = 0.02$); in 1996, there were about 28 MB/person ($GDSP_{1996} = 28$); and by 2000, there may have been 472 MB/person ($GDSP_{2000} = 472$). These values are summarized in Figure 1.

GDSP values signal the amount of storage possibly available to record all the events for each person throughout the year. Here is an analogy that helps explain the storage space implied by these values. In 1985, $GDSP_{1985} = 0.02$ is similar to reserving a small file that could reside on a diskette. In 1996, which is the knee of the curve in Figure 1, $GDSP_{1996} = 28$ makes roughly a 30MB hard drive (or 20 diskettes) available to store information. By 2000, with $GDSP_{2000} = 472$, most of a CD (or 338 diskettes) could be used to record information collected on a single person. Clearly, the amount of storage space potentially available to store information on each person is growing rapidly.

I attempt to estimate how much of an adult's life could be documented on a single piece of letter-size paper (8.5 × 11 inches) and then stored in computers as the technology progressed from 1983 to 2000. Assume a printed page of text contains 54 lines by 60 characters; this information can be stored in 3,240 bytes with no compression. I can use these GDSP figures to compute the amount of a person's time that can be documented on such a page. In 1983, a page could be used to document two months of a person's life. Actual recordings in 1983 did include itemized long distance phone calls, credit card purchases, the volume of electricity used, and so forth. In 1996, a page could be used to document each hour of a person's life. Recordings in 1996 did expand in both size and number. Examples of new collections included items purchased at the grocery store, websites visited,

[1] Data obtained using FAOSTAT, a database produced by the Food and Agriculture Organization (FAO) of the United Nations. Rome, Italy: FAO, 1997.

Figure 1. Characterizing Computer Storage Available for Recording Person-Specific Information

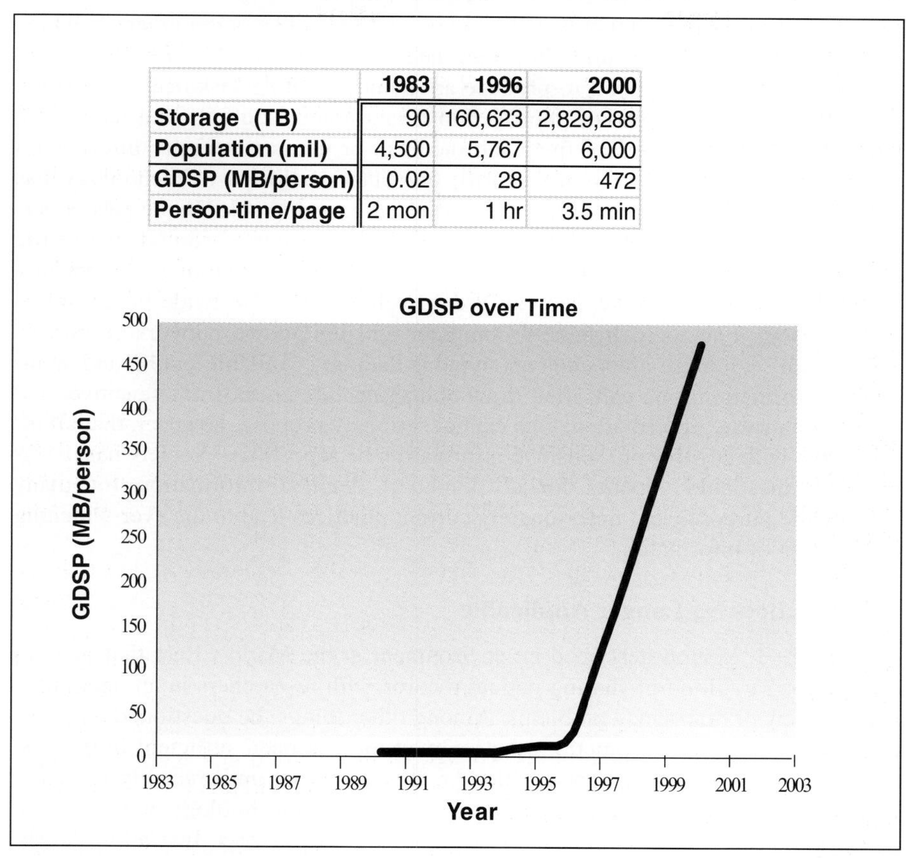

	1983	1996	2000
Storage (TB)	90	160,623	2,829,288
Population (mil)	4,500	5,767	6,000
GDSP (MB/person)	0.02	28	472
Person-time/page	2 mon	1 hr	3.5 min

and in some locations the date and time a car proceeded through a tollbooth. By 2000, with 20 gigabyte drives leading the industry, it is projected that a page could be used to document every 3.5 minutes of a person's life. Collections are expanding to include visual data, surveillance information, and genetic and biometric information such as heart rate, pulse, and temperature. These values are included in the last row of the chart on Figure 1. So GDSP provides a way of characterizing how much space could be used to record daily events for each person in the world population using only new rigid disk drives sold that year.

8. Discussion

In summary, there is no doubt that we are moving toward an environment in which society could collect and store almost all data on all persons. As a result, it is becoming increasingly difficult to produce anonymous and declassified information in today's globally networked society. Most data holders do not even realize the jeopardy at which they place financial, medical, or national security information when they erroneously rely on security practices of the past. Technology has eroded previous protections, leaving the information vulnerable. In the past, a person seeking to reconstruct private information was limited to visiting disparate file rooms and engaging in labor-intensive review of printed material in geographically distributed locations. Today, one can access voluminous worldwide public information using a standard handheld computer and ubiquitous network resources. Thus, from seemingly innocuous anonymous data and available public and semi-public information, one can often draw damaging inferences from sensitive, and heretofore private, information. One cannot seriously propose, however, that all information with any links to sensitive information be suppressed. Society has developed an insatiable appetite for all kinds of detailed information for many worthwhile purposes, but unfortunately current practices lead to the ever-widening distribution of information.

Past Practices No Longer Applicable

In 1997, L. J. Melton described an environment at the Mayo Clinic that had enjoyed a long tradition of sharing patient records with researchers in an open manner with few or no privacy problems. Among other things, he questioned why old, established data-sharing practices that seemed to have been sufficient in the past were no longer considered acceptable. An answer is that until recently there existed natural limits that protected patient privacy, which technology now erodes at an alarming rate. It was not old practices that protected privacy. Instead, it the absence of current technology that provided the protection.

For example, in an earlier time, if one wanted to receive research information from the Mayo Clinic's records, one would have to take time off from work, take a plane to Rochester, Minnesota, and then have access to the files only during the times in which the records room was open. One could leave with only the information one could write down during that time (assuming the absence of copiers). The physical labor involved in manually reviewing records, coupled with the physical restrictions on entering the records room, provided economic boundaries that restricted the dissemination of person-specific data. Now consider what would be involved today if all of the records of the Mayo Clinic were available electronically, assuming the same access practices. One could access all of this information from anywhere in the world using a standard handheld computer and ubiquitous network resources. One could have an exact copy in a matter of seconds and could further distribute it widely to others, around the world, in a matter of minutes. Today's technology does pose unparalleled threats to patient privacy.

Today's technology also makes access to the information easier within the Mayo Clinic itself. As a result, more data tend to be shared internally than ever before.

Data Sharing and Risk

Many details about our lives are documented on computers, and when this information is linked together, the resulting profiles can be as identifying as fingerprints, even when the information contains no explicit identifiers such as name and address. The increase in the availability of detailed data, as well as inexpensive technology to process it, is having a dramatic impact on research. Having more information available will probably lead to even more studies because additional data can often help ensure the validity and generalizability of study results. These studies will likely result in continued increases in data collected and shared. Therefore, the time is right to seriously examine data collection and sharing practices. Most person-specific data are autonomously controlled, and much of the information is replicated across collections. Coherent and comprehensive approaches are needed.

Risk and Liability

U.S. citizens are largely unaware of the loss of privacy and the resulting ramifications that stem from having so much person-specific information available. Clearly a loss of dignity and income can result when personal medical information is widely and publicly distributed. Yet data holders make data-sharing decisions to benefit themselves and minimize their own risk. Doing so does not always provide desirable protections for the persons whose information is contained in these databases.

The idea of 'risk' concerns the likelihood of experiencing loss or damage. Risk becomes an obligation the data holder has to the subjects whose information is contained within the data, and to society. So both the data holder and the subjects want no harm to result from the sharing of data, but from the data holder's perspective, harm appears as legal liability. As a result, actions the data holder may take to 'protect' data may not be the same as actions that would 'protect' the identities of the subjects. Instead, such actions are aimed at limiting the data holder's liability, regardless of their efficiency in protecting subjects. Such self-serving actions include (1) making it difficult to identify the data holder as the source of shared information; (2) making it difficult for society to know what is collected and to whom copies are given; and (3) making it legally difficult for a recipient of the data to publicly admit to being able to identify subjects in data that the data holder asserts are anonymous. These kinds of actions help protect the data holder, but they do not protect the identities of the subjects. Protections for subjects are limited almost entirely to those that can be made available through policies, regulations, and laws. Therefore, it is essential that measurements of risk and characterizations of access policies be based on society's perspective.

Tension Between Privacy and Secondary Uses of Data

The following is an empirically proven claim: Data collected for one reason tends to get used for another. This informal claim gives rise to additional concerns over privacy, because decisions that led to the inclusion of information in the primary data collection typically did not also consider secondary uses of the data. This happens because the demand from secondary uses typically appears after the data are collected. Even in cases where there appears to have been meaningful discussion of secondary uses beforehand, such as when a consent form is used, care is not always taken to ensure that the resulting decision was not coerced or made with little or no understanding of the ramifications. In the next few paragraphs, I look at different ways in which society has made decisions about secondary uses of data, and I provide a way to reason about these findings.

Quality Versus Anonymity. There is a natural tension between the quality of data and the techniques that provide anonymity protection. Consider a continuum that characterizes possible data releases. At one end of the continuum are person-specific data that are fully identified. At the other end are anonymous data that are derived from the original person-specific data, but in which no person can be identified. Between these two endpoints is a finite partial ordering of data releases, where each release is derived from the original data but for which privacy protection is less than fully anonymous (see Figure 2).

The first realization is that any attempt to provide some anonymity protection, no matter how minimal, involves modifying the data and thereby distorting its contents. So, as Figure 2 shows, movement along the continuum from the fully identified data toward the anonymous data adds more privacy protection but renders the resulting data less useful. That is, some tasks exist for which the original data could be used, but those tasks are not possible with the released data because the data have been distorted.

So the original fully identified data and the derived anonymous data are diametrically opposed. The entire continuum describes the domain of possible releases. Framed in this way, a goal of this work is to produce an optimal release of data so that for a given task, the data remain practically useful yet render minimal invasion of privacy.

Figure 2. Optimal Releases of Data

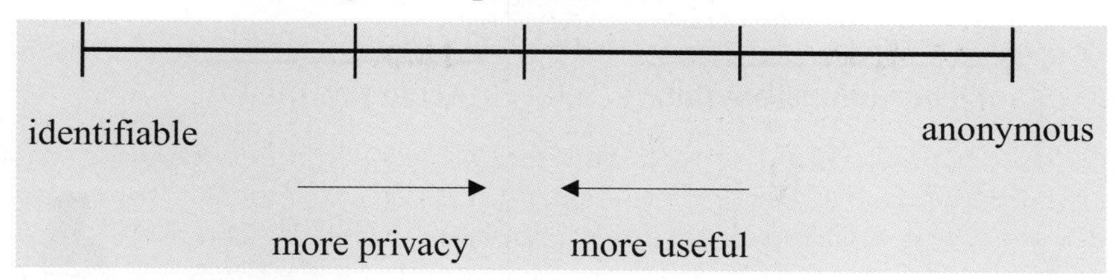

Tug-of-War Between Data Holders and Recipients. The second realization that emerges from Figure 2 is that the usefulness of data is determined by the task to which the recipient puts the data. That is, given a particular task, there exists a point on the continuum in Figure 2 that is as close to anonymous as possible, yet the data remain useful for the task. A release of data associated with that point on the continuum is considered optimal. In the next paragraphs, I provide a skeletal depiction of current practices that determine who gets access to what data. I show that the result can be characterized as a tug-of-war between data holders and data recipients.

In general, the practices of data holders and related policies do not examine tasks in a vacuum. Instead, the combination of task and recipient together are weighed against privacy concerns. This situation can be modeled as a tug-of-war between the data holder and societal expectations for privacy on one side, and the recipient and the recipient's use for the data on the other. In some cases, such as public health legislation, the recipient's need for the data may overshadow privacy protections, allowing the recipient (a public health agent) to get the original, fully identified health data. A tug-of-war is modeled in Figure 3. The privacy constraints on the data holder versus the recipient's demand for the data are graphically depicted by the sizes of the images shown. In the case illustrated, the recipient receives the original, fully identified data.

Figure 3. Recipient's Needs Overpowering Privacy Concerns

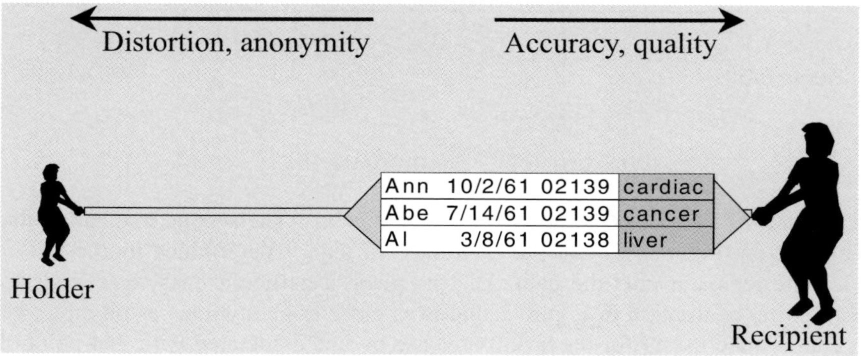

Distortion, anonymity
Accuracy, quality
Ann 10/2/61 02139 cardiac
Abe 7/14/61 02139 cancer
Al 3/8/61 02138 liver
Holder
Recipient

Figure 4 demonstrates the opposite outcome to that of Figure 3. In Figure 4, the data holder and the need to protect the confidentiality or privacy of the information overshadow the recipient and the recipient's use for the data, so the data are completely suppressed and not released at all. Data collected and associated with national security concerns provide an example. The recipient may be a news-reporting agent. Over time, the data may eventually be declassified and a release that is deemed sufficiently anonymous provided to the press, but the original result is as shown in Figure 4, in which no data are released at all.

Figure 4. Data Holder and Privacy Concerns Overpowering Outside Uses of the Data

Distortion, anonymity
Accuracy, quality
Holder
Recipient

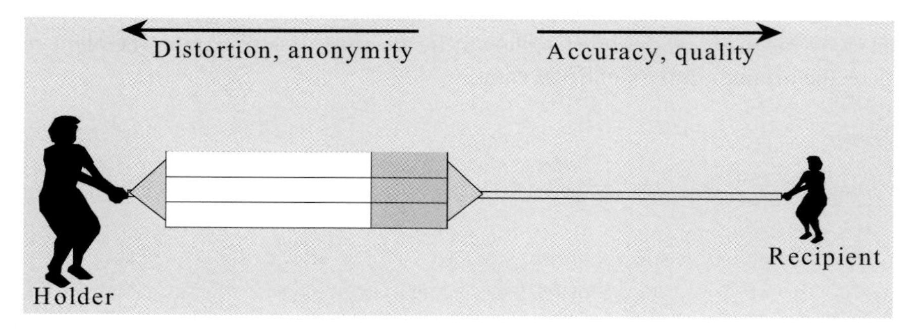

Figure 5. Need for an Optimal Balance Between Privacy Concerns and Uses of the Data

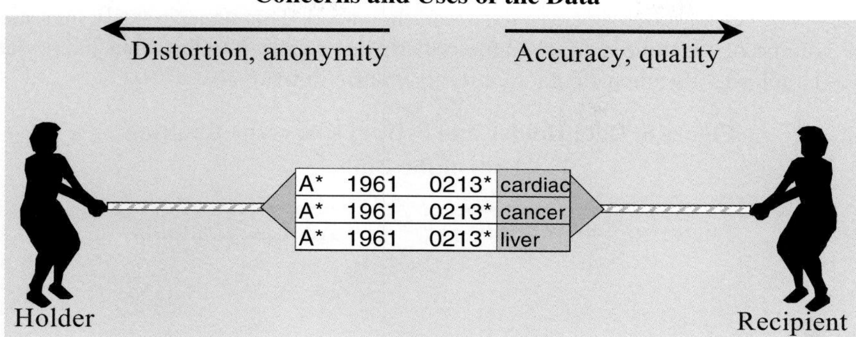

Figures 3 and 4 depict situations in which society has made explicit decisions based on the needs of society as a whole. But secondary uses of medical data—for example, by marketing firms, pharmaceutical companies, epidemiological researchers, and others—do not in general lend themselves to such an explicit itemization. Figure 5 depicts situations in which the needs for privacy are weighed equally against the demand for the data itself. In such situations, a balance should be found in which the data are rendered sufficiently anonymous yet remain practically useful. As an example, this situation often occurs with requests by researchers for patient-specific medical records so that they can undertake clinical outcomes research or administrative research that could possibly benefit society. In most decisions today, the recipient receives either the original patient data or no data at all. Attempts to provide something in between typically result in data with poor anonymity protection or data that are overly distorted. This work seeks to find ways for the recipient to get data that have adequate privacy protection, thereby striking an optimal balance between privacy protection and the data's fitness for a particular task.

At present, data holders often make decisions arbitrarily or by *ad hoc* means. Figure 6 portrays the situation in which some state and federal agencies find themselves when they seek to produce public use files for general use. Over the past few years, there has been a tremendous effort to make more data that government agencies collect available over the World Wide Web. In these situations, protecting the reputation of the agency, and the guarantees for privacy protection by which some agencies are legally bound, outweighs the demands of the recipient. In these cases, a strongly distorted version of the data is often released; the released data are typically produced with little or no consideration of the tasks required. Conversely, some state and federal agencies release poorly protected data. In these cases, the individuals contained in the data can be easily reidentified.

Neither way of releasing data yields optimal results. When strongly distorted data are released, many researchers cannot use the data, or have to seek special permission to get far more sensitive data than they need. This unnecessarily increases the volume of sensitive data available outside the agency. On the other hand, data that do not provide adequate anonymity may harm individuals.

Figure 6. Data Holder and Privacy Concerns Limiting Uses of the Data

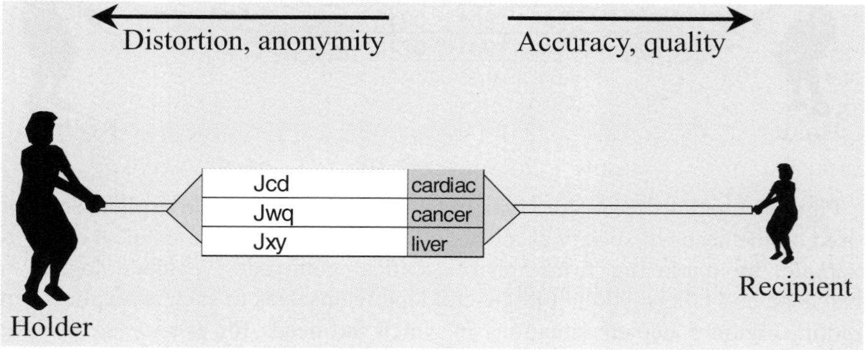

In examining the different struggles between privacy and secondary uses of person-specific data, I make the following informal claims:

- Current policies and practices support crude decisions. A recipient today tends to receive the sensitive data, no data at all, overly distorted data that are of little or no use, or poorly protected data in which individuals can be reidentified. Emerging technology provides many more options.
- Ultimately, the data holder must be held responsible for enforcing privacy protection because the data holder typically reaps a benefit and controls both data collection and dissemination.

While the informal claims above are independent of the content of data, the study of secondary uses of medical data in particular provides a natural incentive to find optimal solutions between researchers and data holders. After all, there are no legislative guidelines to empower one party so that it can overwhelm the other, as was shown in Figures 3 and 4. Also, state and federal agencies tend to be small in number and highly visible in comparison with the dramatic number of holders of medical data. Because there are so many data holders, it is hard to scrutinize their actions, and the resulting damage to individuals can be devastating, yet hard to prove. Strong financial incentives exist not to provide adequate protection in medical data. On the other hand, research from data may lower healthcare costs or save lives. These reasons provide motivation for finding optimal releases of data and for integrating technology with policy for maximal benefit.

A Harris-Equifax poll (Louis Harris and Associates 1994) implies that the public would be willing to share information for research provided researchers and others could not identify any person included in the released data. Rendering data sufficiently anonymous would allow data to be shared more freely.

9. Future Work

Historically, many person-specific pieces of data were limited to few collections in isolated locations (*e.g.,* birth certificate information). Today, however, there are many possible sources of such information independent of the original collecting organization (*e.g.,* health data). Inferences about an individual's health can be found in prescription data, a log of visitors to websites offering information about particular diseases, mailing lists, warranty data, and so forth. Similarly, information about an individual's birth or criminal record can be inferred from different kinds of data, each having different quality issues.

A number of proposed projects of varying degrees of difficulty and skill requirements could extend this work:

1) Track the growth and expansion of common person-specific data collections. While an example provided was the birth certificate data, that of similar expansions in collection and sharing exist in other collections.

2) Document the growth as well as the rationale behind the growth and make predictions about future collection and sharing of the information.

3) Select a piece of information and document how many different ways that information could be inferred through public and semipublic sources, particularly commenting on the quality of the information provided from different sources.

References

'Bargains at a Price: Shopper's Privacy' (1998) *Washington Post*, December 31, p.A01.

'Chapter 100 Immunization Registry' (1998) *Texas Register*, (23)50, December 11.

'Disk/Trend Report 1983'(1983) *Computer Week*, (46) November 11.

Genesis Systems, Inc. (1999) History of Genesis Systems and Electronic Birth Certificate, (Accessed August 14, 2001) http://www.genesisinfo.com/website-v1/prodebc.html and http://www.genesisinfo.com/website-v1/history.html.

General Accounting Office, Income Security Issues (1994) *Child Support Enforcement: Families Could Benefit From Stronger Enforcement Program*. Washington, D.C.: General Accounting Office (HEHS-95-24) December.

'Get to Know Your Customer—Catalina Marketing Lets Retailers Find Out Who's Really Buying' (1998) *Information Week* 691, July 13.

Louis Harris and Associates (1994) *The Equifax-Harris Consumer Privacy Survey*, Atlanta: Equifax.

Massachusetts Department of Public Health, Bureau of Health Statistics, Research and Evaluation (1999) 'Advance Data: Births 1997', (Accessed August 14, 2001) http://www.state.ma.us/dph/birth97/ab97xsum.htm.

Massachusetts Department of Public Health, Registry of Vital Records and Statistics (1999) 'Massachusetts Vital Record Content', (Accessed August 14, 2001) http://www.mass-doc.com/massachusetts_vital_records_content.htm.

Melton, L.J. (1997) 'The Threat to Medical-records Research,' *New England Journal of Medicine,* 337, pp.1466-1469.

National Association of Health Data Organizations (1996) *A Guide to State-Level Ambulatory Care Data Collection Activities,* Falls Church, Va.: National Association of Health Data Organizations.

'Rigid Disk Drive Sales to Top $34 Billion in 1997'(1997) *Disk/Trend News*. Mountain View, Calif.: Disk/Trend, Inc.

State of Illinois Health Care Cost Containment Council (1998) *Data Release Overview,* Springfield, Ill: State of Illinois Health Care Cost Containment Council.

'Students Need Two Measles Shots' (1997) *Oregon Daily Emerald*, November 10.

'Uncle Sam Has All Your Numbers' (1999) *Washington Post*, June 27, p.A01.

U.S. Department of Health and Human Services, Office of Child Support Enforcement (1997) *National Directory of New Hires: NDNH Guide for Data Submission*. Washington, D.C.: U.S. Department of Health and Human Services (HHS-OCSE-1996-1012).

Washington State Department of Public Health, Washington State Vital Statistics (1999) 'Information Sources for Vital Statistics', (Accessed August 14, 2001) http://www.doh.wa.gov/EHSPHL/chs/sub1.htm.

Confidentiality, Disclosure, and Data Access: Theory and Practical Application for Statistical Agencies
Pat Doyle, Julia I. Lane, Jules J.M. Theeuwes and Laura M. Zayatz (Eds)

Chapter 4
Disclosure Risk Assessment

Mark Elliot
University of Minnesota

1. Introduction

Before a statistical agency can release personal data into the public domain, the identity of respondents must be suitably protected. This process has two key components. First, direct identifiers such as name and address must be removed, and second, the information must be checked to make sure that its level of detail is not sufficient to allow *statistical disclosure.*

Statistical disclosure occurs when information on an individual, household, or business is disclosed through release of a dataset that allows an individual's identity to become known even though direct identifiers have been removed. This kind of identification disclosure happens when identification information held by a *data intruder* is linked to data, held in a file that has been cleaned of direct identifiers, through *key* (or common) variables that allow the intruder to derive information about particular individuals that the intruder does not already know.

Accurate assessment of statistical disclosure risk is crucial in data release policy. If data are released too liberally, the risk of statistical disclosure becomes significant. Conversely, if risk assessments are too conservative, released data will be of suboptimal quality and utility.

Marsh *et al.* (1991) elucidated a key distinction in understanding disclosure risk assessment:

- pr(attempt): the probability of an attempt (to disclose) being made.
- pr(identification|attempt): the probability of such an attempt being successful if it is made.

By derivation, pr(identification) = pr(attempt) * pr(identification|attempt).

There is now a substantial body of research examining how pr(identification|attempt) should be measured. However, the complexity of the social, psychological, and political factors that affect the pr(attempt) has meant that little work has been done on the overall likelihood of a successful disclosure attempt. Such analysis is a crucial first step in any project to investigate the risk of statistical disclosure. The psychological components of an attempt (*e.g.*, goals and motives) will influence the type of attack, the strategy, and the probability of identification or disclosure given an attempt.

Section 2 of this chapter examines work on *scenarios of attack* as a means of analyzing this element of the overall disclosure risk. The concept of attack scenarios leads to Section 3, where the concept of key variable is discussed and the resulting framework for risk measurement is elaborated. Sections 4 through 6 describe three approaches to risk measurement: matching experiments, file-level risk metrics, and record-level risk metrics. Section 6 also considers how the file and record levels of risk might be unified and how intermediate levels of risk analysis might also be important. Sections 7 and 8 describe key factors that determine risk (including sample fraction, level of variable detail, and degree of data divergence). The discussion so far assumes that all data are effectively microdata, which is true in many but certainly not all respects. Section 9 discusses special issues that pertain only to tabular data as commonly released. Section 10 considers special factors related to attribute disclosure and secondary differentiation. Section 11 summarizes the chapter.

2. Scenarios of Attack

Marsh *et al.* (1991) argued for the release of samples of so-called anonymized records (SARs) from the 1991 British census, arguing that the probability of a disclosure attempt being successful was negligible. To make the argument, they 1) used the estimated probability of a specified individual in the SAR being correctly matched to records in an external database, and of that match being verified, and 2) classified the risk of a breach in confidentiality as the total probability of an individual being identified from the SAR. This was defined formally as

pr(identification) = pr(identification|attempt) * pr(attempt).

The authors acknowledged the difficulty of assessing the value of pr(attempt) and, assuming that an attempt might be made, concentrated their efforts on a theoretical analysis of pr(identification|attempt), showing that it was negligible. However, they assumed the probability of an attempt to be low because the chances of identification, given an attempt, would be low. That is, the low probability of success would dissuade anybody from attempting it in the first place. However, the value of pr(identification|attempt) in fact depends on the method of identification actually used, and this depends on the motivations of those making the attempt, as I show below.

Marsh *et al*'s analysis was also based on a disclosure attempt made by a large database holder that wished to enhance its database. To draw general conclusions from this analysis presupposes that the matching procedure analyzed is the optimum strategy for breaking confidentiality and that holders of large databases are the main threat to confidentiality. These two points are related, and I show below that there are other possible uses for correctly matched SAR records besides the enhancement of databases. Some of these other uses indicate a different approach to attempts at identification and produce different values for pr(identification|attempt).

Marsh *et al.* identify two main potential *data intruders* (sources of identification attempts):

- Holders of databases, such as credit reference agencies, that wish to update their information on individuals by using census data.
- Journalists or computer hackers who wish to identify people solely to discredit the census.

These two sources represent a dichotomy between attempts where the disclosure of information is the primary goal and attempts where the consequences (such as decline in public confidence) are the primary concern. I suggest that this list should be extended to include political parties and other political organizations, government departments or law enforcement agencies, and individuals attempting to steal another's identity. This is not to imply that such groups are likely to make an attack, but simply to ensure that all possible types of attack are included in the analysis.

The dichotomy between types of attempt highlights a further important point. To establish what a potential data intruder would do, one needs to break down each attempt into its psychological and pragmatic components. Centrally, one needs to understand the motivations of the data intruder, without which it is impossible to clarify the nature of the attack.

To do this, a system of categorizing attempts is needed. Several analysts have used the concept of the disclosure attempt scenario to work toward such a system. Paass (1988) was the first to do so. He carried out a population uniqueness analysis using sets of key variables constructed according to a commonsense conceptualization of the data to which a particular data intruder might have access. Willenborg *et al.* (1990) and Muller *et al.* (1992) adopted similar approaches.

Paass's scheme was essentially *ad hoc.* The first attempt to derive a systematic set of scenarios for a dataset was conducted by Elliot (1996; Elliot and Dale 1999). Elliot carried out extensive empirical work to establish the kinds of identification information available outside a given file (the 1991 British SARs). This led to development of a 12-point taxonomic structure for conducting a scenario analysis:

- *Motivation*—what the intruder is trying to achieve by the intrusion (which may be different from what the data provider fears may happen).
- *Means*—whether the intruder has the necessary resources and skills to carry out the intrusion.
- *Opportunity*—the ease with which the intruder has access (legitimate or otherwise) to the target dataset.
- *Attack type*—the type of statistical/computational procedure involved in the attack.

- *Key/matching variables*—the information available to the intruder with which to conduct the attack.
- *Target variables*—the particular information (if any) that the intruder is seeking to disclose.
- *Effect of data divergence*—the impact of data divergence between the intruder's information and the target file.
- *Likelihood of success*—an estimate of the intruder's likelihood of achieving his or her goals.
- *Goals achievable by other means?*—a query about whether the intruder could achieve his or her goals as effectively by other (perhaps legitimate) means.
- *Consequences of attempt*—the consequences of the attempt being made (whether it is successful or not).
- *Likelihood of attempt*—a rationalistic estimate of whether an attempt of the given sort is likely to be made on the given database.
- *Effect of variations in database structure*—the generalizing concept. Elliot's scheme is essentially pragmatic and is therefore best understood as a tool for analyzing particular target datasets. However, generality can be addressed by considering how variations in the variable structure of the database affect each of the other elements in this scheme.

Development of this scheme had two central purposes. The first was to provide the rationalistic assessment of pr(attempt) and its relationship with pr(identification|attempt). The second was to provide a framework for the empirical validation of key variable selection.

The analytical framework as applied to British census microdata indicated that the crucial scenarios were not (as Marsh and others have proposed) those of a large database holder attempting to enhance or verify existing information, but rather a smaller scale intrusion by an individual or organization with political or other secondary goals.

3. What Is a Key Variable?

The term *key variable* is used widely throughout the statistical disclosure literature. In general the term refers to a variable the value of which is, for a given *target individual,* (1) known to a *data intruder* and (2) present in a *target database* to which the intruder has access. A key variable thus provides the basis for identification of a target individual.

But this simple definition disguises the complexity of defining key variables from a risk assessment viewpoint. Elliot and Dale (1999) identify key availability and key quality as two important properties of variables in determining their suitability as keys. These concepts provide structure for the attack scenarios described in Section 2.

There are three types of analysis one might conduct to identify key variables from a risk assessment point of view:

Scenario analyses. A scenario analysis uses a set of keys that have been defined as representative of those held by a data intruder under each of a set of scenarios.

The set of scenarios is constructed through research and used to generate the key variable sets. This method is empirically based and has the benefit of covering a range of pragmatic possibilities. A scenario analysis is central to any disclosure risk assessment. It provides a pragmatically based framework for conducting analyses. Because it is pragmatically based, however, it will not cover all the possible combination of keys, nor is it the ideal format for considering the systematic variations of codings of variables in risk-driven file construction.

Additional impact analyses. In additional impact analyses, each variable within a dataset is added to a base key. Base keys typically consist of a small number of common variables. All sets are analyzed and summary measures are used to compare the risk to a baseline. If used with a small base key, this method is very good for estimating the maximum risk impact of alternative codings of the same variable, or to compare various constructions of the same file (by using an appropriately weighted mean or median)—see Tranmer *et al.* (2001).

Full file analyses. With some metrics it is meaningful to measure the risk associated with a key that consists of all the variables within the file. Doing so enables one to obtain an *absolute* worst case measure of a file's risk profile; it is also useful as a check on an additional impact analysis when that format is used to assess variation in whole file construction[1].

All three analyses have their value. Scenario analyses give a realistically based selection of key variables. Additional impact analyses are more systematic and provide a detailed view of how each variable contributes to the topology of risk. Full file analyses avoid possible biases caused by key variable selection. Best practice risk analyses will use a combination of these approaches. See Tranmer *et al.* (2001) for an example.

4. The Use of Matching Experiments

Matching experiments are attempts to match records from an identification file with those on the target microdata file using the same methods that a would-be data intruder would use. This method has the obvious advantage that the results are generated by empirical data, rather than coming from theoretical values provided by the uniqueness statistics.

---

[1] This actually corresponds to Elliot and Dale's (1998a) political demonstration scenarios, where a political organization colludes with survey or census respondents to generate a nondivergent dataset (through copies of their census returns, for example).

Muller *et al.* (1992) used the 1987 North Rhine-Westphalia microcensus as an experimental target file and a handbook of German scientists and academics as the identification file. This gave them 10 key variables. They found, using simple matching techniques, that only four records (representing 0.05 percent of the identification file) were correctly matched.

Elliot and Dale (1998b) attempted to match the British General Household Survey to the SARs from the 1991 British census. Using a 17-variable key, they discovered 6 unambiguous true matches but also 227 unambiguous false matches.

These types of matching experiments are extremely useful in understanding the processes an intruder must go through to achieve a correct matching. Both studies found that extensive work was required to optimize the number of matches. The studies also emphasize that two factors provide natural protection against a data intruder: (1) the amount of data divergence between any two files and (2) the degree of clustering on demographic variables, meaning that the rate of statistical twining is very high.

There are, however, two serious disadvantages to using studies such as these to assess disclosure risk.

First, one cannot be sure that a particular identification dataset will provide a valid measure of the level of disclosure risk associated with the target microdata set; the results will be *ad hoc* with respect to the identification dataset and the particular experiments conducted. A different dataset with different data divergence from the target microdata set might produce substantially different results.

Second, setting up matching experiments is time-consuming. Considerable effort is usually required to arrive at comparable coding for the two datasets. Complicated arrangements are also necessary to verify the accuracy of matches while maintaining confidentiality within the procedure[2].

So, although matching experiments are useful excercises, they do not provide a general assessment of disclosure risk for a data file.

5. File-Level Risk Metrics

In order to generalize beyond the *ad hoc* results provided by matching experiments, it is necessary to generate risk metrics. A large proportion of statistical disclosure control research has been concerned with the development and testing of such metrics.

[2] This does reflect on the difficulty of the intruder's task, which in itself would tend to indicate a lower value for pr(attempt).

Population Uniqueness

Population uniqueness was developed early in the discussions of disclosure risk (see, *e.g.*, Dalenius 1986; Greenberg 1990). If an individual has unique values on a set of key variables within a population, that individual is said to be a population unique. Population uniqueness is the proportion of such individuals in the population.

The main advantages of population uniqueness are the simplicity of the concept and its seemingly intuitive relationship with disclosure risk (if an individual is known to be population unique and a record matching that individual is found within a dataset, then identification disclosure has occurred at high level of probability). However, population uniqueness has several disadvantages:

- The assessment of risk requires access to population data. These data are rarely available outside national statistical institutes (NSIs) except in a limited and incomplete form (*e.g.*, electoral registers). Even census data within NSI data are not always completely coded for the entire population. For example, the 1991 British census was fully coded only for a 10 percent sample.
- Population uniqueness is independent of sampling fraction. The sample fraction of a sample dataset is known to affect the disclosure risk for an individual record[3].
- There is no obvious method of incorporating *data divergence* into the estimate of risk. Data divergence was a term used by Elliot (1996) to encapsulate all possible sources of difference between a record of an individual and the individual in the world (data-world divergence) or between records in different datasets for the same individual (data-data divergence). These sources include data aging, missing and erroneous responses, data entry errors, data coding errors, differences in coding schemes, the effects of imputation, and the various disclosure limitation methods employed to limit statistical disclosure. Clearly, such factors affect risk levels, but there is no meaningful way to adjust the uniqueness statistic to incorporate their effect.

Union Uniques

Union uniques are records which, for a given sample dataset, are unique in both the sample and the population. Various risk metrics are based on union uniques, the most important being the 'UUSU ratio', a term coined by Elliot *et al.* (1998) to express the following:

$$\frac{\text{UnionUniques}}{SampleUniques}$$

[3] Simply multiplying the sampling fraction by population uniqueness level as some authors have done (see, for example, Marsh *et al.* 1991) is suitable only as an approximation method, because it fails to address the likely interaction of the effects of the sampling fraction with other factors affecting uniqueness levels.

Thus UUSU expresses the probability of an individual being unique in the population given that the person's record is unique in the sample. The logic behind this metric is that an intruder is unlikely to have access to population data. Target files are themselves typically sample files, and possible identification data are either samples or have a very limited number of key variables. Without access to population data, an intruder will need to use information in the sample[4]. Therefore, the ability to assess population uniqueness from sample uniqueness is a potentially important device.

The UUSU is now accepted as a more useful measure of disclosure risk than population uniqueness because it is sensitive to changes in sampling fraction in a monotonic way. However, it shares some problems with population uniqueness:

- Access to population data is required to calculate it.
- Although the UUSU ratio is sensitive to sampling fraction, it does not explicitly incorporate the sampling fraction within its calculation. Clearly, there is a change in absolute risk that is directly proportional to the probability of an individual being in the sample dataset.
- The UUSU ratio and population uniqueness counts share an overarching problem: Neither of them corresponds in a direct way to what an intruder will need to do in order to break into a microdata file. What is needed is an approach that combines the directness of the matching experiments with the ease of use of the uniqueness metrics. The data intrusion simulation methodology provides such an approach.

Data Intrusion Simulation

The concept behind the data intrusion simulation (DIS) method derives from concerns expressed by Elliot (1996) regarding the need to examine statistical disclosure risk from the viewpoint of the data intruder rather than that of the data themselves. A rational intruder will be indifferent to whether a record was sample or population unique. The intruder will know such statuses are unreliable and will have more pragmatic concerns (*e.g.*, 'Is *this* match correct?'). The DIS method simulates this intruder perspective by focusing on the probability of a unique match being a correct one—pr(cm|um).

The basic principle of the DIS method is to remove records from the target microdata file and then resample those records according to the original sampling fraction. This creates two files: a new, slightly truncated target file and a file of the removed records, which is then matched against the target file. The method has two computational forms: the *special form,* where the sampling is actually done, and the *general form*, where the sampling is not actually done but the effect is derived using the partition structure[5] of the microdata file and sampling fraction.

---

[4] And possibly baseline demographic information to assist him/her in identifying rare records.

[5] A partition structure is a frequency of frequencies and a partition class is a cell count size.

DIS: The Special Form

The special DIS method uses a technique similar to the Briggs (1992) technique:

- Take a sample microdata file (A) with sampling fraction S.
- Remove a small random number of records (B) from A, to make a new file (A').
- Copy back a random number of the records in B to A', with each record having a probability of being copied back equal to S.

The result of this procedure is that B will now represent a fragment of an outside database (an identification file) with an overlap with the A' file equivalent to that which will exist between the microdata file and an arbitrary identification file with zero data divergence (*i.e.*, with no differing values for the same individual).

- Match a simulated fragment of the identification file with the target microdata file.
- Generate the probability of a correct match given a unique match, pr(cm|um), for the fragment.
- Iterate through the first four stages until the estimate stabilizes.

DIS: The General Form. A more general method can be derived from the above procedure. Imagine that the removed fragment (B) is just a single record. Clearly there are six possible outcomes depending on whether the record is resampled or not and whether it was a unique, in a pair, or in a larger partition class. One can thus derive the estimated probability of a correct match given a unique match from:

$$\frac{U^*S/100}{(U^*S/100 + P^*(1 - S/100))},$$

where U is the number of sample uniques, P is the number of records in pairs, and S is the sampling percentage. The denominator of the equation provides an estimate of the number of unique matches and the numerator the number of those that are correct.

Elliot (1998, 2000) has tested this method empirically and found that the estimator is unbiased and accurate. Skinner and Elliot (forthcoming) provide a proof that the estimator is unbiased. Recent work by Elliot (2001b) has shown how a variation of the special method can be used to estimate the impact of disclosure control techniques on the disclosure risk level.

6. Record-Level Risk Metrics

The methods outlined in Section 5 specify risk metrics at the file level. Recently there has been a move toward assessing risk at the record level (Benedetti *et al.* 1998; Elliot 2001b; Fienberg and Makov 1998; Skinner and Holmes 1998). The main rationale behind this approach is that risk is not homogeneous within a data file. Some records, categories, or combinations of categories are intrinsically rare and are recognizable as such. For example, a 16-year-old widow will be a rare person in any coding scheme and in combination with a small number of other variables is highly likely to be a population unique. If an intruder knew of such a person and found a corresponding record in a target dataset, the intruder could infer population uniqueness with a high probability of being correct. Elliot *et al.* (1998) refer to this as the special uniques problem. That is, in any dataset there will be individual records that contain a combination of attributes that are intrinsically rare by virtue of being epidemiologically peculiar. This distinguishes them from random unique records that are unique simply by virtue of the way the coding scheme has been constructed.

Some progress has been made in identifying such records. Skinner and Holmes (1998) provide a method that appears on a small range of key variables to classify sample uniques by their individual's probability of population uniqueness. Elliot (2001b) has demonstrated the principle of persistence across aggregation: that sample uniques whose uniqueness is maintained despite aggregation are considerably more likely to be population unique than those whose uniqueness is not.

7. Risk Factors

Sampling Fraction

All risk metrics are affected in a monotonic and fairly consistent fashion by the sampling fraction of the file. Elliot (2000a) found that both file-level (DIS) and record-level measures (Special Uniques Identification) relate monotonically to sampling fraction.

Level of Detail on Variables

Increasing the detail on a key variable increases the risk reported by all risk metrics. However, additional impact analyses conducted by Elliot (2000a) show that the amount of increase is not proportionate to the increase in detail. Also, the impact of the increase in detail depends on the size of the key. The larger the key variable set, the smaller the impact of increasing the detail of one of the variables in that key variable set.

Level of Geographical Detail

The level of geographical detail is an oddity. Unlike other variables, geography has a nonmonotonic relationship with risk. Elliot *et al.* (1998) discovered that for intermediate-level geographies (in the range 30,000 to 500,000 population unit size) there was no increase in risk as measured by UUSU. Elliot (2000a) confirmed this result using the probability of a correct match given a unique match.

Size of Key (Number of Key Variables)

As with level of detail, an increase in key size leads to increased risk. However, the size of this increase becomes smaller as the key size increases. So, all other factors equal, the impact of adding an additional variable to a key decreases as the key size increases. This fact is particularly important to consider in the context of data divergence.

8. Data Divergence

Data divergence, also referred to as *data incompatibilities* (De Waal and Willenborg 1994), can be thought of as the sum of all differences between a target microdata file and the identification file used by the hypothetical intruder. It is important to note that divergence is not the same thing as error because (1) an error could be consistently made on both the target file and the identification file, in which case the files would show no divergence and (2) error is not the only source of divergence. The full list of sources of divergence is as follows:

- Response error on either file.
- Data coding error on either file.
- Data entry error on either file.
- Data aging.
- Differences in coding regimes employed.
- Differences in variable constructs.
- Missing data imputed on one file (and/or imputed differently on the other file).
- Effects of disclosure control techniques.

The combination of these sources has quite a substantial effect on the probability of successful matching. Elliot and Dale (1998b) found that the probability of a correct match given a unique match in the matching experiment reported above was 0.026. The theoretical level estimated by the DIS system was 0.083. This indicates a record for record divergence rate of 0.068. That is, 68 percent of individuals who have records in both the target file and the identification file differ in how their values on one or more variables have been recorded. This was true despite a substantial amount of work done to reduce the impact of this divergence. The results of Muller *et al.* (1992) were consistent with this finding.

One important aspect of data divergence on this scale is the impact on optimum key size. An underlying assumption of much disclosure control work is that, from the intruder's viewpoint, the bigger the key the better. In most situations, however, that probably is not the case. Elliot and Dale (1999) found that the payoff between the data divergence and the greater differentiation provided by a larger key meant that the optimum key size was less than the 17 they used. Precise calculation of the optimum size is impossible, because it will vary according to the particular target and identification files and the particular key variables under consideration. Nevertheless, the general principle that each key variable added to a key has a diminishing effect on risk, combined with the fact that each variable will increase divergence, means that for any given pair of files the optimum key set for effective matching will be a small subset of the full set.

9. Tabular Data—Special Issues

Thus far we have considered all data as if they were microdata. To a large extent this assumption is reasonable since for many purposes microdata and tabular data can be considered as interchangeable. But certain issues pertain only to tabular data as commonly released.

Small Key Size. For disclosure to occur, one or more variables within a given dataset must be unknown to the intruder. For tabular data, which typically consist of two to five variables (including panel variables), this means a correspondingly small key size.

Sample Size. Whereas microdata are typically released as a sample no bigger than 5 percent of the population from which they were drawn, tabular data are often released as population data (for example, standard census output). This means that all metrics that rely on sampling are meaningless (both UUSU and pr(cm|um)) by definition, give probabilities of 1 with population data.

Geographical Detail. Some tabular data are released at a much finer level of geographical detail than would ever be considered for microdata. This means that, from a data intruder's point of view, attempting to identify individuals within tabular data is a different proposition from matching individuals in microdata. The form of such identification is the same, however, providing that one is considering only a single table. Where more than one table is available for the same individuals, as is usually the case, the possibility of *table linkage* needs to be considered.

Table linkage can happen when the set of variables used on one table overlaps with those used on another. This presents a problem because, in effect, it opens the possibility of a supertable being created through the linking of the tables. Even if it is not possible to do this completely, the partial linking of some of the records in tables has the potential to cause problems. Thus far, no comprehensive analysis of the risk presented by table linkage, or indeed of the related problem of linkage of tables to microdata (when they are drawn from the same source data), has been carried out. Such research is urgently required.

A second problem is that of *table differencing*. With increasing user demand for customized output from censuses, this issue has become more pressing. Different users have different demands in terms of variable constructions, particularly in the construction of geographies. Duke-Williams and Rees (1998) demonstrate how this is possible. Where two different geographies overlap, small fragments may be constructed through subtracting tables generated from one geography from tables generated by the other geography. If an area in the first geography falls inside the area in the other, the subtraction may result in a small area in which direct identification might occur. Even if areas in the two geographies overlap it may still be possible to achieve subtraction by estimating proportions of each cell in each of the subdivided areas. Duke-Williams and Rees conclude that a number of factors affect the degree to which differencing can be done—for example, 'differences in area shape' and the 'use of uniform regional boundaries'. If these factors are unfavorable, then the 'simultaneous publication of data for two geographies poses a risk' (p.379).

10. Secondary Intrusion Techniques

The concepts of secondary differentiation and attribute disclosure are two sides of the same coin for an intruder who goes beyond primary identification disclosure as a means of disclosing information about individuals.

Secondary differentiation is the use of characteristics other than those used in the original matching key to differentiate between statistical twins. So, for example, if an intruder finds a pair on the target file that matches an individual identification record, the intruder can attempt to gather further information that distinguishes the members of a pair, effectively extending the key in an *ad hoc* way and, if successful, creating another matched unique[6].

Attribute disclosure is the flip side of secondary differentiation. Where all members of a matched equivalence class share an attribute that is not part of the original key, that attribute is disclosed about the individual if there are no individuals within the population who match the identification record and who do not have that attribute.

The investigation of these possibilities is notoriously difficult, and very little work has been done. The main defense of this neglect is that even if these secondary intrusion techniques present a real problem (which is an empirical question), the effective use of them would be equally difficult for an intruder.

Consequently, the potential threat posed by these secondary techniques is very unclear. They are clearly subject to the same problems of data divergence and inferencing sampling that affect primary key matching. So, for example, if all records within a sample with attributes {a, b, c} also have attribute {y}, at what probability can you infer that an arbitrary record from the population that has attributes {a, b, c} also has attribute {y}?

One possibility to bear in mind is that the two techniques could be used in unison. Thus, after a primary matching, small but non-unique partitions could be

[6] This type of technique is only really likely in small-scale demonstrative scenarios.

broken down by secondary differentiation until unique matches or common attribute twins are reached.

As stated above, the actual value of these techniques to an intruder is far from clear. An intruder would need to be very sophisticated in order to use them at all, and they may not in the end yield better results than would be achieved by primary matching.

Secondary intrusion techniques are relatively underresearched. They need to be analyzed in a formal scenario framework.

11. The Overall Picture

The picture painted in this chapter is understandably complex. Disclosure risk assessment is still a young research area. Many ideas and concepts overlap and inform one another, but as a whole the field is still pre-theoretical.

It is impossible to quantify disclosure risk in an absolute sense, even for a particular dataset. Two of the limitations on such an endeavor are the impossibility of arriving at any meaningful figure for either the probability of an attempt taking place or the relationship of that probability with what is actually attempted. Another limitation is the apparently substantial but inconsistent impact of data divergence.

Mechanisms are in place to allow us to move toward effective *relative risk assessments,* however. This is in itself a major step forward. Elliot (2001b) has indicated how relative risk assessment can be used to evaluate the real impact of disclosure control techniques. This is the first step in the coherent linking of risk assessment and practical disclosure control.

Another example of this practice is use of a benchmark dataset. In the United Kingdom the 1991 individual SAR has become a safety benchmark after extensive detailed analysis of that dataset. This approach has made possible the straightforward evaluation of proposals for the 2001 British census—for example, in determining the specifications for small area microdata from the U.K. census (a proposed microdata sample with low geographical threshold); see Tranmer *et al.* (2001).

What is also clear from the above is the importance of a multilevel risk analysis—that is, a risk analysis based on an appropriate scenario and key variable analysis, which examines risk at the whole-file, intermediate, and record levels. A further extension of this would be to apply disclosure risk assessment at the file construction stage, so that data files could be flexibly constructed according to actual risk assessment, rather than the orthodox threshold and test approach.

To summarize:

- A scenario analysis provides the best way of framing key variable selection and allows a pragmatic assessment of the likely sources of data intrusion.

- Analysis of U.K. data indicates that the most likely source of data intrusion is not likely to be a large database holder but a small-scale intruder whose intent is tangential to the content of the information disclosed.
- Data divergence has a substantial impact on matching probabilities.
- The disclosure risk impact of adding more detail or additional variables to a key decreases as the key size increases.
- Secondary intrusion techniques are relatively underresearched. Work needs to be done analyzing them in a formal scenario framework.
- For many purposes tabular data can be considered microdata. However, because tabular data are typically 100 percent data, some risk metrics become meaningless.
- Tabular data are also affected by the problem of table linkages and table differencing; both of these topics need more research.
- The Data Intrusion Simulation methodology is the most effective file-level risk assessment technique available.
- Absolute risk assessment is not an empirical possibility. However, a good deal can be achieved using relative risk assessment. The use of benchmark files that have been exhaustively tested for their risk profiles and the testing of the impact of disclosure control techniques on matching probabilities are methods that are available now.
- Risk assessment needs to be multilevel, taking account of file level, record level, and probably intermediate levels.

References

Benedetti, R., L. Franconi, and F. Piesimoni (1998) 'Per-Record Risk of Disclosure in Dependent Data', in *Proceedings of the Third International Conference on Statistical Data Protection,* Lisbon, March, pp.287-91.

Blien, U., H. Wirth, and M. Muller (1992) 'Disclosure Risk for Microdata Stemming from Official Statistics', *Statistica Neerlandica* 46, pp.69-82.

Briggs, M. (1992) 'Estimation of the Proportion of Unique Population Elements Using a Sample', Ottawa: Statistics Canada Working Paper.

Dalenius, T. (1986) 'Finding a Needle in a Haystack', *Journal of Official Statistics*, 2(3), pp.329-36.

De Waal, T. and L. Willenborg (1994) 'Principles of Statistical Disclosure Control', Ottawa: Statistics Netherlands Working Paper.

Duke-Williams, O., and P. Rees (1998) 'Factors Affecting Confidentiality Risks Involved in Releasing Census Data for Small Areas', in *Proceedings of the Third International Conference on Statistical Data Protection,* Lisbon, March, pp.369-79.

Elliot, M.J. (1996) 'Attacks on Confidentiality Using the Samples of Anonymized Records', paper presented to the Third International Seminar on Statistical Confidentiality, Bled, Slovenia, October.

Elliot, M.J. (1998) 'DIS: Data Intrusion Simulation—A Method of Estimating the Worst Case Disclosure Risk for a Microdata File', in *Proceedings of the International Symposium on Linked Employee-Employer Records*, Washington, D.C., May.

Elliot, M.J. (2000a) 'Data Intrusion Simulation Project', Report to the Economic and Social Research Council R000 22 2852.

Elliot, M.J. (2000b) 'DIS: A New Approach to the Measurement of Statistical Disclosure Risk', *International Journal of Risk Management,* 2(4), pp.39-48.

Elliot, M.J. (2001a) 'Data Intrusion Simulation: 'Advances and a Vision for the Future of Disclosure Control', paper presented to the Second UNECE Work Session on Statistical Data Confidentiality, Skopje, March.

Elliot, M.J. (2001b) 'The Identification of Special Uniques', paper presented to the Second UNECE Work Session on Statistical Data Confidentiality, Skopje, March.

Elliot, M.J. and A. Dale (1998a) 'What is a Key Variable?' Report to the European Union on Disclosure Risk for Microdata ESP/204 62/DG111; DM1.1.

Elliot, M.J. and A. Dale (1998b) 'A Disclosure Risk Matching Experiment', Report to the European Union on Disclosure Risk for Microdata EPS/204 62/DG111; DM1.5

Elliot, M.J., and A. Dale (1999) 'Scenarios of Attack: The Data Intruder's Perspective on Statistical Disclosure Risk', *Netherlands Official Statistics,* Vol. 14, Spring.

Elliot, M.J., C.J. Skinner, and A. Dale (1998) 'Special Uniques, Random Uniques and Sticky Populations: Some Counterintuitive Effects of Geographical Detail on Disclosure Risk', *Research in Official Statistics* 1(2), pp.53-68.

Fienberg, S.E., and U.E. Makov (1998) 'Confidentiality, Uniqueness and Disclosure Limitation for Categorical Data', *Journal of Official Statistics* 14, pp.385-97.

Greenberg, B. (1990) 'Disclosure Avoidance Research at the Census Bureau', in *Proceedings of U.S. Bureau of the Census Annual Research Conference,* Washington, DC, pp.144-66.

Marsh, C., C. Skinner, S. Arber, P. Penhale, S. Openshaw, J. Hobcraft, D. Lievesley, and N. Walford (1991) 'The Case for a Sample of Anonymized Records from the 1991 Census', *Journal of the Royal Statistical Society,* Series A, 154, pp.305-40.

Paass, G. (1988) 'Disclosure Risk and Disclosure Avoidance for Microdata', *Journal of Business and Economic Statistics,* 6(4), pp.487-500.

Skinner, C.J., and M.J. Elliot (2001) 'A Measure of Disclosure Risk for Microdata', paper forthcoming, *Journal of the Royal Statistical Society,* Series B.

Skinner, C.J., and D.J. Holmes (1998) 'Estimating the Re-Identification Risk per Record in Microdata', *Journal of Official Statistics,* 14(4), pp.361-72.

Tranmer, M., E. Fieldhouse, M.J. Elliot, A. Dale, and M. Brown (2001) *Proposals for Small Area Microdata,* Centre for Census and Survey Research Occasional Paper submitted to *Journal of the Royal Statistical Society,* Series A.

Willenborg, L., R.J. Mokken, and J. Pannokoek (1990) 'Microdata and Disclosure Risks', in *Proceedings of the 1990 Annual Research Conference of the Bureau of the Census,* Washington, D.C., pp.167-80.

Confidentiality, Disclosure, and Data Access: Theory and Practical Application for Statistical Agencies
Pat Doyle, Julia I. Lane, Jules J.M. Theeuwes and Laura M. Zayatz (Eds)

Chapter 5

Disclosure Control Methods and Information Loss for Microdata*

Josep Domingo-Ferrer
Universitat Rovira i Virgili

Vicenç Torra
Institut d'Investigació en Intel•ligència Artificial

1. Introduction

Statistical disclosure control (SDC) seeks to modify statistical data so that they can be published without giving away confidential information that can be linked to specific respondents. The challenge for SDC is to achieve this modification with minimum loss of the detail and accuracy sought by database users. SDC methods for microdata are usually known as *masking methods,* of which there is a wide range. From the point of view of their operational principles, current masking methods fall into the following two categories (Willenborg and De Waal 2001):

- *Perturbative.* The microdata set is distorted before publication. In this way, unique combinations of scores in the original dataset may disappear and new unique combinations may appear in the perturbed dataset; such confusion is beneficial for preserving statistical confidentiality. The perturbation method used should be such that statistics computed on the perturbed dataset do not differ significantly from the statistics that would be obtained on the original dataset.
- *Nonperturbative.* Nonperturbative methods do not alter data; rather, they produce partial suppressions or reductions of detail on the original dataset. Global recoding, local suppression, and sampling are examples of nonperturbative masking.

* Some of the work reported in this chapter was funded in part by the U.S. Bureau of the Census under Contracts No. OBLIG-2000-29158-0-0 and OBLIG-2000-29144-0-0. Thanks go to Laura Zayatz for providing information on rank swapping. We would also like to thank Josep M. Mateo-Sanz and Francesc Sebé for their help in working out some of the examples in this chapter. The comments of the editors and several reviewers are gratefully acknowledged as well.

From the point of view of the data to which they are applied, a second classification of masking methods can be established:

- *Continuous.* This class of methods is suitable for masking data corresponding to continuous variables. A variable is considered continuous if it is numerical and arithmetic operations can be performed with it. Examples are income and age. Note that a numerical variable does not necessarily have an infinite range: for example, the range of age is finite (alas!).
- *Categorical.* This class of methods is suitable for masking data corresponding to categorical variables. A variable is considered categorical when it takes values over a finite set and standard arithmetic operations do not make sense. Examples are day of the week and eye color.

Structure of This Chapter

Section 2 introduces the notation used throughout the chapter. Section 3 describes masking methods that have been proposed for microdata protection. Section 4 proposes several methods to measure the information loss attributable to masking in both continuous and categorical cases. Section 5 concludes the chapter in a way that leads into Chapter 6, which compares SDC alternatives in terms of the information loss/confidentiality trade-off.

2. Notation and Variable Types

We assume that the information in a microdata file can be represented as a two-dimensional table where one dimension is the set of records (*i.e.,* elements, individuals, persons) and the other is the set of variables.

The microdata file contains a value for each record-variable pair, so that it can be modeled as a function:

$$\boldsymbol{V}{:}\boldsymbol{O} \rightarrow D(V_1) \times D(V_2) \times \ldots \times D(V_m)$$

where $\boldsymbol{O}$ denotes the set of records, $V_1, V_2, \ldots, V_m$ denote the variables, and $D(V_i)$ refers to the range of variable V_i.

Without loss of generality, the m-dimensional function V can be assumed to be of the form:

$$V(O) = (V_1(O), V_2(O), \ldots, V_m(O))$$

where $V_i(\bullet){:}\boldsymbol{O} \rightarrow D(V_i)$ is a one-dimensional function assigning a value for variable V_i to a given record.

3. Masking Methods for Microdata Protection

Using the notation of Section 2, we can state the purpose of SDC more formally by saying that its goal is to supply the user with a masked microdata file V' similar to the original V (*i.e.,* with low information loss) in such a way that

(1) Disclosure risk (*i.e.,* risk of identification of an individual) is low.

(2) User analyses (regressions, means, *etc.*) on V' and on V yield the same or at least similar results.

This section describes masking methods that can be used to produce V' from V. Subsections discuss perturbative methods and nonperturbative methods. Complementary reviews of the literature on masking methods can be found in Adam and Wortmann (1989) and Willenborg and De Waal (2001).

Perturbative Methods

Perturbative methods allow for release of the entire microdata set, although perturbed values rather than exact values are given. Not all perturbative methods are designed for continuous data, a distinction that is addressed further below for each method.

Most perturbative methods reviewed below (including additive noise, data swapping, microaggregation, and post-randomization—PRAM) are special cases of matrix masking. If the original microdata set is V, then the masked microdata set V' is computed as

$$V' = AVB + C$$

where A is a record-transforming mask, B is a variable-transforming mask, and C is a displacing mask (noise) (Duncan and Pearson 1991).

Table 1 lists the perturbative methods described below. For each method, the table indicates whether it is suitable for continuous and/or categorical data.

Table 1. Perturbative Methods Versus Data Types

Method	Continuous data	Categorical data
Additive noise	X	
Data distort. by probability distribution	X	X
Microaggregation	X	
Resampling	X	
Lossy compression	X	
Multiple imputation	X	
Camouflage	X	
PRAM		X
Rank swapping	X	X
Rounding	X	

Additive Noise. Additive noise (Kim 1986; Little 1993; Sullivan and Fuller 1989, 1990) consists of adding random noise with the same correlation structure as the original unmasked data. It is currently the only method that can preserve correlations.

Let $v_{ij} = V_i(o_j)$ be the unmasked value of variable V_i for individual o_j. Let $e_{ij} = E_i(o_j)$ be the noise added to v_{ij}, and let $v'_{ij} = v_{ij} + e_{ij}$. Further, let $V = \{V_{ij}\}$ be the matrix having v_{ij} as elements, and similarly $E = \{e_{ij}\}$ and $V' = \{v'_{ij}\}$. It is assumed that the expected value of the noise is $E(E) = 0$ and its variance is $Var(E) = cVar(V)$ for some constant c. The variance of the masked data is $Var(V') = (1+c)Var(V)$. The variance of unmasked variables can be recovered as $Var(V')/(1+c)$.

White (*i.e.,* Gaussian) noise is most frequently used, even though it may be subject to the *bias problem* (Matloff 1986): if V_i is a continuous positive variable with a strictly decreasing density function (*e.g.,* the exponential density) to which a perturbation that is symmetrical around 0 has been added (*e.g.,* Gaussian noise), Matloff shows that

$$E(V_i | V'_i = w) < w$$

where V'_i is the perturbed version of V_i.

Thus, the constant c is the only parameter that can be tuned. It alters (increases or decreases) the random noise being inoculated.

The nature of additive noise makes it unsuitable for categorical data. But it is well suited for continuous data, for the following reasons:

- It makes no assumptions about the range of possible values for V_i (which may be infinite).
- The noise being added is typically continuous and with mean zero, which suits continuous original data well.
- No exact matching is possible with external files. Depending on the amount of noise added, approximate (interval) matching might be possible.

Data Distortion by Probability Distribution. This method, also called Probability Distortion (Liew *et al.* 1985), is suitable for both categorical and continuous variables. Three steps are needed to compute the distorted version of a confidential original dataset:

(1) Identification of the underlying density function of each of the confidential variables in the dataset and estimation of the parameters associated with the density function.

(2) Generation of a distorted series for each confidential variable from the estimated density function.

(3) Mapping and replacement of the distorted series in place of the confidential series.

In the identification and estimation stage, the original series of the confidential variable (*e.g.,* salary) is screened to determine which of a set of predetermined density functions fits the data best. Goodness of fit can be tested with the Kolmogorov-Smirnov test. An example set of predetermined density functions could include Poisson, exponential, normal, gamma, Weibull, log-normal, uniform, triangular, chi-square. If several density functions are acceptable at a given significance level, selecting the one yielding the smallest value for the Kolmogorov-Smirnov statistics is recommended. If no density in the predetermined set fits the data, the frequency imposed distortion method can be used. With the latter method, the original series is divided into several intervals (somewhere between 8 and 20). The frequencies within the interval are counted for the original series and become a guideline to generate the distorted series. By using a uniform random number generating subroutine, a distorted series is generated until its frequencies become the same as the frequencies of the original series. If the frequencies in some intervals overflow, they are simply discarded.

Once the best-fit density function has been selected, the generation stage feeds the estimated distribution parameters to a random value-generating routine to produce the distorted series.

The final stage, mapping and replacement, is needed only if the distorted variables are to be used jointly with other nondistorted variables. Mapping consists of ranking the distorted series and the original series in the same order and replacing each element of the original series with the corresponding distorted element.

It must be stressed here that the approach described in Liew *et al.* (1985) is for one variable at a time. One could imagine a generalization of the method using multivariate density functions. However, this is not a trivial undertaking. It requires multivariate ranking/mapping and can lead to very poor fits.

The example below shows how distribution fitting can be used to construct a masked dataset from an original dataset.

Example 1

A distribution fitting software (Crystal Ball 2001) was used on the original (ranked) dataset 186, 693, 830, 1177, 1219, 1428, 1902, 1903, 2496, 3406. Continuous distributions tried were normal, triangular, exponential, log-normal, Weibull, uniform, beta, gamma, logistic, Pareto, and extreme value; discrete distributions tried were binomial, Poisson, geometric, and hypergeometric. The software allowed for three fitting criteria to be used: Kolmogorov-Smirnov, χ^2, and Anderson-Darling. According to the first criterion, the best fit occurred for the extreme value distribution with modal and scale parameters 1105.78 and 732.43, respectively; the Kolmogorov statistic for this fit was 0.1138. Using the fitted distribution, the following masked dataset was generated and used to replace the original one: 425.60, 660.97, 843.43, 855.76, 880.68, 895.73, 1086.25, 1102.57, 1485.37, 2035.34. To assess the disclosure risk associated with this masking, reidentification experiments such as those described in Chapter 6 can be conducted.

Microaggregation. Records are clustered into small aggregates or groups of size at least k. Rather than publishing an original variable V_i for a given record, the average of the values of V_i over the group to which the record belongs is published. The rationale behind microaggregation is that confidentiality rules permit publication of microdata sets if the records correspond to groups of k or more individuals, where no individual dominates (*i. e.,* contributes too much to) the group and k is a threshold value. To minimize information loss, groups should be as homogeneous as possible.

Classical microaggregation (Defays and Nanopoulos 1993) required that all groups except perhaps one be of size k; allowing groups to be of size $\geq k$ depending on the structure of data can be termed *data-oriented microaggregation* (Domingo-Ferrer and Mateo-Sanz 2002; Mateo-Sanz and Domingo-Ferrer 1999). Figure 1 illustrates the advantages of variable-sized groups. If classical fixed-size microaggregation with $k = 3$ is used, we obtain a partition of the data into three groups, which looks rather unnatural for the data distribution given. If, in contrast, variable-sized groups are allowed, then the five pieces of data on the left can be kept in a single group and the four on the right in another group; such a variable-size grouping yields more homogeneous groups, which implies lower information loss.

Figure 1. Variable-Sized Groups Versus Fixed-Sized Groups

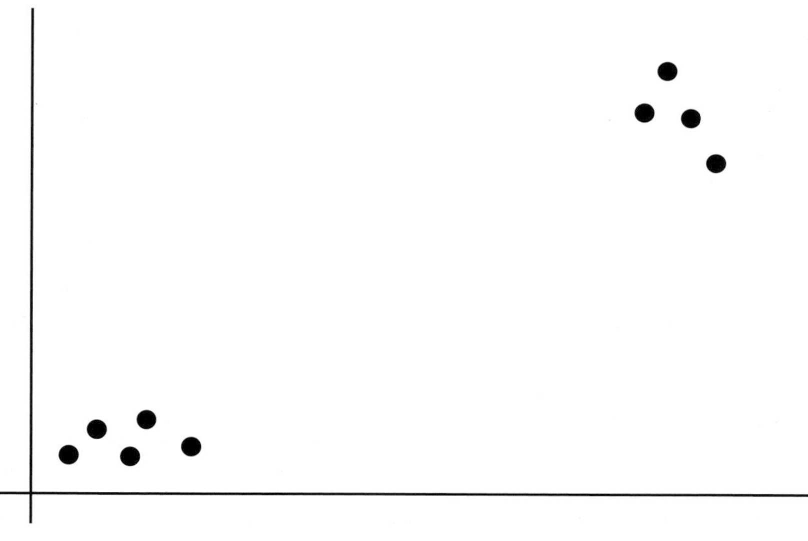

Exactly solving the microaggregation problem—that is, finding a grouping where groups have maximal homogeneity and size at least k—was recently shown to be NP-hard[1] (Oganian and Domingo-Ferrer 2001). Methods in the literature are heuristic and can be univariate or multivariate:

- Univariate methods deal with multivariate datasets by microaggregating one variable at a time—that is, variables are sequentially and independently microaggregated. This approach is known as individual ranking or blurring (Defays and Nanopoulos 1993) and, while it causes low information loss, it can lead to rather high disclosure risks (see Chapter 6).
- Multivariate methods either rank multivariate data by projecting them onto a single axis—for example, using the first principal component[2] or the sum of z-scores[3] (Defays and Nanopoulos 1993)—or dealing directly with unprojected data (Domingo-Ferrer and Mateo-Sanz 2002; Mateo-Sanz and Domingo-Ferrer 1999). When working with unprojected data, one can microaggregate all variables of the dataset at a time, or independently microaggregate groups of two variables at a time, three variables at a time, and so on.

Resampling. Originally proposed for protecting tabular data (Domingo-Ferrer and Mateo-Sanz 1999; Heer 1993), resampling can also be used for microdata. Let V be an original variable in a dataset with n records. Take with replacement t independent samples $X_1, \ldots, X_t$ of size n of the values of V. Independently rank each sample (using the same ranking criterion for all samples). Finally, for $j = 1$ to n, compute the j-th value v'_j of the masked variable V' as the average of the j-th ranked values in $X_1, \ldots, X_t$.

Lossy Compression. This method is new and is recommended for continuous data. The idea is to regard a numerical microdata file as an image (with records being rows, variables being columns, and values being pixels[4]). Lossy compression—for example, JPEG (Joint Photographic Experts Group 2001)—is then used on the image, and the compressed image is interpreted as a masked microdata file. Depending on the lossy compression algorithm used, appropriate mappings be-

---

[1] A problem is NP-hard if it cannot be solved in time polynomial in the input size unless P = NP, where P is the class of problems solvable in polynomial time and NP is the class of problems for which correctness of a solution can be verified in polynomial time. It is conjectured that P is strictly included in NP (Garey and Johnson 1979).

[2] Principal component analysis aims to transform the observed variables into a new set of variables which are uncorrelated and arranged in decreasing order of importance. The principal aim is to reduce the dimensionality of the dataset to make it easier to understand. In particular, the first principal component can be interpreted as a one-dimensional 'summary' of the dataset.

[3] The sum of z-scores is an alternative way to obtain a one-dimensional 'summary' of a dataset. Each variable is standardized, and for each record the standardized values of all variables are added up.

[4] Pixel stands for 'picture element' and corresponds to a dot in a digital image. For a black-and-white image, the value of a pixel is a grayscale level. If b bits are used to encode a pixel, then 2^b grayscale levels can taken by the pixel. Similarly, for a color image, if b bits are used, 2^b combinations of the basic colors (red, green, and blue) are possible for the pixel.

tween variable ranges and color scales will be needed. The example below illustrates the use of lossy compression to obtain a masked dataset from an original dataset.

Example 2

The upper part of Table 2 shows an original dataset with eight variables and eight records; the lower part of the table shows the disclosure protected version of the dataset. The algorithm used is as follows. First, original values have been scaled to integers in the interval *[0,255]*, that is, pixel grayscale values; this gives a grayscale image. Second, JPEG compression with 80 percent quality has been used on the image. Third, the compressed image has been unscaled (using the inverse of the scaling transformation of the first step) to get the masked dataset. The disclosure risk associated to this masking could be estimated in an empirical way, as suggested for the masking of Example 1.

Table 2. Example of SDC Through Lossy Compression

Top, original file; bottom, protected file with JPEG 80%

4173	4621	4527	1428	27	27	3480	4550
2639	6045	4208	1902	1008	808	3136	4100
3315	4765	5645	1903	485	485	4284	5600
1619	3932	2380	1177	700	700	1750	2288
4604	4349	2151	1219	751	1	1606	2100
3433	2463	3217	830	167	50	2448	3200
824	372	8730	186	1030	22	589	7700
4145	629	2500	693	1	1	1912	2500
4100	4710	4498	1451	1	10	3718	4296
2751	5889	4369	1862	1021	808	3052	4098
3314	4465	5763	1903	448	507	4168	5657
1683	4154	2151	1243	654	706	1791	2385
4604	4154	2305	1222	779	1	1559	2100
3536	2374	3363	812	93	124	2443	3263
824	460	8730	199	1030	1	589	7590
4174	572	2538	670	1	23	1965	2605

Multiple Imputation. This method (Rubin 1993) relies on releasing simulated continuous microdata created by multiple imputation techniques based on the original microdata. A way to perform multiple imputation is on a variable-by-variable basis, using a randomized regression (with normal errors) to impute missing values of each continuous variable (Kennickell 1998).

Camouflage. Vector camouflage (Gopal *et al.* 1998) is a method for giving unlimited, correct numerical responses to *ad hoc* queries to a database while not compromising confidential numerical data. No probabilistic assumptions are made, and optimization techniques are used to camouflage the sensitive record (exact answer) in an infinite set of records, thus providing an interval answer. The information loss is the transformation of a point answer into an interval answer. Because of its nature, this method is suitable only for continuous data.

PRAM. The Post-Randomization Method (PRAM) (Gouweleeuw *et al.* 1997) is a probabilistic, perturbative method for disclosure protection of categorical variables in microdata files. In the masked file, the scores on some categorical variables for certain records in the original file are changed to a different score according to a prescribed probability mechanism, namely a Markov matrix. The Markov approach makes PRAM very general, because it encompasses noise addition, data suppression, and data recoding.

PRAM information loss and disclosure risk depend largely on the choice of the Markov matrix (De Wolf *et al.* 1999) and are still open research topics.

The PRAM matrix contains a row for each possible value of each variable to be protected. This rules out using the method for continuous variables, unless these are converted into discrete form (in the same way discussed below in the case of global recoding).

Rank Swapping. Although originally described only for ordinal variables (Moore 1996), this method can be used for any numerical variable. First, values of variable V_i are ranked in ascending order. Then, each ranked value of V_i is swapped with another ranked value randomly chosen within a restricted range (*e.g.*, the rank of two swapped values cannot differ by more than p percent of the total number of records). The use of rank swapping is illustrated below.

Example 3

In Table 3, we can see an original microdata file on the left and its rank swapped version on the right. There are four variables and ten records in the original file; the second variable is alphanumeric, and the standard alphabetic order has been used to rank it. A value of $p = 15$ percent has been used for all variables. The rank swapped data file has been sorted by its first variable.

Table 3. Example of Rank Swapping

Left four columns, original file; right four columns, rank swapped file

1	K	3.7	4.4	1	H	3.0	4.8
2	L	3.8	3.4	2	L	4.5	3.2
3	N	3.0	4.8	3	M	3.7	4.4
4	M	4.5	5.0	4	N	5.0	6.0
5	L	5.0	6.0	5	L	4.5	5.0
6	H	6.0	7.5	6	F	6.7	9.5
7	H	4.5	10.0	7	K	3.8	11.0
8	F	6.7	11.0	8	H	6.0	10.0
9	D	8.0	9.5	9	C	10.0	7.5
10	C	10.0	3.2	10	D	8.0	3.4

Rounding. Rounding methods replace original values of variables with rounded values. For a given variable V_i, rounded values are chosen among a set of rounding points defining a *rounding set.* In a multivariate original dataset, rounding is usually performed one variable at a time (*univariate* rounding); however, multivariate rounding is also possible (Willenborg and De Waal 2001). The operating principle of rounding makes it suitable only for continuous data.

Example 4

Assume a continuous variable V. Then we have to determine a set of rounding points $\{p_1, \ldots, p_r\}$. One possibility is to take rounding points as multiples of a base value b, that is, $p_i = b \cdot i$ for $i = 1, \ldots, r$. The set of attraction for each rounding point p_i is defined as the interval $[p_i - b/2, p_i + b/2]$, $(i = 2, \ldots, r-1)$; for p_1 and p_r, respectively, the sets of attraction are $[0, p_1 + b/2]$ and $[p_r - b/2, V_{max}]$, where V_{max} is the largest possible value for variable V. Now an original value v of V is replaced with the rounding point corresponding to the set of attraction where v lies.

Nonperturbative Methods

Nonperturbative methods do not rely on distortion of the original data but on partial suppressions or reductions of detail. Some of the methods are usable on both categorical and continuous data, but others are not suitable for continuous data. Table 4 lists the nonperturbative methods described below. For each method, the table indicates whether it is suitable for continuous and/or categorical data.

Table 4. Nonperturbative Methods Versus Data Types

Method	Continuous data	Categorical data
Sampling		X
Global recoding	X	X
Top and bottom coding	X	X
Local suppression		X

Sampling. Instead of publishing the original microdata file

$$\boldsymbol{V}:\boldsymbol{O} \rightarrow D(V_1) \times D(V_2) \times \ldots \times D(V_m)$$

what is published is

$$\boldsymbol{V'}:\boldsymbol{S} \rightarrow D(V_1) \times D(V_2) \times \ldots \times D(V_m)$$

where $\boldsymbol{S} \subset \boldsymbol{O}$ is a sample of the original set of records and $\boldsymbol{V'}$ stands for the original function $\boldsymbol{V}$ restricted to $\boldsymbol{S}$.

Sampling is suitable for categorical microdata, but its adequacy for continuous microdata is less clear in a general disclosure scenario. The reason is that the method leaves a continuous variable $V_i(\bullet)$ unperturbed for all records in S. Thus, if variable V_i is present in an external administrative public file, unique matches with V' are very likely, because for a continuous variable (even one truncated due to digital representation) it is unlikely that V_i takes the same value for two different records (*i.e.*, $V_i(o_1) = V_i(o_2)$ if $(o_i \neq o_2)$.

If, for a continuous identifying variable, the score of a respondent is only approximately known by an attacker (as assumed in Willenborg and De Waal 1996), it might still make sense to use sampling to protect that variable. However, assumptions about attacker resources are perilous and may prove too optimistic if good quality external administrative files are at hand. For the purpose of illustration, Example 5 gives the technical specifications of a real-world application of sampling.

Example 5

In 1995, Statistics Catalonia released a sample from the 1991 population census of Catalonia (IDESCAT-Statistics Catalonia 1995). The information released corresponded to 36 categorical variables (including the recoded versions of initially continuous variables); some of the variables were related to the individual person and some to the household. The technical specifications of the sample were as follows:

- *Sampling algorithm:* Simple random sampling.
- *Sampling unit:* Individuals in the population whose residence was in Catalonia as of March 1, 1991.
- *Population size:* 6,059,494 inhabitants.
- *Sample size:* 245,944 individual records.
- *Sampling factor:* 0.0406.

With the above sampling fraction, the maximum absolute error for estimating a maximum-variance proportion is 0.2 percent.

Global Recoding. For a categorical variable V_i, several categories are combined to form new (less specific) categories, thus resulting in a new V'_i with $|D(V'_i)| < |D(V_i)|$ where $|\bullet|$ is the cardinality operator. For a continuous variable V_i, global recoding means replacing V_i by another variable V'_i which is a discretized version of V_i. In other words, a potentially infinite range $D(V_i)$ is mapped onto a finite range $D(V'_i)$. This is a technique used in the μ-Argus SDC package (Hundepool *et al.* 1998).

This technique is more appropriate for categorical microdata, where it helps disguise records with strange combinations of categorical variables. Global recoding is used heavily by statistical offices.

Example 6

If there is a record with 'Marital status = Widow/er' and 'Age = 17', global recoding could be applied to 'Marital status' to create a broader category 'Widow/er or divorced', so that the probability of the above record being unique would diminish.

Global recoding can also be used on a continuous variable, but the inherent discretization very often leads to an unaffordable loss of information. Also, arithmetical operations that were straightforward on the original V_i are no longer easy or intuitive on the discretized V'_i.

Top- and Bottom-Coding. Top- and bottom-coding is a special case of global recoding that can be used on variables that can be ranked, that is, continuous or categorical ordinal. The idea is that top values (those above a certain threshold) are lumped together to form a new category. The same is done for bottom values (those below a certain threshold) (see Hundepool and Willenborg 1999).

Local Suppression. Certain values of individual variables are suppressed with the aim of increasing the set of records agreeing on a combination of values. Ways to combine local suppression and global recoding are discussed in De Waal and Willenborg (1995) and implemented in the μ-Argus SDC package (Hundepool et al. 1998).

If a continuous variable V_i is part of a set of variables, then each combination of values is probably unique. Because it does not make sense to systematically suppress all the values of V_i, we conclude that local suppression is more oriented to categorical variables.

4. Information Loss Measures

Strictly speaking, information loss depends on the data uses to be supported by the masked (*i.e.,* SDC-protected) data. However, potential data uses are very diverse and it may be hard even to identify them all at the moment of data release by a statistical office. It is thus desirable for the data protector to be able to measure information loss in a generic way that reflects how much harm is being inflicted to the data by a given masking method; the amount of information loss measured in this generic way should roughly correspond to the amount of information loss for a reasonable range of data uses. The approach described here to derive generic information loss measures is based on assessing how different the masked dataset is from the original dataset. We will say there is little information loss if the analytic structure of the masked dataset is very similar to the structure of the original dataset. In fact, the motivation for preserving the structure of the dataset is to ensure that the masked dataset will be analytically valid and interesting. Winkler (1998) has determined that

- A microdata set is *analytically valid* if the following are approximately preserved (some conditions apply only to continuous variables):
  (1) Means and covariances on a small set of subdomains.
  (2) Marginal values for a few tabulations of the data.
  (3) At least one distributional characteristic.
- A microdata file is *analytically interesting* if six variables on important subdomains are provided that can be validly analyzed.

More precise conditions of analytical validity and analytical interest cannot be stated without taking specific data uses into account. As imprecise as it may be, the above definition of analytical validity does shed some light on what preserving the dataset structure means. We can actually try several complementary ways to assess the preservation of the original dataset's structure:

- Compare the data in the original and the masked datasets. The more similar the SDC method to the identity function, the less the impact (but the higher the disclosure risk!).
- Compare some statistics computed on the original and the masked datasets. Little information loss should translate to little differences between the statistics.
- Analyze the behavior of the particular SDC method used to measure its impact on the structure of the original dataset.

Information Loss Measures for Continuous Data

Assume a microdata set with n individuals (records) $I_1, I_2, \ldots, I_n$ and p continuous variables $Z_1, Z_2, \ldots, Z_p$. Let X be the matrix representing the original microdata set (rows are records and columns are variables). Let X' be the matrix representing the masked microdata set. The following tools are useful to characterize the information contained in the dataset:

- Covariance matrices V (on X) and V' (on X').
- Correlation matrices R and R'.
- Correlation matrices RF and RF' between the p variables and the p factors PC_1, ..., PC_p obtained through principal components analysis.
- Commonality between each of the p variables and the first principal component PC_1 (or other PC_i's). Commonality is the percentage of each variable that is explained by PC_1 (or PC_i'). Let C be the vector of commonalities for X and C' the corresponding vector for X'.
- Factor score coefficient matrices F and F'. Matrix F contains the factors that should multiply each variable in X to obtain its projection on each principal component. F' is the corresponding matrix for X'.

There does not seem to be a single quantitative measure that completely reflects those structural differences. Therefore, we propose to measure information loss through the discrepancies between matrices X, V, R, RF, C, and F obtained on the original data and the corresponding X', V', R', RF', C', and F' obtained on the masked dataset. In particular, discrepancy between correlations is related to the information loss for data uses such as regressions and cross tabulations.

Matrix discrepancy can be measured in at least three ways:

- *Mean square error:* Sum of squared componentwise differences between pairs of matrices, divided by the number of cells in either matrix.
- *Mean absolute error:* Sum of absolute componentwise differences between pairs of matrices, divided by the number of cells in either matrix.
- *Mean variation:* Sum of absolute percentage variation of components in the matrix computed on masked data with respect to components in the matrix computed on original data, divided by the number of cells in either matrix. This approach has the advantage of not being affected by scale changes of variables.

Table 5 summarizes the measures proposed. In this table, p is the number of variables and n the number of records. Components of matrices are represented by the corresponding lowercase letters (e.g., x_{ij} is a component of matrix X). Regarding $X - X'$ measures, it makes sense to compute them on the averages of variables rather than on all data (call this variant $\overline{X} - \overline{X}'$). It would also be sensible to use $V - V'$ measures to compare only the variances of the variables, that is, to compare the diagonals of the covariance matrices rather than the whole matrices (call this variant $S - S'$).

Table 5. Information Loss Measures for Continuous Microdata

	Mean square error	Mean abs. error	Mean variation
$X-X'$	$\frac{\sum_{j=1}^{p}\sum_{i=1}^{n}(x_{ij}-x'_{ij})^2}{np}$	$\frac{\sum_{j=1}^{p}\sum_{i=1}^{n}\lvert x_{ij}-x'_{ij}\rvert}{np}$	$\frac{\sum_{j=1}^{p}\sum_{i=1}^{n}\frac{\lvert x_{ij}-x'_{ij}\rvert}{\lvert x_{ij}\rvert}}{np}$
$V-V'$	$\frac{\sum_{j=1}^{p}\sum_{1\le i\le j}(v_{ij}-v'_{ij})^2}{\frac{p(p+1)}{2}}$	$\frac{\sum_{j=1}^{p}\sum_{1\le i\le j}\lvert v_{ij}-v'_{ij}\rvert}{\frac{p(p+1)}{2}}$	$\frac{\sum_{j=1}^{p}\sum_{1\le i\le j}\frac{\lvert v_{ij}-v'_{ij}\rvert}{\lvert v_{ij}\rvert}}{\frac{p(p+1)}{2}}$
$R-R'$	$\frac{\sum_{j=1}^{p}\sum_{1\le i\le j}(r_{ij}-r'_{ij})^2}{\frac{p(p-1)}{2}}$	$\frac{\sum_{j=1}^{p}\sum_{1\le i\le j}\lvert r_{ij}-r'_{ij}\rvert}{\frac{p(p-1)}{2}}$	$\frac{\sum_{j=1}^{p}\sum_{1\le i\le j}\frac{\lvert r_{ij}-r'_{ij}\rvert}{\lvert r_{ij}\rvert}}{\frac{p(p-1)}{2}}$
$RF-RF'$	$\frac{\sum_{j=1}^{p}w_j\sum_{i=1}^{p}(rf_{ij}-rf'_{ij})^2}{p^2}$	$\frac{\sum_{j=1}^{p}w_j\sum_{i=1}^{p}\lvert rf_{ij}-rf'_{ij}\rvert}{p^2}$	$\frac{\sum_{j=1}^{p}w_j\sum_{i=1}^{p}\frac{\lvert rf_{ij}-rf'_{ij}\rvert}{\lvert rf_{ij}\rvert}}{p^2}$
$C-C'$	$\frac{\sum_{i=1}^{p}(c_i-c'_i)^2}{p}$	$\frac{\sum_{i=1}^{p}\lvert c_i-c'_i\rvert}{p}$	$\frac{\sum_{i=1}^{p}\frac{\lvert c_i-c'_i\rvert}{\lvert c_i\rvert}}{p}$
$F-F'$	$\frac{\sum_{j=1}^{p}w_j\sum_{i=1}^{p}(f_{ij}-f'_{ij})^2}{p^2}$	$\frac{\sum_{j=1}^{p}w_j\sum_{i=1}^{p}\lvert f_{ij}-f'_{ij}\rvert}{p^2}$	$\frac{\sum_{j=1}^{p}w_j\sum_{i=1}^{p}\frac{\lvert f_{ij}-f'_{ij}\rvert}{\lvert f_{ij}\rvert}}{p^2}$

Information Loss Measures for Categorical Data

Straightforward computation of measures in Table 5 on categorical data is not possible. The following alternatives have been considered in the literature:

- Direct comparison of categorical values.
- Comparison of contingency tables.
- Entropy-based measures.

Direct Comparison of Categorical Values. Comparison of matrices X and X' for categorical data requires the definition of a distance for categorical variables. Definitions consider only the distances between pairs of categories that can appear when comparing a record and its masked version (possible pairs depend on the particular SDC method being used).

For a nominal variable V (a categorical variable that takes values over an unordered set), the only permitted operation is comparison for equality. This leads to the following distance definition:

$$d_V(c, c') = \begin{cases} 0 \ (\text{if } c = c') \\ 1 \ (\text{if } c \neq c') \end{cases}$$

where c is a category in the original dataset and c' is the category corresponding to c in the masked dataset. Correspondence between pairs of categories is determined by the masking method being used.

For an ordinal variable V (a categorical variable that takes values over a totally ordered set), let $\leq_V$ be the total order operator over the range $D(V)$ of variable V. Define the distance between categories c and c' as the number of categories between the minimum and the maximum of c and c' divided by the cardinality of the range:

$$d_V(c, c') = \frac{|c'':(c, c') \leq_V c'' \leq_V \max(c, c')|}{|D(V)|}$$

Comparison of Contingency Tables. An alternative to directly comparing the values of categorical variables is to compare their contingency tables. Given two datasets F and G (the original and the masked set, respectively) and their corresponding t-dimensional contingency tables for $t \leq K$, we can define a contingency table-based information loss measure *CTBIL* for a subset W of variables as follows:

$$CTBIL(F, G; W, K) = \sum_{\substack{\{V_{j1} \cdots V_{jt}\} \subseteq W \\ |\{V_{j1} \cdots V_{jt}\}| \leq K}} \sum_{i_1 \ldots i_t} |x^F_{i_1 \ldots i_t} - x^G_{i_1 \ldots i_t}|$$

where $x^{file}_{subscripts}$ is the entry of the contingency table of *file* at position given by *subscripts*.

Because the number of contingency tables to be considered depends on the number of variables $|W|$, the number of categories for each variable, and the dimension K, a normalized version of expression (1) may be desirable. This can be obtained by dividing expression (1) by the total number of cells in all considered tables.

Distance between contingency tables generalizes some of the information loss measures used in the literature. For example, μ-Argus (Hundepool *et al.* 1998) measures information loss for local suppression by counting the number of suppressions. The distance between two contingency tables of dimension one returns twice the number of suppressions. This is because, when category A is suppressed for one record, two entries of the contingency table are changed: The count of records with category *A* decreases and the count of records with the 'missing' category increases.

Entropy-Based Measures. In De Waal and Willenborg (1999) and Kooiman *et al.* (1998), the use of Shannon's entropy to measure information loss is discussed for local suppression, global recoding, and PRAM. Entropy is an information-theoretic measure, but it can be used in SDC if the masking process is modeled as the noise that would be added to the original dataset in the event of its being transmitted over a noisy channel.

As noted earlier, PRAM is a method that generalizes noise addition, suppression, and recoding methods. Therefore, our description of the use of entropy will be limited to PRAM.

Let V be a variable in the original dataset and V' be the corresponding variable in the PRAM-masked dataset. Let $P_{V,V'} = \{p(V' = j|V = i)\}$ be the PRAM Markov matrix. Then the conditional uncertainty of V given that $V' = j$ is:

$$H(V|V' = j) = -\sum_{i=1}^{n} p(V = i|V' = j)\log p(V = i|V' = j)$$

The probabilities in expression (2) can be derived from $P_{V,V'}$ using Bayes's formula. Finally, the entropy-based information loss measure (*EBIL*) is obtained by accumulating expression (2) for all individuals r in the masked dataset G

$$EBIL(P_{V,V'},\mathrm{G}) = \Sigma_{r\in G}\ H(V|V'=j_r)$$

where j_r is the value taken by V' in record r.

An Alternative Information Loss Measure. From our point of view, information loss when measured using the EBIL expression presents a drawback: The measure is a function of the masked dataset G but does not depend on the original dataset F. We begin with an example to illustrate this point; we then present an alternative approach that can also be applied to any PRAM-like SDC method.

Example 7

Assume that, in a household survey file, variable V contains the town where the household is located. Now consider that V is masked into a new variable V' where the town has been replaced by the state. Locations like 'New York City' and 'Albany' will be recoded into 'NY'. Living in Albany is more specific and identifying (in the sense of being less anonymous) than living in New York City. The information loss measure should somehow reflect that there is more information loss when a household in 'Albany' becomes a household in 'New York State' than when a household in 'New York City' becomes a household in 'New York State'.

Note that in Example 7

$$P(V = \text{'Albany'} \mid V' = NY) < P(V = \text{'New York City'} \mid V' = NY)$$

According to the U.S. Census Bureau's *American FactFinder* (U.S. Census Bureau 2001), the population of New York State in 2000 was 17,990,455, the population of New York City was 7,322,564, and the population of Albany was 101,082. Thus, the above probabilities are $P(V = \text{'Albany'} \mid V' = NY) = 101{,}082/17{,}990{,}455 = 0.05$ *and* $P(V = \text{'New York City'} \mid V' = NY) = 7{,}322{,}564/17{,}990{,}455 = 0.407$.

More generally, the smaller the conditional probability $P(V = i \mid V' = j)$, the larger the information loss. Based on this, we can define the information loss for a variable V as a function of three elements: the conditional probability, the original category i, and the masked category j. If we use minus the logarithm of $P(V = i \mid V' = j)$, the resulting information loss measure satisfies the monotonicity requirement of increasing as the conditional probability decreases. Thus the per-record information loss when $V = i$ is masked as $V' = j$ can be defined as

$$PRIL(P_{V,V'}, i, j) = -\log P(V = i \mid V' = j)$$

Note that it does not make sense to compute *PRIL* for categories i, j such that $P(V = i \mid V' = j) = 0$, because category i will never be masked as j. So *PRIL* is well defined. The information loss for the entire datasets F,G is

$$IL(P_{V,V'}, F, G) = \Sigma_{r \in G} PRIL(P_{V,V'}, i_r, j_r)$$

where i_r is the value taken by V in record r of F and j_r is the value taken by V' in record r of G.

5. Conclusions

The literature on statistical disclosure control for microdata is becoming increasingly vast. This chapter has presented an overview of current proposals as well as a set of measures to assess the extent to which a method damages the informational content of the data being protected. However, the reader should not forget that there is a trade-off between information loss and disclosure risk. In Chapter 6 we compare SDC methods in terms of this inevitable trade-off.

References

Adam, N.R., and J.C. Wortmann (1989) 'Security-Control Methods for Statistical Databases: A Comparative Study', *ACM Computing Surveys,* 21(4), pp.515-56.

Crystal Ball (2001), http://www.cbpro.com/.

Defays, D., and P. Nanopoulos (1993) 'Panels of Enterprises and Confidentiality: The Small Aggregates Method', in *Proceedings of the 1992 Symposium on Design and Analysis of Longitudinal Surveys*, Ottawa: Statistics Canada, pp.195-204.

De Waal, A.G., and L.C.R.J. Willenborg (1995) 'Global Recodings and Local Suppressions in Microdata Sets', in *Proceedings of Statistics Canada Symposium 95*, Ottawa: Statistics Canada, pp.121-32.

— (1999) 'Information Loss Through Global Recoding and Local Suppression', *Netherlands Official Statistics* (special issue on SDC), 14, pp.17-20.

De Wolf, P.-P., J.M. Gouweleeuw, P. Kooiman, and L.C.R.J. Willenborg (1999) 'Reflections on PRAM', in *Statistical Data Protection*, Luxembourg: Office for Official Publications of the European Communities, pp.337-49.

Domingo-Ferrer, J., and J.M. Mateo-Sanz (1999) 'On Resampling for Statistical Confidentiality in Contingency Tables', *Computers & Mathematics with Applications*, 38, pp.13-32.

— (2002) 'Practical Data-Oriented Microaggregation for Statistical Disclosure Control, *IEEE Transactions on Knowledge and Data Engineering* (forthcoming March 2002).

Duncan, G.T., and R.W. Pearson (1991) 'Enhancing Access to Microdata While Protecting Confidentiality: Prospects for the Future, *Statistical Science*, 6, pp.219-39.

Garey, M.R., and D.S. Johnson (1979) *Computers and Intractability: A Guide to the Theory of NP-Completeness*, New York: Freeman.

Gopal, R., P. Goes, and R. Garfinkel (1998) 'Confidentiality via Camouflage: The CVC Approach to Database Query', in J. Domingo-Ferrer (ed) *Statistical Data Protection*, Luxembourg: Office for Official Publications of the European Communities, pp.19-28.

Gouweleeuw, J.M., P. Kooiman, L.C.R.J. Willenborg, and P.-P. De Wolf (1997) 'Post Randomisation for Statistical Disclosure Control: Theory and Implementation', Research Paper No. 9731, Voorburg: Statistics Netherlands.

Heer, G.R. (1993) 'A Bootstrap Procedure to Preserve Statistical Confidentiality in Contingency Tables', in D. Lievesley (ed) *Proceedings of the International Seminar on Statistical Confidentiality*, Luxembourg: Office for Official Publications of the European Communities, pp.261-71.

Hundepool, A., and L. Willenborg (1999) 'ARGUS: Software From the SDC Project', in *Proceedings of Joint UNECE-Eurostat Work Session on Statistical Data Confidentiality*, Luxembourg: UNECE-Eurostat, pp.87-98.

Hundepool, A., L. Willenborg, A. Wessels, L. Van Gemerden, S. Tiourine, and C. Hurkens (1998) *μ-Argus Users Manual Version 2.5*, Voorburg: Statistics Netherlands, March.

IDESCAT-Statistics Catalonia (1995) *Sample of 1991 Population Census of Catalonia*. Barcelona, 1996, (Accessed August 22, 2001) http://www.idescat.es.

Joint Photographic Experts Group [JPEG] (2001) Standard IS 10918-1 (ITU-T T.81), http://www.jpeg.org.

Kennickell, A. (1998) 'Multiple Imputation and Disclosure Protection: The Case of the 1995 Survey of Consumer Finances', in J. Domingo-Ferrer (ed) *Proceedings of Statistical Data Protection '98*, Luxembourg: Office for Official Publications of the European Communities, pp.381-400.

Kim, J.J. (1986) 'A Method for Limiting Disclosure in Microdata Based on Random Noise and Transformation', in *Proceedings of the ASA Section on Survey Research Methodology*, Alexandria, Va.: American Statistical Association, pp.303-8.

Kooiman, P., L. Willenborg, and J. Gouweleeuw (1998) *PRAM: A Method for Disclosure Limitation of Microdata*, Research Report, Voorburg: Statistics Netherlands.

Liew, C.K., U.J. Choi, and C.J. Liew (1985) 'A Data Distortion by Probability Distribution', *ACM Transactions on Database Systems*, 10, pp.395-411.

Little, R.J.A. (1993) 'Statistical Analysis of Masked Data', *Journal of Official Statistics*, 9, pp.407-26.

Mateo-Sanz, J.M., and J. Domingo-Ferrer (1999) 'A Method for Data-Oriented Multivariate Microaggregation', in *Statistical Data Protection,* Luxembourg: Office for Official Publications of the European Communities, pp.89-99.

Matloff, N.E. (1986) 'Another Look at the Use of Noise Addition for Database Security', in *Proceedings of IEEE Symposium on Security and Privacy*, pp.173-80.

Moore, R. (1996) 'Controlled Data Swapping Techniques for Masking Public Use Microdata Sets', Washington, D.C.: U.S. Bureau of the Census, Statistical Research Division RR96/04.

Oganian, A., and J. Domingo-Ferrer (2001) 'On the Complexity of Microaggregation', in *Second Joint UNECE-Eurostat Work Session on Statistical Data Confidentiality*, Skopje, March.

Rubin, D.B. (1993) 'Satisfying Confidentiality Constraints Through the Use of Synthetic Multiply-Imputed Microdata, *Journal of Official Statistics*, 9, pp.461-8.

Sullivan, G., and W.A. Fuller (1989) 'The Use of Measurement Error to Avoid Disclosure', in *Proceedings of the ASA Section on Survey Research Methodology*, Alexandria, Va.: American Statistical Association, pp.802-7.

— (1990) 'Construction of Masking Error for Categorical Variables', in *Proceedings of the ASA Section on Survey Research Methodology*, Alexandria, Va.: American Statistical Association, pp.435-9.

U.S. Bureau of the Census (2001) *American FactFinder,* http://factfinder.census.gov/.

Willenborg, L., and T. De Waal (1996) *Statistical Disclosure Control in Practice*, Lecture Notes in Statistics 111, New York: Springer-Verlag.

Willenborg, L., and T. De Waal (2001) *Elements of Statistical Disclosure Control*, New York: Springer-Verlag.

Winkler W. (1998) 'Re-Identification Methods for Evaluating the Confidentiality of Analytically Valid Microdata', in *Statistical Data Protection*, Luxembourg: Office for Official Publications of the European Communities, 1999. Journal version (1998) in *Research in Official Statistics*, 1(2), pp.50-69.

Confidentiality, Disclosure, and Data Access: Theory and Practical Application for Statistical Agencies
Pat Doyle, Julia I. Lane, Jules J.M. Theeuwes and Laura M. Zayatz (Eds)

Chapter 6

A Quantitative Comparison of Disclosure Control Methods for Microdata*

Josep Domingo-Ferrer
Universitat Rovira i Virgili

Vicenç Torra
Institut d'Investigació en Intel•ligència Artificial

1. Introduction

As described in Chapter 5, there is a plethora of statistical disclosure control (SDC) methods to protect microdata. This chapter provides guidance in choosing a particular SDC method by comparing some of the methods discussed in Chapter 5 on the basis of both information loss and disclosure risk. Information loss can be readily quantified using analytical measures (either generic or data-use-specific). It is far more difficult to assess disclosure risk in a way that is both analytical and applicable to all methods. For this reason, our approach to disclosure risk evaluation is empirical, based on reidentification experiments carried out using record-linkage algorithms.

Methodology

The methodology we use to compare SDC methods is as follows:

- *Test data collection.* We obtained test data from publicly available microdata files. This guarantees public-domain reproducibility of the experiments reported here. The price paid is that the data we start from are not original but have already undergone some amount of disclosure protection.

* This work was funded in part by the U.S. Bureau of the Census under Contracts No. OBLIG-2000-29158-0-0 and OBLIG-2000-29144-0-0 and by the European Commission under Project 'CASC' IST-2000-25069. We are indebted to William Winkler for providing a U.S. Census Bureau implementation of probabilistic record linkage. Thanks go to Francesc Sebé, Narcís Macià, and Àngel Torres for their help in automating the probabilistic record linkage software and running the experiments. Comments by Josep Maria Mateo-Sanz, the editors, and several reviewers are gratefully acknowledged as well.

- *Information loss metrics.* Information loss actually depends on the data uses to be supported by the masked (*i.e.,* SDC-protected) data. However, potential data uses are so diverse that it is hard even to identify them. An alternative and more pragmatic approach for a general purpose comparison is to use a battery of generic, simple, and easily understandable information loss metrics that try to capture structural differences between the original and the masked data files.
- *Disclosure risk assessment.* This risk is empirically quantified as explained above.
- *Empirical work.* Experiments are conducted to obtain t-uples of the form *(method, parms, risk, loss)*, where *parms* are the input parameters to *method*, *risk* is the percentage of reidentified records in the test dataset, and *loss* is the information loss. Given *risk*, it is possible to find *method* and *parms* such that *risk(loss, method(parms))* is minimal (at least over the set of available t-uples). Given *loss*, it is possible to find *method* and *parms* such that *risk(loss, method(parms))* is minimal (at least over the set of available t-uples).

Structure of This Chapter

In Section 2, SDC methods included in the comparison are briefly reviewed, together with their parameterizations. Empirical ways to assess disclosure risk are presented in Section 3. The comparison rationale and results are presented in Section 4 for continuous data and in Section 5 for categorical data. Conclusions are summarized in Section 6.

2. SDC Methods Included in the Comparison

This chapter compares a subset of the methods described in Chapter 5. In this section, we briefly review the methods to be compared and specify the parameterizations we consider. Methods for continuous microdata and categorical microdata are discussed separately.

SDC Methods for Continuous Microdata

Microdata SDC methods can be classified as perturbative and nonperturbative (Willenborg and De Waal 2001). Perturbative methods distort records before release, which allows release of the whole population microdataset. Nonperturbative methods do not alter data but partially suppress or reduce the detail of the original dataset. We consider only perturbative methods for continuous data because reidentification is very easy for unperturbed continuous variables. The reason is that, if a continuous variable V_i is left unperturbed in the masked file and is present in an external administrative public file, unique matches are very likely, because for a continuous variable (even one truncated due to digital representation) V_i is not likely to take the same value for two different records (*i.e.,* it is unlikely that

$V_i(o_1) = V_i(o_2)$ if $o_1 \neq o_2$). Thus, distortion turns out to be the only effective way to protect continuous microdata. The subset of perturbative methods considered is:

- *Additive noise* (Noise*p* for short, where p is a parameter). Gaussian noise is added to the original data to get the masked data. If the standard deviation of the original variable is s, noise is generated using a $N(0,ps)$. Values of p considered in the experiments below are 0.01, 0.02, 0.04, 0.06, 0.08 up to 0.2 with 0.02 increments.
- *Data distortion by probability distribution* (abbreviated Distr (Liew *et al.* 1985)). For each variable in the original microdataset, the best fitting distribution is found; then the fitted distribution is used to generate the masked dataset. There are no parameters. Crystal Ball software (Crystal Ball 2001) is used to find the best fitting distribution.
- *Resampling.* If n is the number of records in the dataset, take with replacement t independent samples $X_1, \ldots, X_t$ of size n of the values of an original variable V_i. Independently rank each sample (using the same ranking criterion for all samples). Build the masked variable V'_i by taking as first value the average of the first values of the samples, as second value the average of the second values, and so on. Resampling is tested for $t = 1$ (Resamp1 for short) and $t = 3$ (Resamp3 for short).
- *Microaggregation.* Records are clustered into small aggregates or groups of size at least k (Defays and Nanopoulos 1993; Domingo-Ferrer and Mateo-Sanz 2002). Rather than publishing a variable for a given individual, the average of the values of the variable over the group to which the individual belongs is published. Variants of microaggregation considered include individual ranking (abbr. MicIR*k*); microaggregation on projected data using z-scores projection (abbr. MicZ*k*), and principal components projection (abbr. MicPCP*k*); and microaggregation on unprojected multivariate data considering two variables at a time (abbr. Mic2mul*k*), three variables at a time (abbr. Mic3mul*k*), four variables at a time (abbr. Mic4mul*k*), or all variables at a time (abbr. Micmul*k*). Values of k between 3 and 10 have been considered.
- *Lossy compression* (abbr. JPEG*q*, where q is a parameter). This method is new and we propose it for continuous data. The idea is to regard a numerical microdata file as an image (with rows being records and columns being variables). Lossy compression, and more specifically the JPEG algorithm (Joint Photographic Experts Group 2001), is then used on the image, and the compressed image is interpreted as a masked microdata file. Depending on the lossy compression algorithm used, appropriate mappings between variable ranges and color scales will be needed. The JPEG quality q is taken as a parameter with values from 5 percent up to 100 percent with 5 percent increments.
- *Rank swapping* (abbr. Rank*p*, where p is a parameter). Although originally described only for ordinal variables, this method can be used for any numerical variable (Moore 1996). First, values of variable V_i are ranked in ascending order; then each ranked value of V_i is swapped with another ranked value randomly cho-

sen within a restricted range (*e.g.,* the rank of two swapped values cannot differ by more than p percent of the total number of records). We consider values of p from 1 to 20.

SDC Methods for Categorical Microdata

For categorical data we consider both perturbative and nonperturbative methods and parameterizations. Each method depends on a single parameter and the set of variables to be masked, as described below.

- *Top-coding* (abbr. Tpv, where p is a parameter and v is a set of variables). This method is applied to ordinal categorical variables. In this case the last p values of the variable are recoded into a new category. We consider values of p from 1 to 9. For example, T5o corresponds to the experiment with a set of variables "o" with the parameter $p = 5$ (*i.e.,* five categories are top-coded).
- *Bottom-coding* (abbr. Bpv, where p and v are as above). This method is also applied to ordinal categorical variables. In this case the first p values of the variable are recoded into a new category. We consider values of p from 1 to 9. For example, B3s corresponds to the experiment with a set of variables "s" with the parameter $p = 3$ (*i.e.,* three categories are bottom-coded).
- *Global recoding* (abbr. Gpv, where p and v are as above). In global recoding, some of the categories of the variable are recoded into new ones. Our experimentation considers the following parameterization: recode the p lowest frequency categories into a single one. As before, we consider values of p from 1 to 9.
- *Post-Randomization Method or PRAM* (abbr. Ppv, where p and v are as above). The scores of some categorical variables for certain records in the original file are changed to a different score according to a prescribed probability mechanism (a Markov matrix). We select the approach described in Kooiman *et al.* (1998) to define the PRAM matrix, as follows. Let $T_V=(T_V(1), ..., T_V(K))t$ be the vector of frequencies of the K categories of variable V in the original file (assume without loss of generality that $T_V(k) \geq T_V(K) > 0$ for $k < K$) and let θ be such that $0 < \theta < 1$. Then the PRAM matrix for variable V is defined as:

$$p_{kl} = \begin{cases} 1 - \theta T_V(K)/T_V(k) & if\ l = k \\ \theta T_V(K)/((K-1)T_V(k)) & if\ l \neq k \end{cases}$$

Let parameter p be $p: = 10\theta$. For each variable we have built nine matrices generated with p taking integer values between 1 and 9.

3. Disclosure Risk Measures

Chapter 5 discusses ways to measure the information loss caused by SDC methods for microdata. However, the assessment of the quality of an SDC method cannot be limited to information loss; disclosure risk must also be measured. The method that optimizes the trade-off between both magnitudes subject to some user requirements turns out to be the best option. To understand the trade-off, consider the two extreme cases between which SDC methods lie:

- If masking consists of encrypting original data, then no disclosure is possible, but no information at all is released (maximum information loss, minimum disclosure risk).
- If no masking is performed and the original data are released, users can perform fully accurate computations, but disclosure of individual respondent data is very likely, especially for microdata (minimum information loss, maximum disclosure risk).

Literature on disclosure risk basically relates to nonperturbative methods, in which a sample of the original dataset is published. Disclosure risk is measured here as the probability that a sample unique is a population unique (Elliot *et al.* 1998; Skinner *et al.* 1994). Unless the sample size is much smaller than the population size, such a probability can be dangerously high; in that case, an intruder who locates a unique value in the released sample can be almost certain that there is a single individual in the population with that value, which is very likely to lead to that individual's identification.

The uniqueness property as stated above is not relevant for perturbative methods, because, even though the whole microdataset is released, it is released with some distortion. Because there is little in the literature on disclosure risk that can be used for a broad class of perturbative methods—disclosure risk measures tend to be method-specific (Adam and Wortmann 1989 is still up-to-date)—empirical methods, such as record linkage techniques, provide a more unified approach. We briefly describe below two approaches to record linkage that yield empirical disclosure risk measures and one analytical measure based on interval disclosure.

Distance-Based Record Linkage

This approach to record linkage is described in Pagliuca and Seri (1999) for the specific case of microaggregation masking using the Euclidean distance. It can be generalized, however, for any perturbative method provided that a distance between the original and the masked value can be defined. As in any record linkage context, it is assumed that an intruder has an external dataset containing as key variables the same variables present in the released masked dataset. The intruder is assumed to try to link the masked dataset with the external dataset using the key variables.

Linkage then proceeds by computing the distances between records in the original and the masked datasets. The distances used are standardized to avoid scaling problems. For each record in the masked dataset, the distance to every record in the original dataset is computed. Then the 'nearest' and 'second nearest' records in the original dataset are considered. A record in the masked dataset is labeled as 'linked' when the nearest record in the original dataset turns out to be the corresponding original record. A record in the masked dataset is labeled as 'linked to 2nd nearest' when the second nearest record in the original dataset turns out to be the corresponding original record. In all other cases, a record in the masked dataset is labeled as 'not linked'. The percentage of 'linked' and 'linked to 2nd nearest' is a measure of disclosure risk.

The Euclidean method above requires rescaling variables as well as an assumption on the weight of variables when computing a distance. For instance, in the proposal of Pagliuca and Seri (1999), all variables have the same weight.

Probabilistic Record Linkage

In Jaro (1989), a record linkage method was described and illustrated on the 1985 census of Tampa, Florida. The matching algorithm uses the linear sum assignment model to 'pair' records in the two files to be matched (the original file and the masked file in our case). The percentage of correctly paired records is a measure of disclosure risk.

Although less simple than the Euclidean method, this approach is attractive because it does not assume rescaling or weighting of variables and requires the user to provide only two probabilities as input. For that reason, this method of record linkage will be termed probabilistic in what follows.

We use the U.S. Census Bureau implementation of probabilistic record linkage provided by William Winkler (U.S. Bureau of the Census 2000; Winkler 1998) (with some additions) in our experimentation.

Interval Disclosure

For a record in the masked dataset, compute rank intervals as follows: Each variable is independently ranked and a rank interval is defined around the value the variable takes on each record. The ranks of values within the interval for a variable around record r should differ less than p percent of the total number of records and the rank in the center of the interval should correspond to the value of the variable in record r. Then, the proportion of original values that fall into the interval centered around their corresponding masked value is a measure of disclosure risk. A 100 percent proportion means that an attacker is completely sure that the original value lies in the interval around the masked value (interval disclosure).

4. Comparison for Continuous Microdata

This section details the steps of the methodology outlined in the introduction for the case of continuous data.

Test Data Collection

We constructed a microdataset using the Data Extraction System (DES) of the U.S. Census Bureau (http://www.census.gov/DES/www/welcome.html). From the available data sources, we chose the Current Population Survey corresponding to 1995—specifically, the file group 'March Questionnaire Supplement—Person Data Files'. Variables and records were selected as follows:

- *Variable selection.* Our continuous variable selection was based on the requirement that the values of each span a wide range. Thirteen variables were selected: AFNLWGT (Final weight), AGI (Adjusted gross income), EMCONTRB (Employer contribution for health insurance), ERNVAL (Business or farm net earnings), FEDTAX (Federal income tax liability), FICA (Social Security retirement payroll reduction), INTVAL (Amount of interest income), PEARNVAL (Total personal earnings), POTHVAL (Total other persons income), PTOTVAL (Total personal income), STATETAX (State income tax liability), TAXINC (Taxable income amount), WSALVAL (Amount: Total wage and salary).
- *Record selection.* Our selection of 1,080 records was based on the need to keep the number of repeated values for each variable low (in principle, one would not expect repeated values for a continuous variable, but there were repetitions in the dataset).
- The resulting dataset had three properties that were important to our work:

  (1) The number of records was fewer than 1,200, which allowed repeated experimentation with the probabilistic record linkage software in reasonable time.

  (2) Seven variables had no repeated values: FEDTAX, AFNLWGT, AGI, EMCONTRB, PTOTVAL, TAXINC, STATETAX. Because absence of repeated values is a distinguishing feature of really continuous variables, these seven were chosen as key variables for record linkage.

  (3) 1,080 (the number of records) is the largest integer less than 1,200, which is a multiple of 5, 8, and 9. Thus, when the microaggregation SDC method is used, the dataset can be microaggregated with minimal group sizes $k = 3, 4, 5, 6, 8, 9$, and 10 so that all groups have exactly size k.

We used the resulting data to carry out the empirical work described in the 'Empirical Work' section below.

Information Loss Metrics

Let X and X' be the matrices representing the original and the masking datasets, respectively. Let V and R be the covariance matrix and the correlation matrix of X; let $\overline{X}$ be the vector of variable averages for X and let S be the diagonal of V. Define V', R', $\overline{X}'$, and S' analogously from X'. The Information Loss (IL) is computed by averaging the mean variations of $X - X'$, $\overline{X} - \overline{X}'$, $V - V'$, $S - S'$, and the mean absolute error of $R - R'$ and multiplying the resulting average by 100. According to the formulae given in Chapter 5 we obtain:

$$IL = 100*\left(\frac{\sum_{j=1}^{p}\sum_{i=1}^{n}\frac{|x_{ij}-x'_{ij}|}{|x_{ij}|}}{np} + \frac{\sum_{j=1}^{p}\frac{|\overline{x}_j-\overline{x}'_j|}{|\overline{x}_j|}}{p} + \frac{\sum_{j=1}^{p}\sum_{1\le i\le j}\frac{|v_{ij}-v'_{ij}|}{|v_{ij}|}}{\frac{p(p+1)}{2}} + \frac{\sum_{j=1}^{p}\frac{|v_{jj}-v'_{jj}|}{|v_{jj}|}}{p} + \frac{\sum_{j=1}^{p}\sum_{1\le i\le j}|r_{ij}-r'_{ij}|}{\frac{p(p-1)}{2}}\right)/5$$

Term 2 is analogous to term 1 but only with respect to the averages of variables. Term 4 is analogous to term 3 but only with respect to the variances of variables (not covariances).

Disclosure Risk Assessment

The Distance Linkage Disclosure risk (*DLD*) is the average percentage of linked records using distance-based record linkage; the average is computed over the number of key variables that the intruder is assumed to know (we consider knowledge of anywhere from one to seven variables). Similarly, the Probabilistic Linkage Disclosure risk (*PLD*) is the average percentage of correctly paired records using probabilistic linkage. The Interval Disclosure (*ID*) is the average percentage of original values falling in the intervals around their corresponding masked values (averages that have been computed over all parameter values, *i.e.,* 1 percent to 10 percent with 1 percent increments).

Empirical Work

Table 1 contains a ranking of methods described in Section 2 (we try the parameter values described in that section for each method). The table contains columns specifying *IL, DLD, PLD,* and *ID* and also an overall score constructed as follows:

$$Score = 0.5(IL) + 0.125(DLD) + 0.125(PLD) + 0.25\,(ID)$$

The rationale of the above weighting is to give equal weight to information loss (0.5) and to disclosure risk. The 0.5 weight of disclosure risk is equally divided among *ID* (0.25) and record linkage. The 0.25 weight of record linkage is equally divided among both approaches to record linkage. The correlation between *DLD* and *PLD* is actually 0.962, so both approaches are very similar. The (*IL, DLD*), (*IL, PLD*), and (*IL, ID*) correlations are –0.605, –0.551, and –0.807; thus, the lower the information loss, the higher the disclosure risk, as one would expect. The *ILRank*, *DLDRank*, *PLDRank*, and *IDRank* columns contain the ranking of each method with respect to *IL*, *DLD*, *PLD*, and *ID*; the lower the rank, the better a method performs (*i.e.*, lower information loss and disclosure risk).

Table 1: Comparison Results for Continuous Microdata

Method	IL	DLD	PLD	ID	Score	IL-Rank	DLD-Rank	PLD-Rank	ID-Rank
Rank15	19.01	1.19	0.15	35.05	18.44	53	6	7	21
Rank19	22.95	0.93	0.08	28.04	18.61	59	2	2	2
Rank16	20.91	1.39	0.11	32.18	18.69	56	8	5	16
Rank13	16.77	2.17	0.12	40.35	18.76	48	12	6	28
Rank14	19.72	1.92	0.07	37.00	19.36	55	10	1	25
Rank11	14.32	2.43	0.25	47.81	19.45	44	13	14	39
Rank12	16.37	2.50	0.25	43.73	19.46	47	14	11	35
Rank20	25.81	0.69	0.09	26.83	19.71	64	1	3	1
Rank18	25.74	0.95	0.09	29.25	20.31	63	4	4	6
Rank10	13.37	3.90	0.38	53.17	20.51	41	24	17	45
Rank17	25.12	1.52	0.20	30.95	20.51	61	9	9	10
Rank09	11.66	5.01	0.52	57.58	20.91	38	37	29	49
Rank08	11.60	6.07	0.85	63.37	22.51	37	39	39	56
Rank07	9.25	7.51	1.08	68.71	22.87	30	41	43	63
Rank06	7.87	9.02	2.79	73.80	23.86	26	43	56	71
Mic3mul07	11.06	19.34	4.70	72.34	26.62	36	68	65	69
Rank05	6.78	16.80	13.60	78.89	26.91	22	58	70	77
Mic3mul09	13.46	19.22	3.44	69.91	27.04	42	67	60	65
Mic3mul10	14.84	17.99	3.44	68.61	27.25	46	64	59	62
Mic4mul04	12.14	19.76	6.67	71.85	27.33	39	69	68	68
Mic4mul05	14.50	17.43	5.45	69.09	27.39	45	61	66	64
Mic3mul08	13.51	20.81	4.15	70.68	27.54	43	71	63	66
Mic4mul08	18.89	17.78	3.35	62.84	27.80	52	62	58	55
Mic3mul06	10.24	20.41	13.90	74.00	27.91	33	70	71	72
Mic4mul07	19.36	17.10	2.08	64.41	28.18	54	60	53	58
Mic4mul06	17.91	17.82	3.98	66.41	28.28	50	63	62	60
Mic4mul09	21.35	15.93	2.00	61.66	28.33	58	57	52	54
Mic4mul10	22.98	16.85	2.37	60.56	29.03	60	59	55	51
Mic3mul05	9.73	23.78	18.29	76.59	29.27	31	76	73	74
Mic3mul04	7.45	23.49	22.75	79.14	29.29	24	75	75	79
Mic4mul03	10.69	22.88	16.69	76.89	29.51	35	74	72	75
Rank04	5.90	22.77	22.78	84.12	29.67	20	73	76	86
Micmul03	27.67	14.26	1.88	57.23	30.16	65	54	50	47

Table 1: Comparison Results for Continuous Microdata (Continued)

Method	IL	DLD	PLD	ID	Score	IL-Rank	DLD-Rank	PLD-Rank	ID-Rank
Micmul04	31.74	13.72	1.38	52.44	30.86	67	53	48	44
Mic3mul03	6.29	29.70	29.06	82.95	31.23	21	79	80	85
Micmul05	35.12	11.73	1.14	48.43	31.27	70	46	44	41
Micmul07	37.68	13.20	1.20	43.46	31.50	72	52	45	34
Micmul06	38.77	13.00	1.22	45.76	32.60	73	50	46	37
Micmul08	41.53	13.12	0.99	42.66	33.19	75	51	42	32
Rank03	5.07	31.73	36.92	89.53	33.50	18	80	83	93
Mic2mul10	10.68	49.38	27.29	77.43	34.28	34	86	78	76
Micmul10	44.69	14.66	0.50	40.41	34.34	76	55	27	29
Noise0.16	32.56	15.65	4.66	64.39	34.91	68	56	64	57
Micmul09	45.98	12.82	0.85	40.99	34.95	79	49	40	30
Mic2mul09	9.93	51.03	33.04	78.94	35.21	32	87	81	78
Mic2mul08	8.55	54.31	33.70	79.77	35.22	27	88	82	80
Mic2mul07	7.53	54.72	37.41	81.40	35.63	25	89	84	83
Noise0.12	25.24	22.21	22.39	71.58	36.09	62	72	74	67
Noise0.1	21.14	27.70	29.03	75.20	36.46	57	78	79	73
Mic2mul06	7.03	56.38	42.00	82.89	36.54	23	90	86	84
JPEG080	33.97	19.13	6.93	66.35	36.83	69	65	69	59
Noise0.14	35.13	19.21	6.24	67.62	37.65	71	66	67	61
Noise0.18	41.12	11.96	3.52	60.95	37.73	74	47	61	52
Noise0.08	17.43	36.06	39.76	79.84	38.15	49	82	85	81
Rank02	2.90	47.26	57.47	94.56	38.18	11	85	90	96
JPEG070	44.92	9.66	2.34	57.28	38.28	77	44	54	48
Noise0.2	45.97	10.01	0.97	57.63	38.77	78	45	41	50
Mic2mul05	5.88	58.97	56.84	85.40	38.77	19	92	89	88
JPEG085	29.47	23.85	24.48	72.80	38.98	66	77	77	70
Mic2mul04	4.90	61.53	60.69	87.26	39.54	17	94	91	89
JPEG090	18.17	35.37	46.98	80.87	39.60	51	81	87	82
Noise0.06	13.03	45.54	56.22	84.16	40.28	40	84	88	87
Mic2mul03	3.28	66.97	64.79	90.51	40.74	15	95	92	94
Noise0.04	8.93	58.51	65.28	88.95	42.18	28	91	94	90
JPEG075	50.45	12.67	2.90	61.27	42.49	80	48	57	53
JPEG095	9.06	60.11	66.56	89.23	42.67	29	93	96	92
Resamp3	3.15	67.90	67.63	96.81	42.72	14	96	97	97
Rank01	2.34	69.19	66.35	99.54	43.00	9	97	95	106
JPEG065	57.77	7.02	1.90	53.87	43.47	81	40	51	46
Noise0.02	4.24	77.34	71.32	94.42	44.31	16	99	98	95
Resamp1	3.11	75.42	71.85	98.36	44.56	13	98	99	99
MicPCP03	69.62	3.16	0.77	38.41	44.90	84	17	38	26

Table 1: Comparison Results for Continuous Microdata (Continued)

Method	IL	DLD	PLD	ID	Score	IL-Rank	DLD-Rank	PLD-Rank	ID-Rank
JPEG055	63.70	5.57	1.26	49.70	45.13	83	38	47	42
Noise0.01	2.57	85.19	74.13	97.03	45.46	10	100	103	98
JPEG100	3.06	87.14	73.03	99.14	46.34	12	101	100	101
MicIR10	1.19	97.37	74.07	99.12	46.81	8	102	102	100
MicIR08	1.03	97.84	74.07	99.29	46.83	6	108	101	103
MicIR09	1.14	97.96	74.40	99.24	46.93	7	109	104	102
MicIR06	0.87	97.66	75.28	99.51	46.93	5	106	105	105
MicIR05	0.69	97.58	75.99	99.58	46.94	3	104	106	107
MicIR03	0.45	97.39	78.96	99.79	47.22	1	103	107	109
MicIR04	0.64	97.63	79.78	99.67	47.41	2	105	108	108
MicIR07	0.81	97.79	88.06	99.42	48.49	4	107	109	104
MicPCP04	78.84	3.43	0.62	36.00	48.92	87	19	32	23
JPEG050	73.20	4.26	0.67	47.96	49.21	86	31	36	40
JPEG060	71.24	7.66	1.52	51.71	49.69	85	42	49	43
MicPCP05	82.55	3.94	0.69	34.10	50.38	88	25	37	20
MicPCP07	89.28	4.02	0.62	32.56	53.36	91	27	33	17
MicPCP09	90.78	4.54	0.25	31.40	53.84	94	34	12	13
MicPCP06	90.26	3.37	0.50	33.42	53.97	93	18	26	19
MicZ03	90.25	3.16	0.61	35.71	54.52	92	16	31	22
JPEG035	88.80	3.65	0.44	43.20	55.71	90	20	23	33
JPEG045	87.55	4.15	0.67	46.78	56.07	89	30	35	38
MicZ04	94.94	3.70	0.53	33.04	56.26	96	21	30	18
MicPCP08	96.93	3.97	0.34	32.04	57.02	97	26	16	14
MicPCP10	97.82	4.13	0.46	31.19	57.28	98	29	24	11
JPEG040	90.99	3.72	0.66	44.98	57.29	95	22	34	36
MicZ07	102.87	4.27	0.38	30.53	59.65	99	32	20	9
MicZ06	103.92	3.88	0.41	30.43	60.10	100	23	21	8
MicZ05	104.06	4.03	0.42	31.30	60.41	101	28	22	12
MicZ08	107.92	4.55	0.52	29.60	61.99	102	35	28	7
MicZ10	109.79	4.83	0.38	28.20	62.59	103	36	18	3
MicZ09	110.91	4.35	0.38	28.36	63.14	105	33	19	4
Distr	58.62	43.05	64.88	88.98	65.04	82	83	93	91
JPEG030	110.48	3.02	0.48	41.79	66.12	104	15	25	31
JPEG025	155.15	2.13	0.25	38.76	87.56	106	11	13	27
JPEG020	164.91	1.36	0.29	36.11	91.69	107	7	15	24
JPEG015	202.66	1.10	0.15	32.06	109.50	108	5	8	15
JPEG010	269.38	0.93	0.22	28.44	141.94	109	3	10	5

Because publishing the original data without masking yields a score of 50 (*IL* = 0 and *DLD* = *PLD* = *ID* = 100), methods scoring above 50 in Table 1 are of no use.

Figure 1. Comparison of SDC Methods for Continuous Microdata With Best Parameter Choice

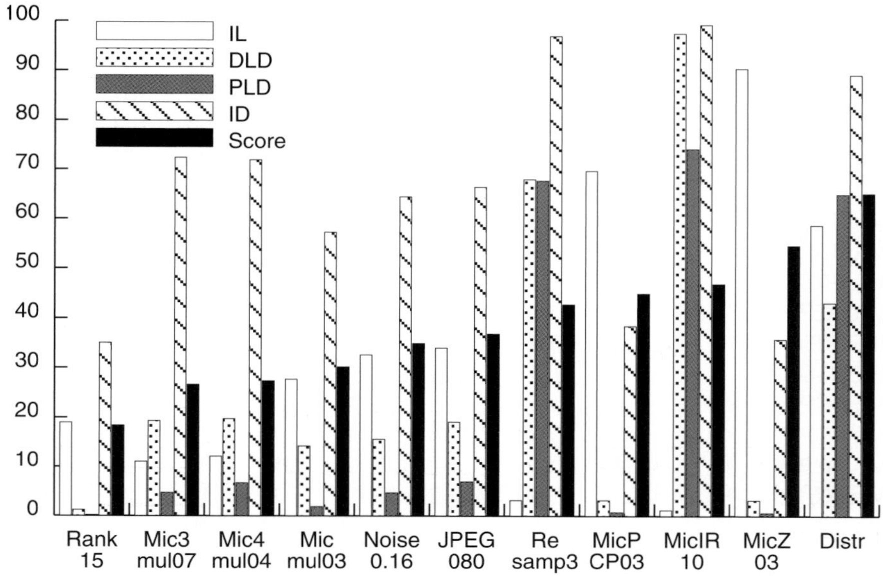

Figure 1 gives a comparison of SDC methods for continuous microdata when the best parameter choice is made. Rank swapping with parameter 15 (Rank15) stands out as the best performer. For multivariate microaggregation, taking three variables at a time and groups of size at least 7 (Mic3mul07) turns out to be the best parameter choice. See Section 6 for expanded conclusions.

5. Comparison for Categorical Microdata

This section details steps of the methodology outlined in the introduction for the case of categorical data.

Test Data Collection

In order to compare masking methods for categorical microdata, we used data from the *American Housing Survey 1993* (also obtained from the U.S. Census Bureau using the DES). We selected the variables BUILT (year structure was built), DEGREE (long-term average degree days), GRADE1 (highest school grade), METRO (metropolitan areas), SCH (schools adequate), SHP (shopping facilities adequate),

TRAN1 (principal means of transportation to work), WHYMOVE (primary reason for moving), WHYTOH (main reason for choice of house), WHYTON (main reason for choosing this neighborhood). We took the first 1,000 records from the corresponding data file (we chose a small dataset for the same reasons mentioned above for continuous microdata).

Five subsets of variables were defined from the set of selected variables, and the same analysis was performed for each of them in the testing process. Three groups were defined by grouping variables with similar number of categories; 's', 'm', and 'l' correspond to groups of variables with small, medium, and large number of categories, respectively. Additionally, 'u' corresponds to the group of variables with medium or large number of categories (union of 'm' and 'l'); 'o' corresponds to the subset of variables defined as ordered. The variables used and the groups of variables are given in Table 2. This table also includes the number of categories for each variable.

Table 2: Variables Used in the Masking Process and in the Reidentification Process

Variables	u	l	s	m	o	N. of Cat.
BUILT	X	X			X	25
DEGREE			X		X	8
GRADE1	X	X			X	21
METRO			X			9
SCH			X			6
SHP			X			6
TRAN1	X			X		12
WHYMOVE	X	X				18
WHYTOH	X			X		13
WHYTON	X			X		13

Information Loss Metrics

For categorical data, we considered three kinds of information loss measures: direct comparison of categorical values, comparison of contingency tables, and entropy-based measures (see Chapter 5 for details).

Direct Comparison of Categorical Values. A distance is defined over the range of categorical variables. When the range of a variable is an ordinal scale, the distance between category a and b is proportional to the number of categories between a and b. When the range of a variable is not ordinal, the distance is 1 if the values are different and 0 if they are not. We denote this information loss measure by Dist.

Comparison of Contingency Tables. For a given subset of variables, contingency tables are computed for a file before and after applying the masking process. The number of differences between the two contingency tables is denoted by *CTBIL*. Because the number of cells in a contingency table depends on the number of categories in the variable, we also consider the normalizing of *CTBIL* by dividing it by the number of cells in all tables. We denote the resulting information loss measure by *ACTBIL*.

Entropy-Based Measures. In Kooiman *et al.* (1998), the use of Shannon's entropy to measure information loss is discussed. The idea is that this information-theoretic measure can be used in SDC, if the masking process is modeled as the noise that would be added to the original dataset in the event of its being transmitted over a noisy channel. Because this measure depends only on the masked dataset and does not account for its relation with the original data, we define a new information loss measure.

Let V be a variable in the original dataset and V' be the corresponding variable in the PRAM-masked dataset (we take PRAM because it is a very general method encompassing the rest of the masking methods considered for categorical data). Then the entropy-based information loss measure *EBIL* is defined as:

$$EBIL(P_{V,V'}, G) = \Sigma_{r \in G} H(V|V' = j_r)$$

where j_r is the value taken by V' in record r, and

$$H(V|V' = j) = -\sum_{i=1}^{n} p(V = i|V' = j) \log p(V = i|V' = j)$$

$P_{V,V'} = \{p(V' = j|(V = i)\}$ being the PRAM Markov matrix.

The new information loss measure taking original data into account is:

$$IL(P_{V,V'}, F, G) = \Sigma_{r \in G} PRIL(P_{V,V'}, i_r, j_r)$$

where i_r is the value taken by V in record r of F and j_r is, as before, the value taken by V' in record r of G and

$$PRIL(P_{V,V'}, i, j) = -\log P(V = i|V' = j)$$

We compute EBIL and IL using two different data files to estimate probabilities: the same file masked (EBILMF and ILMF) and a reference file (EBILRF and ILMF).

Disclosure Risk Assessment

The Probabilistic Linkage Disclosure risk (*PLD*) is the number of correctly paired records using probabilistic linkage. *PLDRank* is the ranking of the method with respect to *PLD* (normalized with maximal value 100). The inability to use the Euclidean distance for categorical data makes computation of *DLD* and *ID* more cumbersome than for continuous data. Although distance definitions exist for categorical variables (as the ones used for the information loss measures), difficulties arise in ordering pairs of records.

Empirical Work

Table 3 contains a ranking of methods described in the section 'SDC Methods for Categorical Microdata' (the parameter values described in that section were tried for each method). The table contains columns with information loss measures (*Dist*, *CTBIL*, *ACTBIL*, *EBILRF*, *ILRF*, *EBILMF*, and *ILMF*), a reidentification measure (*PLD*), and a score of the methods (*Score* and *Ave. Score*) defined as an average between the ranks of disclosure risk and information loss measures. The *(Dist, PLD)*, *(CTBIL, PLD)*, *(ACTBIL, PLD)*, *(EBILRF, PLD)*, *(ILRF, PLD)*, *(EBILFM, PLD)*, and *(ILFM, PLD)* correlations are, respectively, -0.4898, -0.4156, -0.5520, -0.345, -0.288, -0.3368 and -0.408. Here again, as expected, the lower the information loss, the higher the disclosure risk.

Table 3: Comparison Results for Categorical Microdata

Method	PLD	Dist	CTBIL	ACTBIL	EBILRF	ILRF	EBILMF	ILMF	PLD Rank	AIL Rank	Score	Ave. Score
T5u	235	2716	31830	11.10	2892.10	3030.30	2952.60	2952.60	5.60	72.82	39.19	41.60
T5s	428	2372	14788	27.00	2453.40	2564.10	2506.90	2506.90	23.90	70.19	47.04	41.60
T5o	347	1093	6148	5.20	1661.00	1727.40	1687.50	1687.50	15.00	56.76	35.88	41.60
T5l	853	319	1820	1.00	486.80	481.40	474.90	474.90	58.30	32.36	45.35	41.60
T5m	769	344	1874	2.70	438.70	466.20	445.70	445.70	45.00	36.11	40.56	41.60
T6u	200	3152	35786	12.50	3764.10	3900.70	3819.50	3819.50	5.00	80.23	42.62	42.19
T6s	457	2789	16400	30.00	3256.00	3357.30	3299.80	3299.80	26.70	74.81	50.74	42.19
T6o	287	1242	6886	5.80	2106.70	2171.90	2130.50	2130.50	10.00	59.77	34.88	42.19
T6l	789	391	2220	1.30	671.20	663.60	655.50	655.50	48.90	37.18	43.03	42.19
T6m	751	363	1966	2.90	508.10	543.40	519.70	519.70	41.70	37.64	39.65	42.19
T3u	288	1081	14154	4.90	942.10	963.80	946.40	946.40	10.60	55.65	33.10	42.90
T3s	748	819	6150	11.20	691.20	717.40	706.70	706.70	41.10	51.62	46.37	42.90
T3o	758	559	3258	2.80	576.40	588.40	579.10	579.10	42.80	41.90	42.34	42.90
T3l	943	177	1028	0.60	179.40	178.90	173.30	173.30	78.90	23.38	51.13	42.90
T3m	825	262	1442	2.10	250.90	246.50	239.70	239.70	52.80	30.37	41.57	42.90
T4u	254	2475	29456	10.30	2289.30	2357.70	2311.40	2311.40	7.20	69.07	38.15	43.15

Table 3: Comparison Results for Categorical Microdata (Continued)

Method	PLD	Dist	CTBIL	ACTBIL	EBILRF	ILRF	EBILMF	ILMF	PLD Rank	AIL Rank	Score	Ave. Score
T4s	492	2144	13756	25.10	1896.40	1943.90	1917.60	1917.60	28.30	67.27	47.80	43.15
T4o	529	854	4896	4.10	1117.70	1135.10	1124.90	1124.90	30.00	50.65	40.32	43.15
T4l	908	233	1348	0.80	304.90	303.10	297.10	297.10	71.10	27.31	49.21	43.15
T4m	775	331	1812	2.60	392.90	413.80	393.80	393.80	46.10	34.40	40.25	43.15
G1u	423	55	770	0.30	0.00	0.00	0.00	0.00	23.30	12.41	17.87	43.25
G1s	1000	51	408	0.90	0.00	0.00	0.00	0.00	98.30	13.80	56.06	43.25
G1o	998	35	210	0.20	0.00	0.00	0.00	0.00	95.60	6.48	51.02	43.25
G1l	998	4	24	0.00	0.00	0.00	0.00	0.00	96.10	0.65	48.38	43.25
G1m	962	4	24	0.00	0.00	0.00	0.00	0.00	84.40	1.39	42.92	43.25
T7u	265	3453	38344	13.40	4389.90	4564.00	4475.60	4475.60	7.80	82.64	45.21	44.15
T7s	687	3067	17418	31.80	3801.90	3943.70	3879.90	3879.90	35.00	80.09	57.55	44.15
T7o	307	1314	7172	6.10	2281.90	2345.20	2303.70	2303.70	12.80	61.76	37.27	44.15
T7l	719	483	2736	1.60	903.20	893.80	885.50	885.50	38.30	41.71	40.02	44.15
T7m	751	386	2072	3.00	588.00	620.30	595.70	595.70	42.20	39.17	40.69	44.15
T9u	429	4163	43976	15.40	4921.40	5230.90	5133.50	5133.50	24.40	88.15	56.30	44.18
T9s	187	3695	19294	35.30	4097.70	4368.10	4305.50	4305.60	4.40	83.56	44.00	44.18
T9o	297	1604	8162	6.90	2925.90	2917.30	2834.90	2834.90	12.20	66.99	39.61	44.18
T9l	534	812	4392	2.50	1665.60	1585.90	1536.00	1536.00	30.60	49.35	39.95	44.18
T9m	725	468	2442	3.50	823.70	862.70	828.00	828.00	38.90	43.19	41.04	44.18
T8u	510	4167	43940	15.30	4957.20	5244.20	5140.20	5140.20	28.90	88.33	58.61	44.75
T8s	44	3774	19386	35.40	4337.20	4583.70	4511.20	4511.20	0.60	84.44	42.50	44.75
T8o	329	1494	7772	6.60	2635.50	2622.50	2542.50	2542.60	14.40	64.81	39.63	44.75
T8l	643	679	3726	2.10	1308.40	1222.40	1175.50	1175.50	33.30	47.08	40.21	44.75
T8m	774	393	2108	3.10	620.00	660.50	629.00	629.00	45.60	40.00	42.78	44.75
G4u	388	1269	16606	6.90	1246.10	1272.40	1241.80	1241.80	16.70	60.88	38.77	45.35
G4s	729	1215	8690	19.80	1183.50	1203.20	1175.60	1175.60	39.40	61.48	50.46	45.35
G4o	810	424	2508	2.40	560.10	597.20	517.50	517.50	51.10	39.21	45.16	45.35
G4l	983	71	420	0.30	81.90	144.90	76.10	76.10	90.00	14.63	52.31	45.35
G4m	893	54	316	0.50	62.60	69.30	66.10	66.10	66.10	13.94	40.02	45.35
T2u	324	655	8802	3.10	406.50	400.40	388.90	388.90	13.90	44.35	29.12	45.36
T2s	890	462	3598	6.60	301.40	299.50	293.30	293.30	65.60	39.72	52.64	45.36
T2o	903	330	1960	1.70	218.00	210.50	205.90	205.90	68.30	30.79	49.56	45.36
T2l	965	132	780	0.40	83.20	79.40	76.10	76.10	85.00	19.49	52.25	45.36
T2m	873	193	1058	1.50	105.00	101.00	95.60	95.60	61.10	25.37	43.24	45.36
T1u	397	222	3058	1.10	0.00	0.00	0.00	0.00	19.40	25.00	22.22	46.88
T1s	1000	172	1362	2.50	0.00	0.00	0.00	0.00	98.90	23.70	61.30	46.88
T1o	1000	109	654	0.60	0.00	0.00	0.00	0.00	99.40	15.56	57.50	46.88
T1l	996	40	240	0.10	0.00	0.00	0.00	0.00	93.90	6.30	50.09	46.88
T1m	933	50	272	0.40	0.00	0.00	0.00	0.00	76.70	9.91	43.29	46.88
G3u	389	546	7384	3.10	440.40	445.50	432.80	432.80	17.20	43.56	30.39	47.03
G3s	916	513	3872	8.80	410.20	412.90	402.20	402.20	73.30	43.80	58.56	47.03

Table 3: Comparison Results for Categorical Microdata (Continued)

Method	PLD	Dist	CTBIL	ACTBIL	EBILRF	ILRF	EBILMF	ILMF	PLD Rank	AIL Rank	Score	Ave. Score
G3o	943	234	1384	1.30	249.60	246.70	225.90	225.90	79.40	28.24	53.84	47.03
G3l	992	40	236	0.10	39.30	45.40	27.60	27.60	91.70	9.35	50.51	47.03
G3m	918	33	194	0.30	30.20	32.60	30.60	30.60	73.90	9.81	41.85	47.03
G2u	411	211	2858	1.20	118.70	123.70	118.10	118.10	21.10	28.19	24.65	47.47
G2s	1000	194	1466	3.30	110.30	115.40	110.10	110.10	100.00	29.07	64.54	47.47
G2o	992	117	702	0.70	79.30	74.40	71.00	71.00	92.80	19.40	56.09	47.47
G2l	996	11	66	0.00	5.50	7.60	7.00	7.00	94.40	4.63	49.54	47.47
G2m	940	17	100	0.20	8.50	8.30	8.00	8.00	78.30	6.76	42.55	47.47
B9u	591	5061	49270	17.20	6282.70	6359.20	6228.20	6228.20	32.80	95.79	64.28	47.92
B9s	50	4000	20000	36.60	5199.20	5375.00	5291.70	5291.80	1.70	89.40	45.53	47.92
B9o	284	1498	7934	6.70	2553.10	2661.80	2468.10	2468.10	8.90	65.32	37.11	47.92
B9l	733	655	3676	2.10	1046.90	1103.30	931.80	931.80	40.00	45.46	42.73	47.92
B9m	760	1061	5270	7.60	1083.50	984.20	936.40	936.40	43.90	55.97	49.93	47.92
B8u	571	4990	49068	17.10	6018.90	6139.00	6020.20	6020.30	31.70	94.91	63.29	48.66
B8s	50	4000	20000	36.60	5199.20	5375.00	5291.70	5291.80	1.10	89.95	45.53	48.66
B8o	296	1465	7810	6.60	2438.60	2535.20	2351.40	2351.40	11.70	63.70	37.69	48.66
B8l	777	599	3380	1.90	869.70	908.70	749.70	749.70	46.70	42.87	44.77	48.66
B8m	813	990	5068	7.40	819.70	763.90	728.50	728.50	51.70	52.41	52.04	48.66
G9u	778	4746	48140	20.10	6035.90	6223.70	6120.40	6120.50	47.20	94.91	71.06	48.98
G9s	50	4000	20000	45.70	5199.20	5375.00	5291.70	5291.80	2.80	90.79	46.78	48.98
G9o	285	1313	7214	6.80	2336.50	2818.20	2404.20	2404.20	9.40	63.47	36.46	48.98
G9l	764	411	2356	1.50	735.20	1172.60	795.10	795.10	44.40	39.72	42.08	48.98
G9m	779	746	4140	6.90	836.60	848.70	828.70	828.70	47.80	49.21	48.50	48.98
G5u	395	2899	34516	14.40	2933.20	3140.60	3073.50	3073.50	18.90	74.58	46.74	49.22
G5s	834	2809	16918	38.60	2820.80	3007.50	2947.30	2947.30	55.00	74.81	64.91	49.22
G5o	648	631	3698	3.50	886.80	1080.00	914.10	914.10	33.90	46.34	40.12	49.22
G5l	972	116	680	0.40	152.80	285.00	152.00	152.00	87.20	19.77	53.50	49.22
G5m	876	90	502	0.80	112.40	133.10	126.20	126.20	62.20	19.44	40.83	49.22
G8u	858	4283	45544	19.00	5680.40	5891.10	5792.00	5792.00	59.40	93.06	76.25	49.33
G8s	50	4000	20000	45.70	5199.20	5375.00	5291.70	5291.80	2.20	91.34	46.78	49.33
G8o	253	1244	6952	6.60	2195.50	2588.70	2239.80	2239.80	6.70	61.06	33.87	49.33
G8l	825	322	1856	1.20	543.60	893.30	581.10	581.10	53.30	34.40	43.87	49.33
G8m	849	283	1544	2.60	481.20	516.10	500.20	500.20	57.20	34.58	45.90	49.33
G6u	464	3288	37942	15.90	3734.20	3940.40	3854.40	3854.40	27.20	81.34	54.28	49.49
G6s	694	3152	17984	41.10	3547.30	3716.70	3641.70	3641.70	36.70	81.25	58.96	49.49
G6o	471	875	5064	4.80	1394.00	1674.80	1427.80	1427.80	27.80	52.87	40.32	49.49
G6l	925	176	1026	0.60	249.40	475.20	262.50	262.50	75.60	25.23	50.39	49.49
G6m	876	136	766	1.30	186.90	223.80	212.60	212.60	62.80	24.21	43.50	49.49
B7u	667	4886	48578	17.00	5678.90	5814.10	5707.10	5707.10	34.40	93.52	63.98	49.58
B7s	78	3967	19934	36.40	5028.60	5226.80	5146.70	5146.80	3.30	88.06	45.69	49.58
B7o	289	1368	7430	6.30	2220.10	2300.80	2129.80	2129.90	11.10	61.94	36.53	49.58

Table 3: Comparison Results for Categorical Microdata (Continued)

Method	PLD	Dist	CTBIL	ACTBIL	EBILRF	ILRF	EBILMF	ILMF	PLD Rank	AIL Rank	Score	Ave. Score
B7l	854	478	2720	1.60	593.70	607.40	465.20	465.20	58.90	38.52	48.70	49.58
B7m	842	919	4842	7.00	650.20	587.30	560.30	560.30	55.60	50.46	53.01	49.58
B6u	688	4761	47670	16.60	5139.10	5323.30	5230.60	5230.70	35.60	90.42	62.99	49.76
B6s	160	3855	19644	35.90	4552.60	4795.10	4728.20	4728.20	3.90	85.93	44.91	49.76
B6o	320	1287	7044	6.00	1922.40	2011.50	1852.40	1852.40	13.30	59.49	36.41	49.76
B6l	899	420	2406	1.40	428.60	432.00	299.30	299.30	67.80	35.05	51.41	49.76
B6m	847	906	4796	7.00	586.50	528.20	502.50	502.50	56.70	49.49	53.08	49.76
G7u	880	3867	42628	17.80	4839.90	5078.10	4985.60	4985.70	63.90	87.08	75.49	50.25
G7s	354	3681	19362	44.20	4546.80	4745.60	4665.60	4665.70	15.60	85.19	50.37	50.25
G7o	246	1181	6710	6.30	2062.70	2430.60	2101.80	2101.80	6.10	59.54	32.82	50.25
G7l	881	243	1408	0.90	371.50	695.60	403.50	403.50	64.40	30.23	47.34	50.25
G7m	875	186	1044	1.70	293.10	332.40	320.00	320.00	61.70	28.80	45.23	50.25
B5u	691	4581	46568	16.30	4524.60	4697.50	4608.60	4608.60	36.10	87.08	61.60	50.36
B5s	278	3694	19288	35.30	4007.60	4247.40	4181.10	4181.10	8.30	83.01	45.67	50.36
B5o	393	1207	6634	5.60	1633.60	1724.00	1576.50	1576.40	17.80	57.31	37.55	50.36
B5l	915	394	2280	1.30	351.40	363.60	241.90	241.90	72.80	33.43	53.10	50.36
B5m	861	887	4728	6.90	517.00	450.20	427.50	427.50	60.00	47.78	53.89	50.36
B4u	824	3516	37970	13.30	2937.10	2998.90	2937.50	2937.50	52.20	78.94	65.58	52.55
B4s	446	2650	15784	28.90	2502.60	2625.10	2585.60	2585.60	25.60	71.20	48.38	52.55
B4o	522	818	4780	4.00	1065.10	1149.10	1032.30	1032.30	29.40	49.77	39.61	52.55
B4l	974	119	704	0.40	132.80	188.10	90.10	90.10	87.80	19.44	53.61	52.55
B4m	882	866	4640	6.70	434.50	373.80	351.90	351.90	65.00	46.11	55.56	52.55
B3u	825	3150	34974	12.20	2053.00	2066.90	2017.70	2017.70	53.90	73.61	63.75	52.71
B3s	576	2294	14336	26.20	1661.50	1742.00	1713.70	1713.70	32.20	67.13	49.68	52.71
B3o	710	606	3550	3.00	604.40	668.50	587.20	587.20	37.80	43.29	40.53	52.71
B3l	985	102	604	0.30	86.50	128.50	64.60	64.60	90.60	16.85	53.70	52.71
B3m	896	856	4598	6.70	391.50	324.80	304.00	304.00	66.70	45.09	55.88	52.71
P9u	394	6086.3	1858	0.90	5678.30	5499.00	6146.30	5676.60	18.30	78.70	48.52	54.83
P9s	733	3086.3	886	2.60	3125.20	3839.40	3243.10	3738.60	40.60	65.28	52.92	54.83
P9o	800	98.462	662	0.70	5318.10	6890.70	4916.00	4718.10	50.00	44.72	47.36	54.83
P9l	947	1012	248	0.20	5305.10	6112.10	5005.10	4164.70	81.10	51.67	66.39	54.83
P9m	907	3000	142	0.30	2553.10	1659.70	2903.20	1938.00	70.60	47.36	58.96	54.83
B2u	795	2738	31508	11.00	1197.40	1205.40	1185.20	1185.20	49.40	66.20	57.82	55.20
B2s	698	1949	12786	23.40	997.90	1042.30	1028.10	1028.10	37.20	63.47	50.35	55.20
B2o	921	336	2006	1.70	193.80	248.90	214.10	214.10	75.00	31.81	53.40	55.20
B2l	992	81	484	0.30	48.60	74.80	46.70	46.70	92.20	14.72	53.47	55.20
B2m	945	789	4346	6.30	199.50	163.10	157.10	157.10	80.60	41.34	60.95	55.20
P8u	399	6077	1748	0.90	5624.80	5464.30	6091.80	5641.80	20.00	77.73	48.87	55.42
P8s	758	3077	804	2.40	3077.30	3808.70	3193.40	3707.00	43.30	64.17	53.75	55.42
P8o	828	87.916	604	0.60	5276.60	6879.40	4871.70	4696.90	54.40	43.33	48.89	55.42
P8l	960	1010.9	230	0.20	5290.60	6110.10	4990.30	4155.40	83.90	50.74	67.31	55.42

Table 3: Comparison Results for Categorical Microdata (Continued)

Method	PLD	Dist	CTBIL	ACTBIL	EBILRF	ILRF	EBILMF	ILMF	PLD Rank	AIL Rank	Score	Ave. Score
P8l	904	3000	140	0.30	2547.50	1655.60	2898.40	1934.80	69.40	47.08	58.26	55.42
P7u	387	6068.8	1544	0.80	5564.20	5426.50	6031.00	5605.60	16.10	76.99	46.55	55.44
P7s	788	3068.8	706	2.10	3024.80	3779.00	3139.80	3678.10	48.30	62.96	55.65	55.44
P7o	850	78.787	522	0.60	5236.10	6861.30	4829.40	4671.00	57.80	42.08	49.93	55.44
P7l	958	1009.9	224	0.20	5272.20	6102.00	4974.00	4139.50	83.30	49.95	66.64	55.44
P7m	905	3000	134	0.30	2539.30	1647.50	2891.30	1927.40	70.00	46.81	58.40	55.44
P6u	410	6059.3	1426	0.70	5503.80	5367.20	5971.60	5544.80	20.60	75.74	48.15	56.37
P6s	802	3059.3	668	2.00	2971.80	3736.80	3086.70	3633.40	50.60	61.90	56.23	56.37
P6o	862	68.418	480	0.50	5188.40	6845.20	4782.90	4645.10	60.60	40.60	50.58	56.37
P6l	966	1009	186	0.10	5255.80	6107.80	4962.60	4137.00	86.10	49.03	67.57	56.37
P6m	910	3000	118	0.20	2532.00	1630.50	2884.90	1911.40	72.20	46.39	59.31	56.37
B1u	537	1950	23888	8.30	0.00	0.00	0.00	0.00	31.10	51.76	41.44	57.54
B1s	1000	1302	9198	16.80	0.00	0.00	0.00	0.00	96.70	49.72	73.19	57.54
B1o	1000	146	876	0.70	0.00	0.00	0.00	0.00	97.20	18.80	58.01	57.54
B1l	1000	59	354	0.20	0.00	0.00	0.00	0.00	97.80	9.63	53.70	57.54
B1m	970	648	3770	5.50	0.00	0.00	0.00	0.00	86.70	36.02	61.34	57.54
P5u	421	6050.2	1228	0.60	5445.30	5288.20	5912.90	5466.10	22.20	74.40	48.31	58.09
P5s	842	3050.2	578	1.70	2922.40	3677.50	3035.10	3575.20	56.10	60.46	58.29	58.09
P5o	903	58.17	432	0.50	5143.50	6815.80	4739.30	4610.10	68.90	39.58	54.24	58.09
P5l	975	1007.9	182	0.10	5234.60	6106.00	4946.10	4130.80	88.90	48.38	68.63	58.09
P5m	925	3000	98	0.20	2522.90	1610.70	2877.90	1891.00	76.10	45.83	60.97	58.09
P4u	429	6041.6	954	0.50	5379.50	5225.40	5848.40	5402.80	25.00	72.78	48.89	58.78
P4s	880	3041.6	468	1.40	2866.00	3629.60	2978.10	3526.80	63.30	58.94	61.13	58.78
P4o	908	48.183	376	0.40	5092.30	6785.80	4690.50	4571.30	71.70	37.45	54.56	58.78
P4l	974	1006.5	164	0.10	5214.80	6105.30	4932.20	4122.90	88.30	47.64	67.99	58.78
P4m	939	3000	66	0.10	2513.50	1595.80	2870.30	1876.00	77.80	44.91	61.34	58.78
P3u	446	6034.1	842	0.40	5316.40	5186.40	5786.00	5363.30	26.10	71.76	48.94	59.63
P3s	896	3034.1	406	1.20	2812.70	3592.20	2923.50	3488.70	67.20	57.45	62.34	59.63
P3o	933	39.313	322	0.30	5041.30	6753.90	4641.70	4531.20	77.20	35.93	56.57	59.63
P3l	979	1005.1	142	0.10	5191.00	6091.00	4915.40	4101.60	89.40	46.71	68.08	59.63
P3m	944	3000	60	0.10	2503.70	1594.20	2862.50	1874.60	80.00	44.40	62.20	59.63
P2u	419	6026.2	706	0.40	5248.30	5142.60	5718.40	5320.70	21.70	70.23	45.95	60.30
P2s	920	3026.2	340	1.00	2754.30	3557.10	2863.80	3454.40	74.40	56.06	65.25	60.30
P2o	951	29.846	242	0.30	4986.80	6724.90	4588.80	4494.50	81.70	33.94	57.80	60.30
P2l	992	1003.5	106	0.10	5165.30	6082.30	4896.60	4084.50	93.30	45.60	69.47	60.30
P2m	953	3000	42	0.10	2494.00	1585.50	2854.60	1866.20	82.20	43.84	63.03	60.30
P1u	422	6013.3	360	0.20	5172.10	5048.10	5642.10	5226.80	22.80	65.79	44.28	61.36
P1s	965	3013.3	178	0.50	2687.40	3463.20	2794.80	3360.60	85.60	52.45	69.00	61.36
P1o	987	15.666	126	0.10	4920.50	6659.70	4526.30	4422.00	91.10	30.74	60.93	61.36
P1l	996	1002.2	50	0.00	5133.50	6071.40	4874.00	4067.70	95.00	44.31	69.65	61.36
P1m	956	3000	16	0.00	2484.70	1584.90	2847.40	1866.20	82.80	43.10	62.94	61.36

In this table, *Dist*, *CTBIL*, *ACTBIL*, *EBILRF*, *ILRF*, *EBILMF*, and *ILMF* are the information loss measures described in Subsection 5.2. *AILRank* is an average rank defined from the ranks of these measures (*DistRank*, *CTBILRank*, *ACTBILRank*, *EBILRFRank*, *ILRFRank*, *EBILMFRank*, *ILMFRank*—not displayed) that gives the same weight to the three classes of information loss measures (distance, contingency table, and entropy-based measures). Within a class, all measures also have the same importance. Therefore, *AILRank* is defined as:

$$AILRank = \frac{DistRank + \frac{CTBILRank + ACTBILRank}{2} + \frac{EBILRFRank + ILRFRank + EBILMFRank + ILMFRank}{4}}{3}$$

PLD is the number of correctly paired records using probabilistic linkage, and *PLDRank* is the ranking of the method with respect to *PLD* (normalized with maximal value 100).

The score is defined as

$$Score = (PLDRank + AILRank)/2$$

to give the same importance to disclosure risk and information loss, and an average score is computed for each pair *(masking method, parameter)*. The average is defined over the different choices of variables considered. This is the column labeled *Ave. Score*. Figure 2 is a graphical comparison of the selected SDC methods when the best parameter choice is made (average values over the considered groups of variables—'m', 'u', 's', 'o', and 'l'—are displayed). Top-coding with several parameterizations turns out to be the best performer (the best parameter choice being 5). See Section 6 for more details.

The results presented here are based on ranks instead of on the values of the information loss measures, because the ranges of the latter are different and difficult to compare with one another.

Figure 2. Comparison of SDC Methods for Categorical Microdata With Best Parameter Choice

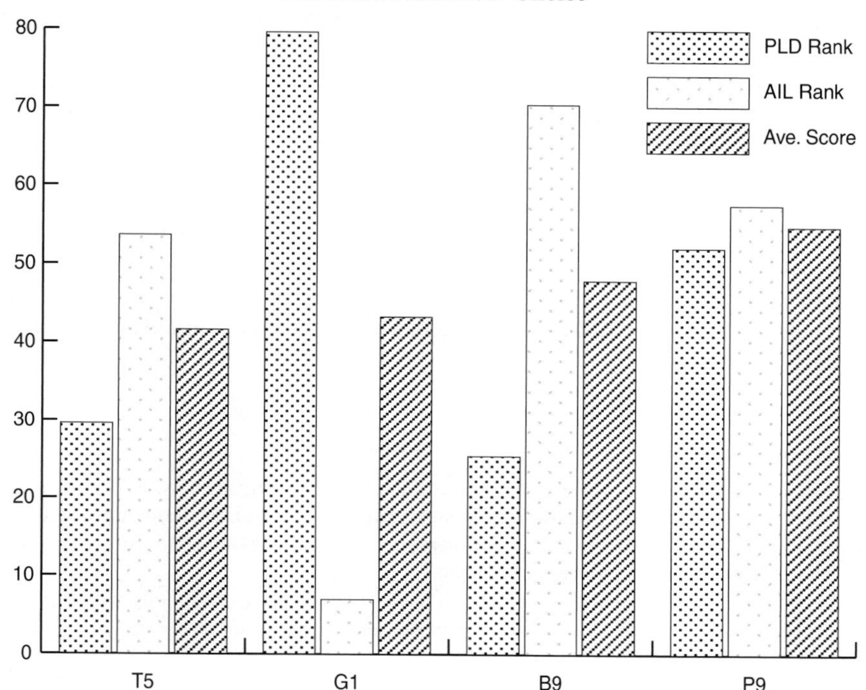

6. Conclusions

There is a rich array of methods for microdata disclosure control. A set of proposals for continuous and categorical microdata has been identified and described in this chapter. Measures for assessing information loss have also been described.

Regarding methods for masking continuous microdata (Table 1), Distr, MicZ*k*, MicIR*k*, MicPCP*k*, and resampling score around or above 50 for all tried *k*, so their use is not recommended. Multivariate microaggregation on unprojected data, proposed in Domingo-Ferrer and Mateo-Sanz (2002), is the only form of microaggregation scoring well. Taking three variables at a time seems the best strategy, but even in this case, microaggregation is second to rank swapping. Rank swapping outperforms the best microaggregation and is the best performer (for p around 15 percent). However, being a stochastic method, it is not reproducible and this may lead to disclosure in on-line databases allowing repeated queries. Microaggregation on unprojected data can still be a good option in such cases. Lossy compression (JPEG) in its current form is not excellent, but the general approach is promising; it is a new proposal, so there is room for improvement (such as using other compression algorithms). It is worth noting that random noise masking as implemented to carry out this experimentation is a simple algorithm based on univariate Gaussian noise generation; refinements using multivariate Gaussian noise (Kim 1986) should perform much better.

Regarding methods for categorical microdata, parameterizations of top-coding are the best rated, while the PRAM methods (with the parameters described above) are poorly rated. In general, for the former methods the reidentification risk is low while the information loss is moderate. For the PRAM methods, both reidentification risk and information loss are high. However, the results show that the information loss and the number of reidentifications are highly dependent on the set of variables and the number of categories in each variable. In this sense, the selected parameterizations for PRAM seem particularly ill-suited for variables with a large number of categories.

References

Adam, N.R., and J.C. Wortmann (1989) 'Security-Control Methods for Statistical Databases: A Comparative Study', *ACM Computing Surveys*, 21(4), pp.515-56.

Crystal Ball (2001), http://www.cbpro.com.

Defays, D., and P. Nanopoulos (1993) 'Panels of Enterprises and Confidentiality: The Small Aggregates Method', in *Proceedings of the 1992 Symposium on Design and Analysis of Longitudinal Surveys*, Ottawa: Statistics Canada, pp.195-204.

Domingo-Ferrer, J., and J.M. Mateo-Sanz (2002) 'Practical Data-Oriented Microaggregation for Statistical Disclosure Control', *IEEE Transactions on Knowledge and Data Engineering* (forthcoming), (Accessed August 22, 2001) http://www.computer.org/tkde/statistical_database.htm.

Elliot, M.J., C.J. Skinner, and A. Dale (1998) 'Special Uniques, Random Uniques and Sticky Populations: Some Counterintuitive Effects of Geographical Detail on Disclosure Risk', *Research in Official Statistics*, 1(2), pp.53-67.

Jaro, M.A. (1989) 'Advances in Record-Linkage Methodology as Applied to Matching the 1985 Census of Tampa, Florida', *Journal of the American Statistical Association*, 84, pp.414-20.

Joint Photographic Experts Group (2001) Standard IS 10918-1 (ITU-T T.81), http://www.jpeg.org.

Kim, J.J. (1986) 'A Method for Limiting Disclosure in Microdata Based on Random Noise and Transformation', in *Proceedings of the ASA Section on Survey Research Methodology*, pp.303-8.

Kooiman, P., L. Willenborg, and J. Gouweleeuw (1998) *PRAM: A Method for Disclosure Limitation of Microdata*, Research Report, Voorburg: Statistics Netherlands.

Liew, C.K., U.J. Choi, and C.J. Liew (1985) 'A Data Distortion by Probability Distribution', *ACM Transactions on Database Systems*, 10, pp.395-411.

Moore, R. (1996) 'Controlled Data Swapping Techniques for Masking Public Use Microdata Sets', U.S. Bureau of the Census (unpublished manuscript).

Pagliuca, D., and G. Seri (1999) *Some Results of Individual Ranking Method on the System of Enterprise Accounts Annual Survey*, Esprit SDC Project, Deliverable MI-3/D2.

Skinner, C., C. Marsh, S. Openshaw, and C. Wymer (1994) 'Disclosure Control for Census Microdata', *Journal of Official Statistics*, 10, pp.31-51.

U.S. Bureau of the Census (2000) 'Record Linkage Software: User Documentation'. Available from U. S. Bureau of the Census.

Willenborg, L., and T. De Waal (2001) *Elements of Statistical Disclosure Control*, New York: Springer-Verlag.

Winkler, W. (1998) 'Re-identification Methods for Evaluating the Confidentiality of Analytically Valid Microdata', in *Statistical Data Protection*, Luxembourg: Office for Official Publications of the European Communities, 1999. Journal version (1998) in *Research in Official Statistics*, 1(2), pp.50-69.

Confidentiality, Disclosure, and Data Access: Theory and Practical Application for Statistical Agencies
Pat Doyle, Julia I. Lane, Jules J.M. Theeuwes and Laura M. Zayatz (Eds)

Chapter 7

Disclosure Limitation Methods and Information Loss for Tabular Data*

George T. Duncan
Carnegie Mellon University

Stephen E. Fienberg
Carnegie Mellon University

Ramayya Krishnan
Carnegie Mellon University

Rema Padman
Carnegie Mellon University

Stephen F. Roehrig
Carnegie Mellon University

1. Introduction

Even in the age of electronic dissemination of statistical data, tables are central data products of statistical agencies. Prominent examples are the *American Fact-Finder* and similar documents from the U.S. Bureau of the Census, the United Kingdom's Office of National Statistics, and Statistics Netherlands. Many survey and census data are categorical in nature, and thus the representation of survey results in the form of cross-classifications or tables is a natural device for statistical reporting. But even when they collect measurement data, statistical agencies often represent the information from them in the form of distinct quantities. As a result, tables of counts are a primary unit of reporting and analysis. Sometimes these tables represent simple cross-classifications of the counts of survey and census elements. Other times the sample units are weighted according to probabilities of selection or are interpretable as the numbers of people in the population (based on

* Original research reported in this chapter was supported in part by the National Science Foundation under grant EIA-9876619 to the National Institute of Statistical Sciences.

the sample). In such tables of counts, the occurrence of small values is usually taken to present the possibility of a disclosure risk, because an intruder or data snooper may use data for individuals who are unique in the population in matching against other databases.

Considerable effort has gone into developing disclosure limitation methods for tabular data that effectively lower disclosure risk and provide products with high utility to legitimate data users (Duncan 2001; Duncan *et al.* 1993; Willenborg and de Waal 1996, 2000). These techniques include cell suppression, local suppression, global recoding, rounding, and various forms of perturbation (Federal Committee 1994). Under cell suppression, for example, the values of table cells that pose confidentiality problems are determined and suppressed (as primary suppressions) as are values of additional cells that can be inferred from released table margins (as secondary suppressions) (Cox 1980). Perturbation is used through controlled rounding (Cox 1987), versions of post-randomized response (Gouweleeuw *et al.* 1998), and Markov perturbation approaches, which have been proposed in various forms by Duncan and Fienberg (1999), Fienberg *et al.* (1998), and Fienberg *et al.* (2001).

Many of these methods can be represented in the form of matrix masks (Duncan and Pearson 1991). The computational problems associated with these approaches have been widely explored in recent years through such techniques as (1) network methods by Cox (1995); (2) mathematical programming—linear programming (LP) or integer linear programming (IP)—and graph theory as addressed by Fischetti and Salazar-González (1996, 1998, 1999), Chowdhury *et al.* (1999), and Duncan *et al.* (2001); and (3) branch and bound methods by Dobra (2001) and Dobra and Fienberg (2001).

In Section 2 of this chapter, we describe a framework for simultaneously examining the impact of disclosure limitation techniques on the two attributes of confidentiality protection and information loss. The first attribute is characterized as inverse to disclosure risk, and it measures the extent to which confidentiality is protected from the attacks of a data snooper. The second attribute is characterized as data utility, and it measures the extent to which data users will still find the tabular data product useful even though there may be some information loss. In Section 3, we describe a variety of techniques, some quite new and under development, for limiting disclosure for tabular data. In Section 4, we consider the topic of disclosure auditing for tabular data. Disclosure auditing involves procedures for examining a proposed data product and assessing its vulnerability to attack by an intruder or data snooper. In Section 5, we show how the framework in Section 2 can provide what we call an R-U confidentiality map for evaluating and analyzing disclosure risk and data utility of tabular data. We devote our attention to tables of counts and look at both two-way and multiway tables. Most of the methods we describe, however, are applicable in related form to weighted tables of various kinds.

The methods surveyed in this chapter comprise a substantial part of the working arsenal of disclosure limitation practitioners at statistical agencies. Many of the references at the end of the chapter point to seminal works in the field, and thus will be useful to both practitioners and researchers. The newer techniques described here are indicative of directions currently being pursued, and so will be of interest to researchers wishing to extend the state of the art.

To lend concreteness to our exposition throughout the chapter, we make use of the three-dimensional table presented in Table 1. It illustrates the various issues and disclosure limiting methods. Rows are indexed by i = 1,2,3,4, columns by j = 1,2,3,4, and levels by k = 1,2,3. We refer to the three two-way marginal totals derivable from this table by summing over variables as *IJ*+ (for the row by column totals), *I*+*K* (for the row by layer totals), and +*JK* (for the column by layer totals). Similarly, *I*++, +*J*+, and ++*K* represent the corresponding one-way marginal totals. The tables above the horizontal line are the three (i,j) levels, and the table below it is the *IJ*+ marginal.

If we consider this example a three-way population table, then the six cells with entries of '1' represent individuals who are *unique* in the population and thus pose a confidentiality problem. Most users would also consider the six cells with entries of '2' to pose serious disclosure risk, because one individual recorded in such a cell sees the other as unique. There are no such entries in any of the two-way marginals, however, and this fact might generate a false sense of security on the part of a data administrator, as we show below, if the marginals alone were published.

Table 1. Our Illustration

k = 1

1	4	66	3	74
1	2	0	0	3
0	4	3	1	8
3	0	0	3	6
5	10	69	7	91

k = 2

2	3	2	68	75
0	4	78	3	85
0	0	0	61	61
0	3	1	0	4
2	10	81	132	225

k = 3

0	80	0	1	81
4	2	2	1	9
3	0	4	45	52
61	3	55	4	123
68	85	61	51	265

3	87	68	72	230
5	8	80	4	97
3	4	7	107	121
64	6	56	7	133
75	105	211	190	581

2. A Framework for Disclosure Risk and Information Loss

Some argue that the legitimate objects of inquiry for statistical research are inferences drawn from aggregates over individual records—for example, the proportion of pilots for commercial airlines in the United States who have an alcohol abuse problem—perhaps given a set of additional demographic and occupational characteristics. The statistical agency often seeks to provide users with data that will allow accurate inferences about such population characteristics. Unfortunately, the additional characteristics of interest may well make the resulting cross-classification quite sparse, possibly replete with entries of '1' and '2'. Because of confidentiality promises—whether explicit or implicit—the statistical agency seeks to thwart the data snooper who might seek to use the disseminated data to draw accurate inferences about, say, the alcohol abuse status of a particular pilot for American Airlines. This capability by a data snooper represents a statistical disclosure, but that level of statistical detail may still be of legitimate statistical interest to a careful analyst, who has no interest in using the information about this particular pilot beyond the context of this statistical analysis.

There are two major types of disclosures—identity disclosure and attribute disclosure. *Identity disclosure* occurs with the association of a respondent's identity with a disseminated data record (Paass 1988; Spruill 1983; Strudler *et al.* 1986). *Attribute disclosure* occurs when the respondent can be associated with either an attribute value in the disseminated data or an estimated attribute value based on the disseminated data (Duncan and Lambert 1989; Lambert 1993). In the case of identity disclosure, the association is assumed to be exact. In the case of attribute disclosure, the association can be approximate. Most statistical agencies emphasize limiting the risk of identity disclosure, perhaps because of its substantial equivalence to the inadvertent release of an identified record, a clear administrative slip-up. On the other hand, an attribute disclosure, even though it invades the privacy of a respondent, may not be so easily traceable to actions of the agency.

We introduce a conceptual framework to provide context to our discussion of disclosure limitation methods. We take the data user to be primarily interested in the estimation of a conditional or a joint probability or population proportion based on the tabular data. We assume that an intruder or data snooper has access to external information that will make it likely that the snooper can compromise confidentiality when a cell count is small. Our framework establishes quantitative measures for two basic attributes:

(1) **Disclosure Risk**—a measure of risk to confidentiality that the data trustee, such as a statistical agency, would experience by a data release.

(2) **Data Utility**—a measure of the value of information to a legitimate data user.

Generally, the application of a disclosure limitation method would have the desirable effect of lowering disclosure risk and, concomitantly, have the undesirable effect of lowering data utility.

Disclosure risk generally is determined based on how the agency envisions the data snooper making a disclosure. More simply, it may be based on some measure of the percentage of the population that could be easily compromised because of the uniqueness of their attribute values. Chen and Keller-McNulty (1998), Duncan and Lambert (1986, 1989), and Fienberg and Makov (1998), among others, develop specific disclosure risk models. Finally, the released data will have more or less utility for the user depending on the degree of perturbation from the original data and the intended use of the data. The statistical disclosure limitation problem is to choose a methodology for data release so that disclosure risk is adequately low while statistical information (data utility) in the disseminated data are as high as possible (Duncan and Fienberg 1999). This characterization of the problem is displayed in Figure 1. Data utility is the value of the statistical information that the agency provides to a legitimate user.

Figure 1. The Statistical Disclosure Limitation Problem

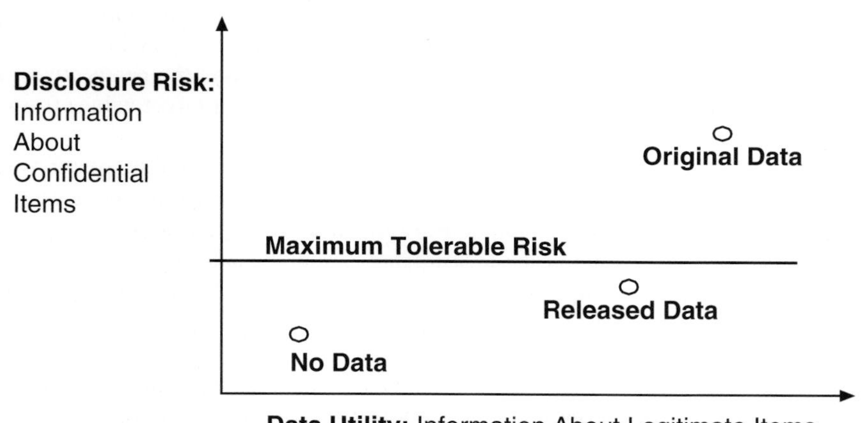

In addressing the statistical disclosure limitation problem, the agency is shoring up the two pillars of its foundation: satisfying the data users who depend on its products and reassuring the respondents who provide it with data. Data snoopers pose a threat to the agency's ability to deliver on its promise of confidentiality to respondents. Thus, Figure 1 is a graphical assessment of how an agency provides data utility to users and lowers disclosure risk in the face of attack. Domingo-Ferrer (1999) terms a data product of high data utility to be both *analytically valid* (key statistical characteristics, such as means and covariances, preserved) and *analytically interesting* (several variables on important subdomains provided) and notes that it involves low information loss; he terms data products with low disclosure risk to be *safe*. Zaslavsky and Horton (1998) use a decision-theoretic approach based on the structure of Figure 1 to derive an optimal disclosure limitation scheme for minimum cell size in tabular data.

Generally, a data snooper has *a priori* knowledge about a target (Duncan and Lambert 1986). Typically, this knowledge would take the form of a database with identified records (Adam and Wortmann 1989). Certain variables may be in common with the subject database. These variables are called *key* or *identifying* (De Waal and Willenborg 1996, 1998). When a single record matches on the key variables, the data snooper has a candidate record for identification. This candidacy is promoted to an actual identification if the data snooper is convinced that the individual is in the target database. This would be the case either if the data snooper has auxiliary information to that effect or if the data snooper is convinced that the individual is unique in the population. The data snooper may find that according to certain key variables, a sample record is unique. The question then arises as to whether the individual is also unique on these key variables in the population. Bethlehem *et al.* (1990) have examined detection of records agreeing on simple combinations of keys based on discrete variables in the files. Record linkage methodologies have been examined by Fuller (1993) and extensively by Winkler (1998), who uses a version of the matching algorithm originally proposed by Fellegi and Sunter (1969). Trottini (2001) presents an even more comprehensive framework for the decision-theoretic trade-offs between the perspectives of the agency and the perspectives of users, in light of the extent to which an intruder is able to infer target values from a released dataset.

Elliot and Dale (1999) and Paass (1988) explore the psyche and motivations of the data snooper. The snooper may or may not be someone with limited access to the data and may or may not be malicious. Prudently, however, the database administrator must assume a worst-case scenario, that is, that the data snooper has access to sophisticated analytical tools, is knowledgeable about the data and has ready access to relevant external data sources, and has the necessary computational power to attempt an attack on the data.

Data utility is a positive expression of information loss. A variety of measures of data utility have been proposed. For example, Özsoyoğlu and Chung (1986) suggested a measure for tabular data under disclosure limitation through cell suppression as simply the percentage of suppressed cells. This particular measure is crude at best. Similarly, De Waal and Willenborg (1998) consider a variety of options for choosing local suppressions (*i.e.*, values for specific variables in specific records) by focusing on the total number of such suppressions, or the number of categories affected by the local suppressions.

More generally, we presume that once the database administrators are able to hold disclosure risk to an adequately low level, they should then seek to maximize data utility. This follows from the perspective that all disclosure limitation methods attempt to maximize data utility for a given user task subject to a constraint on disclosure risk. We examine the fundamental trade-off between data utility and disclosure risk. As information loss increases because of disclosure limitation, an estimate of a conditional probability becomes less precise, and data utility consequently goes down. Simultaneously, disclosure risk also decreases. This conceptual framework can be used to compare alternative disclosure limitation methods. Typically one can control the use of any particular disclosure limitation method by choosing certain parameter values. For example, with noise addition, the parameter is the variance, say τ^2, of the added noise. As τ^2 is changed, the disclosure risk R and the data utility U change. These changes can be presented graphically in what is called an R-U confidentiality map (Duncan and Keller-McNulty 2001; also Section 5 below).

Here we model disclosure risk for any individual table cell as the sum

$$\sum_k r(k)p(k),$$

where *r(k)* is the risk associated with the data snooper obtaining knowledge that a cell entry has true value *k*, and *p(k)* is the probability that the cell value is *k*, given the table and knowledge held by the data snooper about the disclosure-limiting method employed. Thus the sum ranges over the possible true values a cell could have, given the published value.

To illustrate how this model could be applied, consider a table protected by rounding to base 3. For example, if from Table 1 the *IJ+* marginal table (the 4×4 table below the horizontal line) is rounded to base three, Table 2 results. (We discuss table rounding in more detail in Section 3.) A cell with a published value 3 could have true value 1, 2, 3, 4, or 5. The data disseminator would assign $r(k), k \varepsilon \{1, 2, 3, 4, 5\}$ according to the perceived risk associated with an intruder determining that *k* was the true value. Generally, *r*(*k*) would be a decreasing function of *k*, because the disclosure risk would be higher with smaller cell counts.

Table 2: A Table Rounded to Base Three

3	87	69	72	231
6	9	81	3	99
3	3	6	108	120
63	6	57	6	132
75	105	213	189	582

In determining *p(k)* for each *k*, the data disseminator would attempt to assess the probability distribution that the data snooper would use. There are several possible approaches. A conservative approach is to act as though the data snooper places high probability on the lower possible values of *k*. Another approach is to base assessments of $p(k)$ on the actual frequency distribution of cell counts in some reference population. Then the data disseminator can model information loss (the complement of data utility) as, for example, the mean square error in estimating conditional probabilities.

As we noted above, this framework allows us, in principle, to examine the various disclosure limitation schemes with regard to both data utility and disclosure risk. In the following sections, we illustrate this framework with a numerical example—an example that we will use in the evaluation and analysis section to compare some of the disclosure limitation methods and show how it can quantify the utility/risk trade-off.

3. Disclosure Limitation

In this section, we discuss a number of protection schemes that have been proposed for tables of counts. All the methods discussed have appeared in the statistical literature with applications to two-dimensional tables. We will discuss their applicability to three- and higher-dimensional tables, because many desirable properties of statistical disclosure limitation techniques for two-dimensional tables are absent from the corresponding techniques for tables of higher dimension (Cox 1999).

Sampling

One of the surest ways to limit disclosure is to release only part of the data. Thus, releasing a table whose counts are based on a sample of the units in the original table is a way to provide a serious measure of protection for the original reporting units. The statistical agency practice of releasing microdata samples is essentially based on this approach. The virtue of sampling in this context is that if the details of the sampling procedure are available, a user can make valid inferences about the population underlying the sample table, albeit with less precision than would have been possible with the original table.

Cell Suppression

Cell suppression has been used for many years by a large number of statistical agencies (Cox 1980, 1995, 2002; De Carvalho *et al.* 1994; Fischetti and Salazar-González 2000; Giessing 2002; Kelly *et al.* 1992). On the face of it, the idea is simplicity itself. If a table contains an entry that is deemed sensitive, the disseminator simply does not provide a value for it. This has no effect at all on other table entries, so it effectively localizes the distortion of the data to individual cells. A suppressed cell, in isolation, would appear to be able to take on any value whatsoever (and so provide complete protection), but in the context of the table as a whole, there are evident constraints that arise from marginal values and algebraic relationships between cell entries. Furthermore, data disseminators sometimes publish the rules used to determine suppressions, thus providing further clues to their likely values.

Under cell suppression, each sensitive cell in the table is suppressed (*primary* suppressions). If marginal totals or other linked tables are also to be published, it may be necessary to remove additional cell values (*secondary* or *complementary* suppressions) that would allow an intruder to use algebraic or other means to identify the sensitive cell values. In terms of our R-U framework, the goal of cell suppression is to find secondary suppressions that maximize the utility of the resultant table while affording sufficient protection. Often the total number of suppressed cells is taken as the measure of utility, but other measures (*e.g.*, entropy) have been used as well.

Except in special circumstances, the secondary cell suppression problem is computationally NP-hard (Kelly *et al.* 1992), suggesting that any solution procedure will grow exponentially in complexity with increasing problem size. Recent work by Fischetti and Salazar-González (2000) has increased considerably the size of tables that can be protected optimally by suppression, but it is still quite common for heuristics to be used instead of procedures that are provably optimal. In addition, many of the heuristics in current use do not guarantee a specific level of protection. For such heuristics, one of the disclosure auditing techniques described in Section 4 should always be applied before a table is made public.

Many of the published heuristic methods for cell suppression (Cox 1980, 1995; Kelly *et al.* 1992) rely on the simple structure of two-dimensional tables. Algebraic relationships between cells in two-dimensional tables are effectively captured in network flow models, which typically have fast solution routines. Unfortunately, once the transition is made to three- and higher-dimensional tables, the network representation breaks down, voiding many useful theoretical results and algorithms. The exact methods of Fischetti and Salazar-González (2000), however, do not depend on network structure, and so can find optimal suppression patterns for arbitrary n-dimensional tables, at least those of moderate size. When the size of the tables is such that optimal suppressions cannot be computed, meta-heuristic approaches such as tabu search (Glover and Laguna 1997) provide a way of looking for 'good' solutions within a reasonable time. Essentially, tabu search is a neighborhood search method with a built-in mechanism to prevent the algorithm from becoming stuck at local optima. Duncan, Krishnan, *et al.* (2001) apply the tabu search approach in conjunction with the fast disclosure auditing approach of Chowdhury *et al.* (1999) to solve cell suppression problems in three-dimensional tables. The quality of the tabu search solutions was comparable to the quality of the IP solutions.

Unfortunately, there is only a limited set of circumstances where the special structure of the tables to be protected permits an efficient solution algorithm. For example, see Duncan, Krishnan, *et al.* (2001), and the decomposition and reducibility results of Dobra and Fienberg (2000), which characterize an important class of linked tables that give rise to these simplifications. In the absence of these special structures, however, large cell suppression problems (especially ones involving large tables and substantial numbers of primary suppressions) are computationally difficult.

Duncan and Fienberg (1999) and Fienberg (1997, 2001) have criticized cell suppression because it causes unnecessary loss of statistical information. In particular, they note that complementary suppressions destroy data that are not themselves sensitive, and the resulting tables greatly reduce the user's ability to make correct inferences about relationships in the original unsuppressed table. Thus cell suppression achieves disclosure limitation at the expense of elimination of some data.

Rounding

Rather than simply suppress a sensitive cell, one might disguise its true value by modifying it in a principled way. One way to do this is to choose a positive integer b and round table entries to an integer multiple of it. This is usually done for *all* cells in the table. Rounding has the general advantage of providing at least a roughly correct value for every cell (assuming b is small) and thereby helps the data user to avoid badly incorrect inferences about cell values. Cell suppression, on the other hand, does open the possibility that the user may draw false inferences about the suppressed cell values (Duncan and Fienberg 1999). Less positively, with all cell values rounded, many more would typically be changed from their true values than would be the case with cell suppression. Also, when multiple, overlapping tables are rounded individually, a common cell may end up being rounded to two different values.

Rounding can be done with more or less sophistication. Fellegi (1972, 1975) introduced the notion *of controlled rounding,* which insists that the rounded table be additive, meaning that rows, columns, and so on sum to the their respective (rounded) marginals (see also Cox and Ernst 1982). *Zero-restricted* controlled rounding (Kelly, Golden, and Assad 1990) further requires that cell values in the unmodified table that are already multiples of b (in particular, zeros) remain so. Finally, *unbiased* controlled rounding (Causey, *et al.* 1985; Cox 1987) specifies that the expected value of a rounded cell value equal its unrounded value. The requirements imposed by these different rounding methods are all related to attempts to improve the data utility for users while still minimizing disclosure risk in some formal sense.

Simple and efficient polynomial-time algorithms have been devised for all these flavors of rounding, at least for two-dimensional tables. Polynomial-time algorithms have the property that the worst-case solutions time grows only as a polynomial function of the size of the problem (here, the number of cells to be rounded). More difficult computational problems are classified as NP-hard, and for such problems solution times may increase exponentially with size. The controlled rounding problem for three dimensions has been shown to be NP-hard (Kelly, Assad, and Golden 1990), although the heuristic given in Kelly, Golden, and Assad (1990) has proven to be effective when unbiased solutions are not required. Fischetti and Salazar-González (1998) provide advice for implementing controlled rounding approaches empirically in three and more dimensions. The method of Dobra (2001) for bounding cell values implicitly or explicitly generates feasible tables and thus has the potential to identify controlled rounding solutions or near solutions in higher dimensions.

If natural assumptions are made about the distribution of cell values in a table (*i.e.*, that it is nearly uniform across local intervals of length b), it is often easy to specify probabilities for each possible cell value in a rounded table. In many cases, even if the value of the rounding base b is not explicitly announced, it can be easily deduced from the published table itself.

Data Swapping, Confidentiality Edit, and Simulated Tables

Dalenius and Reiss (1978) first proposed a method for swapping observations 'at random' while preserving marginal totals. In Dalenius and Reiss (1982), they illustrate the implementation of a kth-order swap in which all k-dimensional margins of a p-dimensional table are preserved. This is similar to the notion of randomly selecting a replacement table from a restricted set of alternative tables with the same k-dimensional marginal totals. An obvious issue is how to choose k. Dalenius and Reiss illustrate their proposal with $k = 2$. There is also the issue about what fraction of records to swap. Without going into details, we can say that Dalenius and Reiss were unable to come up with a general method for accomplishing data swapping. Nor were they able to assess the increase in variability associated with the added randomness.

The U.S. Census Bureau used a variant of data swapping in the context of the Confidentiality Edit as part of the 1990 decennial census. It wanted to interchange a subset of households in different census blocks that had a number of characteristics, say k, in common. This has the result of holding the corresponding k-dimensional totals for those blocks fixed, as well as the $(p-k)$-dimensional margin adding across blocks and across the variables held constant under swapping. The primary method matched records on $k = 6$ variables. For further details, see Fienberg *et al.* (1996), Griffin *et al.* (1989), and Navarro *et al.* (1988). Again we have a statistical issue about the choice of k, as well as the issue about the fraction of records to be swapped.

To illustrate this alternative notion of data swapping, we consider a 3×2×2 contingency table with entries $\{n_{ijk}\}$ as shown in Table 3.

Table 3: Contingency Table

n_{111}	n_{121}	n_{1+1}	n_{112}	n_{122}	n_{1+2}
n_{211}	n_{221}	n_{2+1}	n_{212}	n_{222}	n_{2+2}
n_{311}	n_{321}	n_{3+1}	n_{312}	n_{322}	n_{3+2}
n_{+11}	n_{+21}	n_{++1}	n_{+12}	n_{+22}	n_{++2}

We want to track what happens when we swap the values for a randomly selected pair of individuals, one in layer 1 and the other in layer 2. Suppose that the individual selected from layer 1 is in the (1,2,1) cell and that we are swapping his/her characteristics with a randomly selected individual in the (3,1,2) cell. The result is shown in Table 4.

Table 4: Result of Swapping Values in Table 3

n_{111}	$n_{121}-1$	$n_{1+1}-1$	n_{112}	$n_{122}+1$	$n_{1+2}+1$
n_{211}	n_{221}	n_{2+1}	n_{212}	n_{222}	n_{2+2}
$n_{311}+1$	n_{321}	$n_{3+1}+1$	$n_{312}-1$	n_{322}	$n_{3+2}-1$
$n+_{11}+1$	$n_{+21}-1$	n_{++1}	$n_{+12}-1$	$n_{+22}+1$	n_{++2}

Note that the two-dimensional total for the first two variables (adding over layers) is unchanged, as is the one-dimensional total for the third variable. This process is now repeated for pairs of randomly selected units in the two layers, thus producing a confidentiality edit that continues to preserve the same marginal totals. One variant of this, used by the Census Bureau for the 2000 census, is to select swapping 'partners' by targeting certain unique records. Moore (1996) describes additional applications of variants on data swapping and contrasts them with the simple matrix-masking technique of adding noise.

We note that data swapping, like cell suppression, is a method of altering cell counts in a multidimensional cross-classification while maintaining fixed marginal totals. This observation led Fienberg, Steele, and Makov (1996) and Fienberg, Makov, and Steele (1998) to propose a more elaborate version of repeated data swapping that allows for a series of moves from one table to another, subject to marginal constraints. Their method utilizes the tool of Gröbner bases described in Section 4 below, and in essence replaces the original table by a random draw from the exact distribution under the log-linear model whose minimal sufficient statistics correspond to the released marginals, subject to those marginals being fixed. Thus probabilistic simulation yields a replacement table with the same marginal totals as the original table. Actually, Fienberg, Makov, and Steele (1998) go further in proposing the retention of the simulated table only if it is consistent with some more complex log-linear model.

The data-swap transformation described above represents one of a subclass of possible 'moves' in a Markov chain algorithm proposed by Diaconis and Sturmfels (1998). Such moves alone, however, do not always suffice to generate the exact distribution. Even in the cases where they do suffice, one needs to run the Markov chain a very long time to simulate the exact distribution as explored by Fienberg, Makov, and Steele (1998). Making a small proportion of swaps, as is done in practice, is not sufficient to rest the methodology on a firm statistical foundation that a user can invoke in order to assess the added uncertainty that results from the alteration of the data.

An extremely important feature of this simulation methodology is that information on the variability it introduces into the data is directly accessible to the user, because anyone can begin with the reported table and information about the margins that are held fixed and then run the Diaconis-Sturmfels Markov chain algorithm to regenerate the full distribution of all possible tables with those margins. This then allows the user to make inference about the added variability in a formal modeling context in a form that is similar to the approach to inference in Gouweleeuw *et al.* (1998). As a consequence, simulation and perturbation methods represent a major improvement from the perspective of access to data over cell suppression and data swapping.

This approach offers the prospect of simultaneously smoothing the original counts *and* providing disclosure limitation protection. But there remain many practical issues regarding the use and efficacy of such methods for generating disclosure-limited public-use samples. For example,

- How effective are such devices for limiting disclosure—that is, protecting against attack by a data snooper?
- What is the data utility (correspondingly, information loss) when we compare actual data with those released?
- How can they be used when the full cross-classification of interest is very sparse, consisting largely of 0s and 1s?
- How can we use models to generate the simulated data when the users have a multiplicity of models and even classes of models that they would like to apply to the released data?
- What if a release involves thousands of tables with overlapping cells?

We discuss some of the implications of using Markov perturbation in more detail below.

Markov Perturbation

The method of simulating from the exact distribution of a table given a set of marginals is intimately related to the notion of Markov perturbation described by

Duncan and Fienberg (1999). Thinking of cell values as counts of entities classified in a particular category, Markov perturbation deliberately misclassifies by selectively moving entities from one classification to another. This is done in such a way that marginal totals are preserved, and the expected values of all cells are unchanged. In employing Markov perturbation, the statistical agency (1) lowers disclosure risk by increasing the uncertainty of a data snooper about the true cell value and (2) gives the legitimate data user a value for analysis, albeit one that is subject to misclassification error—an error process that any good data analyst of categorical data must contend with anyway.

The procedure of Markov perturbation works as follows, described for a two-dimensional table. An *elementary data square* is chosen as a 2×2 submatrix of the table. Then each entity (*i.e.,* each individual contributor to the counts in the four cells in the submatrix) moves to an adjacent cell (up or down, left or right) according to a Markov transition matrix chosen to be stationary. The transition matrix is chosen so that row sums and column sums are unchanged. To protect the entire table, the process is repeated with a random sequence of the possible elementary data squares.

In this section we have surveyed the principal techniques used for disclosure limitation. Because some of these techniques may in practice rely on heuristic algorithms to provide the desired level of protection, the next section discusses methods that test whether this protection has been achieved.

4. Disclosure Auditing

Disclosure auditing is a process of examining a proposed data product to assess its vulnerability to attack by a data snooper. As part of a sensible procedure for evaluating security implementations, protectors should play the role of those who might attempt to compromise data security. They should search for weak points. Prudently, they should assume that the attacker has adequate resources to similarly identify and exploit such weak points. In this section we present methods for disclosure audit that have been available for some time, as well as new methods. Special attention is given to higher-dimensional tables.

A data disseminator might wish to publish an entire, original table. This table may well have cells, which we call sensitive cells, that are deemed to pose unacceptable disclosure risks. It is common to declare a cell in a population table whose value is small, say 1 or 2, as posing an unacceptable disclosure risk and hence sensitive. In that case, the table should not be disseminated in its original form. Instead, it should first be transformed through some disclosure limitation procedure, such as one of the techniques described in Section 3. If the technique used cannot inherently guarantee the requisite level of protection, it is necessary to audit the proposed release, that is, apply a procedure to test the level of protection actually afforded. For example, cell suppression patterns are often determined through heuristic methods that do not provide such guarantees.

Alternatively, for reasons of brevity or protection, the disseminator might wish to publish tables—say one or more of the two-dimensional marginal tables *IJ+, I+K or +JK* in our running example—that are *derived* from the original higher-dimensional table. If the goal is to protect values in the original table or values in an unpublished margin, auditing is again necessary to ensure such protection.

Before an audit can proceed, the disseminator must decide what constitutes a sensitive cell. Sufficient protection exists for a sensitive cell provided that in the released data product the true value of this cell entry is sufficiently ambiguous to a data snooper. A common and useful scheme is to define a protection range and demand that protection be such that any value in the range is potentially the correct cell value.

If such a protection range is given for a cell, then an audit verifies that realizations of the table indeed exist that agree with the published data but have the sensitive cell with a value anywhere in the protection range. For example, Table 1 might be protected by suppressing all cells with values 1 or 2 (and also additional cells if necessary—see Section 3), so that for each sensitive cell, any value in the range [0, 4] is feasible.

Linear and Integer Programming

To verify a protection range for a sensitive cell of a table with published marginal totals, the obvious technique is linear programming or integer linear programming (Zayatz 1993). The published cells are used to form linear constraints on the possible values of the cell. Then the sensitive cell value is both minimized, to obtain the lower bound on the cell value, and maximized, to obtain the upper bound on the cell value. The lower and upper bounds then provide the protection range for that cell. In the case of multiple sensitive cells, the procedure is repeated for each. Because the constraints implied by the published cell values apply to every sensitive cell, the max/min pair for one cell can be calculated independently of that for any other cell.

Several difficulties arise in the use of standard linear programming (LP), and consequently there is considerable interest in finding alternative techniques. This is especially true for implementing procedures on large tables, which can require considerable computational effort, depending on the number of sensitive cells needing protection. For the example given above, 11 cells are sensitive, so an equal number of maxima and minima must be calculated, and this is computationally feasible. For much larger tables and many more suppressions, the task can become daunting. As an example, the Bureau of Labor Statistics quarterly *ES-202 Employment and Wages* has nearly 40,000 cells for some county/state aggregations, more than half of them suppressed. A 'brute force' LP approach makes little sense in such a case.

Fortunately, it is often possible to decompose the overall audit problem into smaller pieces. For example, large tables often have one or more attributes arranged hierarchically (*e.g.*, counties within a state, or four-digit SIC (Standard Industrial Classification) codes 'rolling up' to three-digit codes). In such cases, auditing can be done at each level separately—if due care is given to the fact that margins at one level correspond to internal table entries at the next higher level. With such a decomposition, the number of LP problems to be solved increases, but the size of each problem (as measured by the number of variables plus constraints) decreases. Because the average time requirements of most linear programming algorithms increase roughly as the cube of the problem size, a large net savings in total computation can accrue. As an illustration, a two-level hierarchical table with a total of 8 rows and 8 columns contains 64 internal cells and 16 marginal constraints. If this table can be decomposed hierarchically into four 4×4 tables, then each of the four has 16 internal entries and 8 marginal constraints. Auditing the original table would consume computing resources proportional to $(64 + 16)^3$, while the four smaller ones would need resources proportional to $4 \times (16 + 8)^3$, about a 90 percent savings.

Another way in which LP-based auditing procedures can be improved takes advantage of the linked structure of the table. In the process of maximizing, say, one cell value, a simplex-based LP algorithm will incrementally increase the current value for that cell, subject to the imposed constraints. At each step, the value of every other cell is recorded in a data structure (the 'simplex tableau') that can be easily examined to see if any cell (other than the one that is currently being maximized) is at its guaranteed maximum or minimum. Detecting these occurrences is a simple matter, and it obviates the need to perform a separate optimization on that cell. Empirically, this can speed up the auditing process considerably.

A second potential difficulty with linear programming lies in the nature of some auditing problems. For tables of more than two dimensions, there is no general guarantee that the optima produced by linear programming will be integers. Two-dimensional tables can be represented as a network, with rows and columns as nodes and internal cells as arcs. Because of a special property of networks, optimal solutions to max/min problems are integer, provided the known values (unsuppressed cells, row and column sums) are integer. Unfortunately, three- and higher-dimensional tables can no longer be represented as a network, so the integrality property of optima is lost (Roehrig 1999). For example, using Table 1 as the base table, suppose that we wish to publish the three two-dimensional margins *IJ+*, *I+K,* and *+JK.* If cells in the underlying three-dimensional table are considered sensitive, the LP approach to finding inferable bounds will result in an upper bound of 13.5 for cell (1, 3, 3). This is clearly unobtainable in any table of counts, so it cannot be right. Thus auditing techniques that give sharp *integer* bounds are needed. Integer programming (IP) is the obvious solution, but in general it is extraordinarily more difficult computationally than LP, which permits continuous solutions. In many circumstances, LPs giving non-integer optima can be augmented with additional, legitimate constraints that force integrality—'Gomory cuts', for example (Schrijver 1986). A general theory has yet to be developed, but Roehrig (2001b) has obtained results for special cases. Alternative approaches such as meta-heuristic search (*e.g.*, tabu search described by Glover and Laguna 1997) that have performed well on difficult combinatorial optimization problems can also be applied to these difficult auditing problems. Duncan, Krishnan, *et al.* (2001) report on the application of tabu search to a related disclosure limitation problem.

Alternative Approaches

Because of the difficulties outlined above, other approaches to auditing disclosure in tables of counts have been pursued. We describe three of them here, but this is an active area of research, so the list is growing. The first alternative derives from generalizations of the well-known Fréchet and Bonferroni bounds on joint probability distributions given lower-dimensional marginal distributions. It applies to certain situations in which lower-dimensional marginal totals (themselves tables) are to be published and bounds are needed on entries in the original higher-dimensional table. Dobra and Fienberg (2000) show that when the released marginals form a *decomposable graph*, one can combine information from the subgraphs to realize sharp bounds for entries in the original table. Further, the same structure can be used to break the problem of computing bounds for a large table into sets of much smaller ones corresponding to irreducible components.

In our three-dimensional example, suppose that the margins $IJ+$ and $+JK$ are published, but we wish to protect the underlying table IJK. The published marginals directly associate I and J, and J and K, but the interaction between the dimensions I and K are indirect. This can be visualized by drawing a graph with nodes representing the dimensions and arcs that indicate which nodes are joined by the published marginal tables. In this example, the graph is especially simple, as shown in Figure 2.

Figure 2. Graph Representing Released Marginal Tables

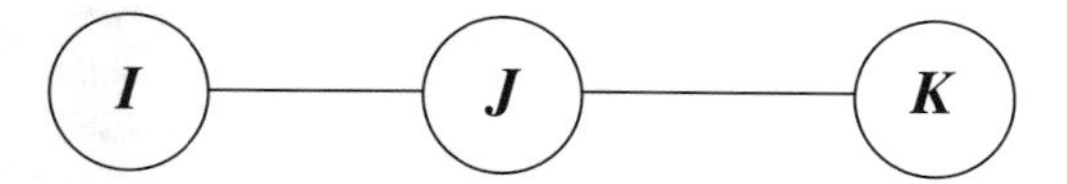

Node J is a 'separator' of nodes I and K. Dobra and Fienberg (2000) show that if a separator (which may be larger than a single node, depending on the released margins) is a clique, then a variant of the normal Fréchet bounds can be used to calculate bounds for entries in the base table (IJK in this case). These results are powerful because the bounds can essentially be 'read off' from the published tables. Dobra and Fienberg give closed-form expressions for these bounds, so that no iterative mechanism is required.

Chowdhury *et al.* (1999) developed an equivalent network-based bounding scheme for another three-dimensional case. Suppose once again that the marginals *IJ*+ and +*JK* from Table 1 are to be released, but now the third two-dimensional marginal *I*+*K* is considered sensitive. Table 5 shows the three two-dimensional marginal tables.

Table 5: Marginal Tables Derived from Table 1

IJ+			
3	87	68	72
5	8	80	4
3	4	7	107
64	6	56	7

+JK		
5	2	68
10	10	85
69	81	61
7	132	51

I+K		
74	75	81
3	85	9
8	61	52
6	4	123

Bounds on *I*+*K* can be quickly determined by either the Dobra and Fienberg or the Chowdhury *et al.* procedure, and compared with the desired bounds from the confidentiality intervals for each sensitive cell. For the marginal table *I*+*K*, Table 6 shows upper bounds of the desired protection range determined as the actual cell value plus 20 percent and the upper bounds of the computed protection range. Note that disclosures occur when the computed upper bound is below the desired upper bound; there are three disclosures—at cells (1,1), (2,2), and (3,3).

Table 6: Upper Bounds of Protection Range

Desired Upper Bounds		
89	90	98
4	102	11
10	74	63
8	5	148

Computed Upper Bounds		
88	152	200
86	94	78
21	120	65
74	71	133

The Dobra-Fienberg approach for the decomposable case has some natural extensions to tables that correspond to reducible graphs, and this class of extensions can reduce substantially the computational demands of the calculation of bounds in large numbers of dimensions—see, for example, Dobra and Fienberg (2001). A third and closely related bounding technique, suggested in Fienberg (1999), elaborated upon by Dobra (2001), and illustrated in Dobra and Fienberg (2001), can also be thought of as a generalization of Buzzigoli and Giusti's (1999) 'shuttle' algorithm. In its basic form, the Dobra procedure starts with loose upper and lower bounds for each cell, then iteratively narrows the bounds by taking advantage of cell relationships inherent in the tabular structure. The resulting bounds are sharp for a well-characterized group of problems, but for the general case the procedure uses a variant of the integer programming technique of implicit enumeration to find a table realization that provably achieves the sharpest bounds. The procedure's especially attractive feature is that it is relatively efficient in computing sharp bounds for the special cases such as when the marginals can be used to describe a decomposable graph, or when the corresponding graph has a reducible structure. The general method also works in the presence of 'structural zeros' and so may be of use in connection with other disclosure limitation approaches such as identifying secondary suppressions.

Using either LP or the methods described above to find bounds on cell entries specifies *extremal* values; the process in and of itself does not give the likelihood for any individual value in the range. To fit into our disclosure risk framework, we need a way to specify or estimate the probability associated with each feasible value. This is possible, at least in principle. Diaconis and Sturmfels (1998) show how one can systematically sample from the set of all tables that agree with the published marginal values. They provide a list of 'moves'—changes to the internal cell entries—that leave the published marginal values unchanged. Such moves (often called 'Gröbner basis' moves because of the method used to generate them) can be described as a set of cell increments and decrements; one example for a 3×3×3 table is shown in Table 7.

Table 7: Example of Gröbner Basis Moves

0	0	0
0	-	+
0	+	-

+	0	-
-	+	0
0	-	+

-	0	+
+	0	-
0	0	0

Applying a move from this list to a feasible table (*i.e.,* one that agrees with the published values) results in a new feasible table. The list is generated in such a way that the set of all feasible tables can be traversed uniformly if one chooses moves randomly with equal probability from the list. Thus an estimate of the probability of a particular cell value can be obtained by moving randomly through the set of feasible tables and tallying the proportion of time that one lands on a table having that cell value. Fienberg, Makov, and Steele (1998) apply this work in the context of disclosure limitation problems and link it to Markov perturbations. Fienberg *et al.* 2001 present an expository treatment of the theory in the context of contingency tables, making explicit links to the theory of log-linear models, and they provide heuristic descriptions of the role of Gröbner bases (the moves described above) when the MCMC (Markov Chain Monte Carlo) procedure approaches the extremal values.

Diaconis and Sturmfels applied this idea to sampling from the space of k-way tables when all (k-1)-way margins are known, and Fienberg *et al.* (2001) make clear how it generalizes to complete k-way tables with any set of marginals fixed, but the idea easily generalizes to other situations, in particular to cell suppression. Cell suppression merely adds some constraints and removes some possible moves, so the basic plan of constructing the Gröbner basis to find legitimate moves still applies.

While elegant in principle, the Gröbner basis idea is limited in practice at present. The difficulty is computational. Currently, the best general-purpose computer programs take many hours to find the Gröbner basis moves for a 3×3×3 table when the three two-way margins are given. Specialized programs can solve the same problem in a fraction of a second, but still take *months* to solve the analogous 5×4×3 problem (Roehrig 2001a). Larger problems are, at least in general, simply out of the question. Recent work (Dobra 2000), however, shows how to calculate a basis quickly for the class of auditing problems whose released marginals form a decomposable graph. Dobra's construction extends to related problems and allows for the combination of Gröbner bases for component subtables in regular graphs.

To apply our framework for disclosure risk, the statistical agency might assume that the data snooper holds a particular probability distribution for the values within the protection range. A reasonable procedure might be to assume a unimodal distribution spanning the interval, with its peak at the center and tails of decreasing probability extending to the endpoints. This is reasonable because although it is possible to have multiple feasible tables achieving the interval endpoint values, interior values allow more freedom for other cells to change, increasing the number of feasible tables possible. Thus one would not expect a distribution uniform over the protection range.

Other Disclosure Limitation Methods

Thus far we have discussed auditing tables that either have not been modified or have been modified using cell suppression. In both cases, the published cells (*i.e.,* the unsuppressed cells) imply constraints that serve to bound the sensitive cells. Some other forms of disclosure limitation may also be audited using LPs. The various forms of rounding (Section 3) all result in cells whose true values are unknown (thus affording protection), yet are known to lie in a well-defined interval. Thus, each published cell value gives rise to a pair of inequality constraints. Just as before, LP can find extreme values for a cell. Of course, the agency needs to invoke LP only to find bounds on quantities not in the rounded table; each rounded cell has obvious bounds. Nonetheless, a data disseminator may still want to know how bounds on sensitive cells in other, linked tables are influenced by the release of the rounded tables. In our example, the tables $IJ+$ and $+JK$ may be released in rounded form, but the disseminator might wish to know the resulting bounds for the table $I+K$.

Tables that have been protected by perturbation, by the addition of random noise, data swapping, or Markov perturbation, rely on the theoretical properties of the method. As an example we now analyze the protection provided by Markov perturbation.

To use the disclosure risk formula given in Section 2, we need a technique for determining the probabilities associated with possible cell values. We begin that process here, by showing the steps necessary to model a data snooper and incorporate that model into the analysis. A procedure along the same lines can be used to find a snooper's beliefs under other disclosure limitation schemes such as cell suppression and rounding.

Consider the elementary data square shown in Table 8, taken as part of a larger two-dimensional table.

Table 8: Elementary Data Square

1	14	15
17	83	100
18	97	115

This square is altered using Markov perturbation, and the resulting (published) square is the snooper's starting point. (The full Markov perturbation process will, as described above, alter cells within this square as other, intersecting, squares are perturbed. We treat the simple single-square case here and defer the full analysis for later research.) The top-left cell has the true value $\omega = 1$. Modifications to the square that leave the margins fixed and the internal entries non-negative are restricted; the masked value M for the top-left cell can be no larger than 15 and no smaller than 0. We can think of this square as being composed of a number of entities (0) that must remain there because of the marginal and non-negativity restrictions, and entities currently classified there but free to move out (1). Similarly, the lower-left cell can be thought of as consisting of unmoving entities (3) and movable ones (14). During the perturbation, some proportion of the top-left cell's movable entities move out, and some of the lower-left cell's movable entities move in. Under moves that are independent and identically distributed for a particular cell, the resulting number of entities in the top-left cell is a random variable that can be expressed as a constant plus the sum of two independent binomial random variables. It has the form

$$M = c + Binom(1, 1 - \theta) + Binom\left(14, \frac{r}{1-r} - \theta\right)$$

where $r = 1/18$ and is included to preserve stationarity (see Duncan and Fienberg 1999 for the details). The parameter $\theta \in [0, \min((1\text{-}r)/r, 1)]$ determines the extent of disclosure limitation: It controls the probability of movement by the entities. If θ is zero, the table remains unchanged, while larger values provide increasing protection.

Suppose that after perturbation, the published data square is as shown in Table 9.

Table 9: Data Square After Perturbation

3	12	15
15	85	100
18	97	115

A data snooper's view of the protected top-left cell is the following:

> I can think of the published cell value M as what I've 'observed', and the true 'state of nature' ω as being the unperturbed value. I want to construct the probabilities of the various true states of nature, given the evidence provided me by the published table. I know enough to construct probabilities for the observed value given the various possible true states of nature. But this is the wrong way around. Yet, with my prior distribution on the possible cell values, I can use Bayes's theorem to 'invert' the conditional probabilities.

Let's see how this can be done.

Because the Markov perturbation process leaves margins fixed, it is easy to determine the range of possible true values ω for the top-left cell (the possible states of nature). For each of these, we first construct *P(Observed | State of Nature)*. For example, one possible state of nature is the value $\omega = 1$ (which happens to be the true state, although unknown to the snooper). There are a number of ways to move from 0 (the count of unmovable entities) to 3 (the published value), each a sum adding to 3 of the two kinds of movement described above. Not all pairs are permissible, however, because the (1,1) cell can 'give up' only one entity. So the feasible set of 'stayer-mover' pairs is $A = \{(0,3), (1,2)\}$. We write $3 = X + Y$, where $X \sim \text{Binom}(1, 1 - \theta)$ and $Y \sim \text{Binom}(14, r/(1-r)\theta)$. Then the likelihood function value at $\omega = 1$ is given by

$$P(Observed\ M = 3 \mid State\ of\ Nature\ \omega = 1) = \sum_A P(X = x)P(Y = y)$$

$$= \sum_A \binom{1}{x}(1-\theta)^x\theta^{1-x} \bullet \binom{14}{y}\left(\frac{r}{1-r}\theta\right)^y\left(1-\frac{r}{1-r}\theta\right)^{14-y}.$$

The general form of the likelihood function is given by

$$L(\omega) = P(Observed\ M \mid State\ of\ Nature\ \omega = 1) =$$

$$\sum_A \binom{\omega}{x}(1-\theta)^x\theta^{\omega-x}\binom{15-\omega}{y}\left(\frac{r}{1-r}\theta\right)^y\left(1-\frac{r}{1-r}\theta\right)^{15-\omega-y}$$

where $A = \{(x, y){:}x + y = M;\ 0 \le x \le \omega; 0 \le y \le 15-\omega\}$ and $r = \frac{\omega}{18}$.

To calculate this likelihood value the data snooper must know, or assume, the value of θ. This would be the case if the agency publicly released the value of this disclosure limitation parameter. If the agency chose not to release the value of θ, then the data snooper would be uncertain about the appropriate likelihood value. The effect of this uncertainty would be to raise the data snooper's perceived chances of error and hence lower the disclosure risk. Thus calculations based on assuming that the data snooper knows the value of θ can be taken to provide upper bounds on the actual disclosure risk. Similarly, without knowing for sure the value of θ, the data user also has increased uncertainty. Based on the value of θ, the snooper can compute the value of r because of the conditioning assumption that the state of nature has value ω, so $r = \omega/18$. Calculating these conditional probabilities for the possible states of nature ω and based on an observed value of M = 3, we get Table 10, whose entries can be interpreted as the likelihood values for each of the given θ values.

Table 10: Likelihood Values

$\theta\backslash\omega$	**0**	**1**	**2**	**3**	**4**	**5**	**6**	**7**	**8**	**9**	**10**	**11**	**12**	**13**	**14**	**15**
0.05	0	0.0007	0.0683	0.7746	0.1482	0.0178	0.0017	0.001	0	0	0	0	0	0	0	0
0.1	0	0.0026	0.1151	0.6193	0.2235	0.515	0.0096	0.0016	0.0002	0	0	0	0	0	0	0
0.2	0	0.0088	0.1655	0.4382	0.2708	0.1121	0.38	0.0114	0.0032	8	0.002	0	0	0	0	0

In keeping with our overall plan, we now need to specify the data snooper's prior beliefs for the various states of nature.

In the simplest case, we might assume that the data snooper holds a uniform distribution over ω. The posterior probabilities *P(State of Nature* = ω | *Observed M* = *3)* are just those in Table 10; the prior distribution is uninformative. Suppose, instead, that the data snooper's prior distribution on ω were as shown in Table 11, anticipating a true state of nature close to 1 (note that the data snooper has prior probability 0 that ω = 0, because true zeroes would be unperturbed).

Table 11: Data Snooper's Prior Distribution on the State of Nature ω

θ / ω	**0**	**1**	**2**	**3**	**4**	**5**	**6**	**7**	**8**	**9**	**10**	**11**	**12**	**13**	**14**
0.05	0	0.001	0.146	0.751	0.096	0.006	0	0	0	0	0	0	0	0	0
0.1	0	0.005	0.259	0.579	0.139	0.016	0.002	0	0	0	0	0	0	0	0
0.2	0	0.015	0.403	0.381	0.157	0.032	0.009	0.002	0	0	0	0	0	0	0

Then the posterior probabilities for ω are as in Table 12.

Table 12: Posterior Probabilities for Prior Distribution Given in Table 11

0	**1**	**2**	**3**	**4**	**5**	**6**	**7**	**8**	**9**	**10**	**11**	**12**	**13**	**14**	**15**	**0**
0	0.3	0.3	0.15	0.1	0.05	0.04	0.03	0.02	0.01	0	0	0	0	0	0	0

A tractable family of distributions for basing the data snooper's prior distribution for ω is the beta-binomial family, conditioned on the known upper and lower bounds for *n*. A special case of the beta-binomial distribution is the discrete uniform distribution discussed above.

The general picture is this. Different choices of the disclosure limitation parameter θ produce differing amounts of 'blurring' of the probabilities of the true state of nature, and so provide varying degrees of disclosure risk. At the same time, these different values of θ cause different amounts of data distortion, and consequently affect data utility. In Section 5 we will illustrate some of these trade-offs.

5. Evaluation and Analysis

We are now in a position to show how two disclosure limitation methods can be compared using the R-U confidentiality map described initially in Section 2. For ease of exposition, we consider the simple 2×2 table with marginal totals of Section 3 and compare cell suppression with Markov perturbation. The technique is extensible to larger tables and different limitation methods.

Because our 2×2 table contains a 1 in cell (1,1), we assume that this cell is a primary suppression. This decision necessitates complementary suppressions, which for this simple case must obviously include the remaining three interior cell entries. Thus under cell suppression, all that can be published are the marginal totals. For Markov perturbation, we assume that the published table is the one discussed in Section 4 above.

To trade off disclosure risk and data utility, we require specific measures. To assess disclosure risk, for this illustration we use the reciprocal of entropy. Specifically, we take $R = 1/(-\Sigma(p_\omega \log p_\omega))$, where p_ω is the snooper's probability that the (1,1) cell value is ω. This measure assumes that disclosure risk is reduced as the snooper's probability function over the possible true cell values ω spreads out. To measure data utility, we use mean squared precision, specifically the reciprocal of the mean squared error based on the probability distribution of ω available to the data user and the fact that the true value of ω is 1.

In the case of Markov perturbation, we assume both the data user and the snooper are aware of the form of disclosure limitation that has been applied to the table, and for convenience we assume that both parties have the same prior distributions on the disguised value ω in cell (1,1). For cell suppression, we measure both disclosure risk and data utility according to the bounds that can be computed for the missing (1,1) cell. As noted in the previous section, the value ω is easily seen to be in the range [0, 15]. For simplicity of illustration, let us suppose that both the data snooper and the data user are interested in the value of ω. In further developments we will take the data snooper to be primarily interested in whether ω can be taken to be 1, and the data user to be interested in inference about the probability of falling in the (1,1) cell according to a probability model.

With these measures, we find the data utilities and disclosure risks shown in Table 13, also depicted in Figure 1.

Table 13: Data Utilities and Disclosure Risks

	Data Utility	**Disclosure Risk**
Markov Perturbation		
$\theta = 0.05$	0.24	1.263
$\theta = 0.1$	0.232	0.969
$\theta = 0.2$	0.216	0.713
Cell suppression	0.159	0.255
Rounding	0.346	0.919
Original Data	∞	2

Included in the table and figure are the results of an identical risk/utility analysis of our simple 2×2 table after rounding to base three. If we assume that the 'No Data' case (as shown in Figure 1) amounts to publishing just the marginal totals, the risk and utility values coincide with those for cell suppression. On the other hand, if 'No Data' means that not even the margins are published, risk and utility are both zero. The relative performance of the various disclosure limitation methods examined in this simple example should not be taken as suggesting that one method is universally superior to another. These and other methods need to be examined in the context of their actual use, with actual data products.

Figure 3. R-U Confidentiality Map for Suppression and Markov Perturbation

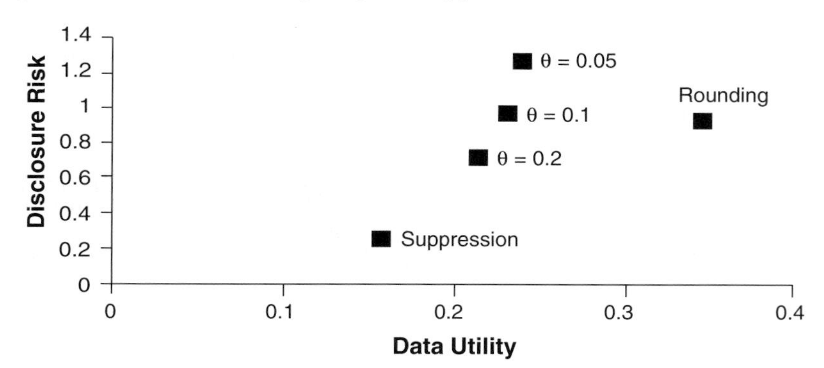

With each data product there is a disclosure threshold, above which the disclosure risk is too great. As Figure 3 illustrates, for different risk thresholds, different disclosure limitation methods, or their parameters, may be preferred. In this example, by varying the Markov exchange parameter θ it is possible to move from no protection ($\theta = 0$) to protection essentially equivalent to that provided by cell suppression, yet with higher data utility. This example is extreme, because it only analyzes an elementary 2×2 table. Nonetheless, it is clear that under reasonable assumptions, very different forms of disclosure limitation can be compared successfully.

There are several ways the R-U confidentiality map might be used within an organization. First, the organization may be unclear on the actual level of disclosure risk that has been borne in the past, and therefore unsure how to proceed in the future. Generating R-U maps for previous data releases would enable it to quantify the risks taken in the past and compare such risks among different data products. Such a program could enable the organization to develop a coherent strategy of dissemination in which comparable (or perhaps justifiably different) risks exist over the various releases. Knowledge of risk in past releases could be further used as a benchmark for the risk associated with a new release, especially one using a new disclosure limitation technique.

Along this same line, a comparison of two competing limitation techniques might result in an R-U confidentiality map like the one in Figure 4. In this figure, the choice of disclosure limitation method depends crucially on the disclosure threshold. As the threshold is raised above the point of intersection of the two curves, method 1 provides considerably more utility for a given threshold increase, suggesting new possibilities for the risk-utility trade-off. Our central theme, that the R-U confidentiality map allows much more informed decision making, is especially apparent in this example.

Figure 4: A Hypothetical R-U Confidentiality Map

Method 2
Method 1
Disclosure Risk
Disclosure Threshold
Data Utility

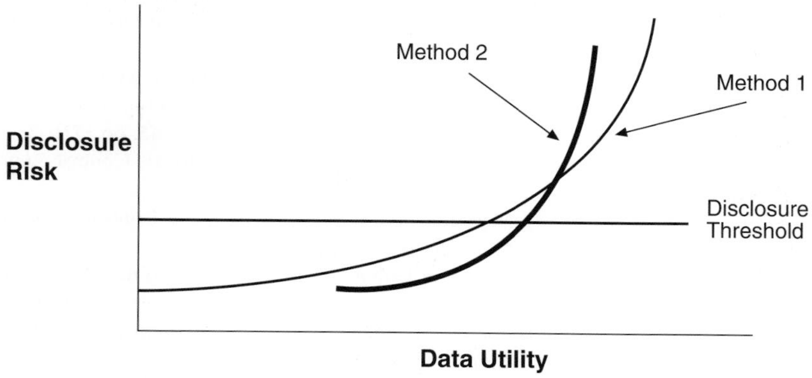

6. Conclusions

Statistical agencies have a variety of disclosure limitation methods available for use in their efforts to protect the confidentiality of tabular data. A systematic way of comparing the merits of these methods is through the R-U confidentiality map. An important further consideration is the computational burden of the procedure.

Perturbation methods are attractive because they offer the prospect of providing users with more data and in a form that allows for proper statistical inferences. We discuss several related versions in this chapter.

Recent advances in computational algorithms and the new statistical perspectives that have been brought to bear on disclosure limitation problems suggest that we may soon be in a position to do a much more thorough job of examining tabular data for possible disclosures and then applying disclosure limitation methods in such a form as to give users greater access to data for analysis.

Toward this end, we see the need for further research to identify procedures that have increased data utility while maintaining low disclosure risk, and more attention to the development of efficient computational algorithms that scale to the high-dimensional tabular problems typical of much statistical agency data.

References

Adam, N.R., and Wortmann, J.C. (1989). 'Security-Control Methods for Statistical Databases: A Comparative Study', *ACM Computing Surveys,* 21, pp.515-56.

Bethlehem, J.G., W.J. Keller, and J. Pannekoek (1990) 'Disclosure Control of Microdata', *Journal of the American Statistical Association,* 85, pp.38-45.

Buzzigoli, L., and A. Giusti (1999) 'An Algorithm to Calculate the Lower and Upper Bounds of the Elements of an Array Given Its Marginals', in *Statistical Data Protection (SDP'98) Proceedings*, pp.131–47, Luxembourg: Eurostat.

Causey, B., L. Cox, and L. Ernst (1985) 'Applications of Transportation Theory to Statistical Problems', *Journal of the American Statistical Association,* 80, pp.903-9.

Chen, G., and S. Keller-McNulty (1998) 'Estimation of Identification Disclosure Risk in Microdata', *Journal of Official Statistics,* 14, pp.79-95.

Chowdhury, S. D., G.T. Duncan, R. Krishnan, S.F. Roehrig, and S. Mukherjee (1999) 'Disclosure Detection in Multivariate Categorical Databases: Auditing Confidentiality Protection Through Two New Matrix Operators', *Management Science,* 45, pp.1710-23.

Cox, L.H. (1980) 'Suppression Methodology and Statistical Disclosure Control', *Journal of the American Statistical Association,* 75, pp.377-85.

—— (1981) 'Linear Sensitivity Measures and Statistical Disclosure Control', *Journal of Statistical Planning and Inference,* 5, pp.153-64.

—— (1987) 'A Constructive Procedure for Unbiased Controlled Rounding', *Journal of the American Statistical Association,* 82, pp.38-45.

—— (1995) 'Network Models for Complementary Cell Suppression', *Journal of the American Statistical Association,* 90, pp.1453-62.

—— (1999) 'On Properties of Multi-Dimensional Statistical Tables', unpublished manuscript.

—— (2002) 'Disclosure Risk for Tabular Economic Data', Chapter 8 of this volume.

Cox, L.H. and L.R. Ernst (1982) 'Controlled Rounding', *INFOR 20,* pp.423-432.

Dalenius, T., and S.P. Reiss (1978) 'Data-Swapping: A Technique for Disclosure Control' (extended abstract), in *Proceedings of the Section on Survey Research Methods,* Alexandria, Va.: American Statistical Association, pp.191-4.

—— (1982) 'Data-Swapping: A Technique for Disclosure Control, *Journal of Statistical Planning and Inference,* 6, pp.73-85.

De Carvalho, F.N. Dellaert, and M. de Sanches Osorio (1994) 'Statistical Disclosure in Two-Dimensional Tables: General Tables', *Journal of the American Statistical Association,* 89, pp.1547-57.

De Vries, R.E. (1993) *Disclosure Control of Tabular Data Using Subtables,* Voorburg: Statistics Netherlands.

De Waal, A.G., and A.J. Pieters (1995) *ARGUS User's Guide Report,* Voorburg: Department of Statistical Methods, Statistics Netherlands.

De Waal, A.G., and L.C.R.J. Willenborg (1994) *Minimizing the Number of Local Suppressions in a Microdata Set,* Voorburg: Statistics Netherlands.

—— (1996) 'A View on Statistical Disclosure for Microdata', *Survey Methodology,* 22, pp.95-103.

—— (1998) 'Optimal Local Suppression in Microdata', *Journal of Official Statistics,* 14, pp.421-35.

Diaconis, P., and B. Sturmfels (1998) 'Algebraic Algorithms for Sampling from Conditional Distributions', *Annals of Statistics,* 26(1), pp.363-97.

Dobra, A. (2000) *Measuring the Disclosure Risk for Multi-Way Tables With Fixed Marginals Corresponding to Decomposable Log-Linear Models*, Technical Report, Pittsburgh, Pa.: Department of Statistics, Carnegie Mellon University.

— (2001) *Computing Sharp Integer Bounds for Entries in Contingency Tables Given a Set of Fixed Marginals*, Technical Report, Pittsburgh, Pa.: Department of Statistics, Carnegie Mellon University.

Dobra, A., and S.E. Fienberg (2000) 'Bounds for Cell Entries in Contingency Tables Given Marginal Totals and Decomposable Graphs', *Proceedings of the National Academy of Sciences,* 97, 11185-92.

— (2001) 'Bounds for Cell Entries in Contingency Tables Induced by Fixed Marginal Totals', paper prepared for Second Joint ECE/Eurostat Work Session on Statistical Data Confidentiality, March 14-16, Skopje.

Domingo-Ferrer, Josep (1999) 'Microdata Masking Methods', Workshop on Confidentiality Research, U.S. Bureau of the Census. Alexandria, Va., May 3-4.

Duncan, G.T. (2001) 'Confidentiality and Statistical Disclosure Limitation', *International Encyclopedia of the Social and Behavioral Sciences,* forthcoming.

Duncan, G.T., and S.E. Fienberg (1999) 'Obtaining Information While Preserving Privacy: A Markov Perturbation Method for Tabular Data', in *Statistical Data Protection (SDP'98) Proceedings*, pp.351-62, Luxembourg: Eurostat.

Duncan, G.T., T.B. Jabine, and V.A. de Wolf (1993) *Private Lives and Public Policies: Confidentiality and Accessibility of Government Statistics,* Panel on Confidentiality and Data Access, Committee on National Statistics, Washington, D.C.: National Academy Press.

Duncan, G.T., and S. Keller-McNulty (2001) *Disclosure Risk vs. Data Utility: The R-U Confidentiality Map*, Technical Report, Statistical Sciences Group, Los Alamos, N.M.: Los Alamos National Laboratory.

Duncan, G.T., R. Krishnan, R. Padman, P. Reuther, and S. Roehrig (2001) 'Exact and Heuristics Methods for Cell Suppression in Multi-Dimensional Linked Tables, *Operations Research*, forthcoming.

Duncan, G.T., and D. Lambert (1986) 'Disclosure-Limited Data Dissemination' (with discussion), *Journal of the American Statistical Association,* 81, pp.10-28.

— (1989) 'The Risk of Disclosure of Microdata', *Journal of Business and Economic Statistics,* 7, 207-17.

Duncan, G.T., and R. Pearson (1991) 'Enhancing Access to Microdata While Protecting Confidentiality: Prospects for the Future' (with discussion), *Statistical Science,* 6, pp.219-39.

Elliot, M., and A. Dale (1999) 'Scenarios of Attack, the Data Intruders' Perspective on Statistical Disclosure Risk', *Netherlands Official Statistics,* 14, pp.6-10.

Ernst, L.R. (1989) *Further Applications of Linear Programming to Sampling Problems,* Technical Report Census/SRD/RR-89-05. Washington, D.C.: Statistical Research Division, U.S. Bureau of the Census.

Federal Committee on Statistical Methodology (1994) *Statistical Policy Working Paper 22: Report on Statistical Disclosure Limitation Methodology,* Washington, D.C.: U.S. Office of Management and Budget.

Fellegi, I.P. (1972) 'On the Question of Statistical Confidentiality', *Journal of the American Statistical Association,* 67, pp.7-18.

— (1975) 'Controlled Random Rounding', *Survey Methodology,* 1, pp.123-133.

Fellegi, I.P., and A.B. Sunter (1969), 'A Theory for Record Linkage', *Journal of the American Statistical Association,* 64, pp.1183-210.

Fienberg, S.E. (1994) 'Conflicts Between the Needs for Access to Statistical Information and Demands for Confidentiality', *Journal of Official Statistics,* 10, pp.115-32.

— (1997) 'Confidentiality and Disclosure Limitation Methodology: Challenges for National Statistics and Statistical Research', paper commissioned by the Committee on National Statistics for presentation at its 25th anniversary meeting.

— (1999) 'Fréchet and Bonferroni Bounds for Multi-Way Tables of Counts with Applications to Disclosure Limitation', in *Statistical Data Protection (SDP'98) Proceedings*, pp.115-29, Luxembourg: Eurostat.

— (2001) 'Statistical Perspectives on Confidentiality and Data Access in Public Health', *Statistics in Medicine,* 20, pp.1347-56.

Fienberg, S.E., and E.U. Makov (1998) 'Confidentiality, Uniqueness, and Disclosure Limitation for Categorical Data, *Journal of Official Statistics,* 14, 385-98.

Fienberg, S.E., E.U. Makov, M.M. Meyer, and R.J. Steele (2001) 'Computing Exact Distribution for a Multi-Way Contingency Table Conditional on Its Marginal Totals', in A. Saleh (ed), *Data Analysis from Statistical Foundations: Papers in Honor of D.A.S. Fraser,* Commack, N.Y.: Nova Science Publishing.

Fienberg, S.E., U.E. Makov, and R.J. Steele (1998) 'Disclosure Limitation Using Perturbation and Related Methods for Categorical Data' (with discussion), *Journal of Official Statistics,* 14, pp.485-512.

Fienberg, S.E., R.J. Steele, and U.E. Makov (1996) 'Statistical Notions of Data Disclosure Avoidance and Their Relationship to Traditional Statistical Methodology: Data Swapping and Log-Linear Models', *Proceedings of Bureau of the Census 1996 Annual Research Conference*, pp.87-105, Washington, D.C.: U.S. Bureau of the Census.

Fischetti, M., and J.J. Salazar-González (1996) 'Models and Algorithms for the Cell Suppression Problem', *Proceedings of the Third International Seminar on Statistical Confidentiality,* pp.114-22, Luxembourg: Eurostat.

— (1998) 'Experiments with Controlled Rounding for Statistical Disclosure Control in Tabular Data with Linear Constraints', *Journal of Official Statistics,* 14, pp.553-66.

____ (1999) 'Models and Solving the Cell Suppression Problem for Linearly Constrained Tabular Data', in *Statistical Data Protection (SDP'98) Proceedings*, pp.401-9, Luxembourg: Eurostat.

____ (2000) 'Models and Algorithms for Optimizing Cell Suppression in Tabular Data with Linear Constraints', *Journal of the American Statistical Association,* 95, pp.916-28.

Fuller, W. (1993) 'Masking Procedures for Microdata Disclosure Limitation', *Journal of Official Statistics,* 9, pp.383-406.

Giessing, S. (2002) 'A Practitioner's Guide to Non-Perturbative Disclosure Control Methods for Tabular Data', Chapter 9 of this volume.

Glover, F., and M. Laguna (1997), *Tabu Search*, Boston: Kluwer Academic Publishers.

Gouweleeuw, J.M., P. Kooiman, L.C.R.J. Willenborg, and P.P. de Wolf (1998) 'Post Randomisation for Statistical Disclosure Control: Theory and Implementation' (with discussion), *Journal of Official Statistics,* 14, pp.463-84.

Griffin, R., A. Navarro, and L. Flores-Baez (1989) 'Disclosure Avoidance for the 1990 Census', *Proceedings of the Section on Survey Research*, Alexandria, Va.: American Statistical Association, pp.516-21.

Kelly, J.P., A.A. Assad, and B.L. Golden (1990) 'The Controlled Rounding Problem: Relaxations and Complexity Issues', *OR Spektrum,* 12, pp.129-38.

Kelly, J., B. Golden, and A. Assad (1990) 'Controlled Rounding of Tabular Data', *Operations Research,* 38, pp.760-72.

— (1992) 'Cell Suppression: Disclosure Protection for Sensitive Tabular Data', *NETWORKS,* 22, pp.397-417.

Lambert, D. (1993) 'Measures of Disclosure Risk and Harm, *Journal of Official Statistics,* 9, pp.313-31.

Moore, R.A. (1996) *Controlled Data Swapping Techniques for Masking Public Use Microdata Sets,* RR 96-05,Washington, D.C.: U.S. Bureau of the Census.

Nargundkar, M.S., and W. Saveland (1972) 'Random Rounding to Prevent Statistical Disclosure', *Proceedings of the American Statistical Association, Social Statistics Section,* pp.382-5.

Navarro, A., L. Flores-Baez, and J. Thompson (1988) 'Results of Data Switching Simulation', presented at the spring meeting of the American Statistical Association and Population Statistics Census Advisory Committees.

Özsoyoğlu G. and J. Chung (1986) 'Information Loss in the Lattice Model of Summary Tables Due to Cell Suppression', *Proceedings of IEEE Symposium on Security and Privacy,* pp.160-73.

Paass, G. (1988) 'Disclosure Risk and Disclosure Avoidance for Microdata', *Journal of Business and Economic Statistics,* 6, pp.487-500.

Roehrig, S.F. (1999) 'Auditing Disclosure in Multiway Tables with Cell Suppression: Simplex and Shuttle Solutions', paper presented at the American Statistical Association Joint Statistical Meetings, Baltimore, August 8.

— (2001a) 'Computing Gröbner Bases for Statistical Disclosure Limitation, paper prepared for presentation at Grostat 2001, New Orleans, September.

— (2001b) *Finding Integer Solutions to Disclosure Limitation Problems Using Strong Inequalities*, Working Paper, Pittsburgh, Pa.: The Heinz School of Public Policy and Management, Carnegie Mellon University.

Schrijver, A. (1986) *Theory of Linear and Integer Programming*. New York: Wiley.

Spruill, N.L. (1983) 'The Confidentiality and Analytic Usefulness of Masked Business Microdata', *Proceedings of the Section on Survey Research Methods,* pp.602-7, Alexandria, Va.: American Statistical Association.

Statistics Netherlands, http://www.cbs.nl/en/figures/keyfigures/index.htm.

Strudler, M., H.L. Oh, and F. Scheuren (1986) 'Protection of Taxpayer Confidentiality with Respect to the Tax Model', *Proceedings of the Section on Survey Research Methods,* Alexandria, Va.: American Statistical Association.

Trottini, M. (2001) 'A Decision-Theoretic Approach to Data Disclosure Problems', paper prepared for Second Joint ECE/Eurostat Work Session on Statistical Data Confidentiality, Skopje, March 14-16.

U.K. Office of National Statistics (the U.K. government site), http://www.statistics.gov.uk.

U.S. Bureau of the Census, American FactFinder, http://factfinder.census.gov/servlet/BasicFactsServlet.

Willenborg, L., and T. de Waal (1996) *Statistical Disclosure Control in Practice,* Lecture Notes in Statistics 111, New York: Springer-Verlag.

—— (2000). *Elements of Statistical Disclosure Control*, Lecture Notes in Statistics 155, New York: Springer-Verlag.

Winkler, W.E. (1998) 'Reidentification Methods for Evaluating the Confidentiality of Analytically Valid Microdata', *Research in Official Statistics,* 1, pp.87-104.

Zaslavsky, A.M., and N.J. Horton (1998) 'Balancing Disclosure Risk Against the Loss of Nonpublication', *Journal of Official Statistics*, 14, pp.411-9.

Zayatz, L. (1993) 'Using Linear Programming Methodology for Disclosure Avoidance Purposes', *Proceedings of the International Seminar on Statistical Confidentiality.* pp.341-51, Luxembourg: Eurostat.

Zayatz, L.V., and S. Rowland (1999) 'Disclosure Limitation for American FactFinder', paper presented at the American Statistical Association Joint Statistical Meetings, Baltimore, August 8.

Confidentiality, Disclosure, and Data Access: Theory and Practical Application for Statistical Agencies
Pat Doyle, Julia I. Lane, Jules J.M. Theeuwes and Laura M. Zayatz (Eds)

Chapter 8

Disclosure Risk for Tabular Economic Data

Lawrence H. Cox

1. Introduction

Much of the early effort (1960s and 1970s) on statistical disclosure limitation (SDL) was focused on SDL for tabular economic data, and strong interest continues today. This interest is fueled by two factors: incentives for a business to discover confidential business information pertaining to a competitor and the highly structured tabular framework used to report economic statistics. These factors combine to increase the risk of disclosure of confidential business data.

This chapter examines disclosure risk in tabular economic data. Its primary objective is to develop and examine quantitative links between risk of disclosing confidential business information, rules for defining statistical disclosure, and methods for limiting statistical disclosure in tabular economic data. These concepts are examined in relation to prior information, assessing the effectiveness of disclosure limitation rules and methods and the effects of disclosure limitation methods on the usefulness of tabular economic data products.

The chapter is organized as follows. Section 2 introduces the tabular framework in which economic statistics are reported and discusses the ways in which this framework poses risks of disclosing confidential respondent information. Section 3 summarizes quantitative rules to define disclosure used by national statistical offices and illustrates the relationship of disclosure rules to actual disclosure risk. Section 4 introduces complementary cell suppression as the primary means of limiting statistical disclosure in tabular economic data. It includes a brief history of cell suppression methodology and automated systems. Topics discussed in this section include computational and operational complexity of complementary cell suppression, vulnerabilities of cell suppression, and effectiveness of cell suppression for limiting respondent disclosure risk. The section describes new approaches such as partial suppression and controlled estimates; examines strengths and weakness of existing methods, in particular the failure of current methods to deal directly with disclosure at the respondent level; and discusses an approach to limiting respondent level disclosure directly. Section 5 contains concluding comments.

2. Tabular Economic Data

Economic statistics typically take the form of aggregations of a quantitative attribute over a well-defined set of business units. Typical quantitative attributes are total sales, receipts or profits for business statistics, total value of shipments or total capital expenditures for manufacturing statistics, income or total acres by crop for agriculture statistics, and construction cost, sales price, or rental receipts for construction statistics. Other economic statistics take the form of count data, such as number of employees, number of units manufactured or sold, or number of business or farms within a particular area. Although these statistics can be incorporated within the framework that follows and can be expressed as aggregates of quantitative attributes, we focus here on quantitative attributes of the first kind.

Well-defined sets of business units are typically determined by geographic location and kind of business activity. The simplest, and a typical, example of geographic classification in the United States begins at the city level. Cities, plus the nonurban remainder of the county (*balance-of-county*), aggregate to the county in which they are located. Counties aggregate to states, which in turn comprise census regions and ultimately the entire United States. This geographic structure is *hierarchical*. Hierarchical aggregation is essentially one-dimensional. It can be modeled in a mathematically straightforward manner, and it presents the fewest problems from the standpoints of both disclosure limitation and computation. Modern economic statistics also include Metropolitan Statistical Areas (MSAs), which can overlap standard geographic areas such as states. Overlaps also result from cities located in multiple counties. Geographic overlaps create a nonhierarchical structure that increases the size and complexity of the system of tabulation equations, which in turn increases computational demands and the potential for disclosure risk.

Business activity is often classified hierarchically. Until recently, the United States has used Standard Industry Code (SIC), a coding with meaningful two-, three-, four- and six-digit hierarchical groupings. The North American Industry Classification System (NAICS) has now been introduced to replace SIC.

A two-dimensional tabulation structure results when a hierarchical geographic classification scheme, such as city-county-state-region-USA, is cross-classified with a hierarchical kind-of-business scheme, such as SIC. Each geography-by-SIC classification defines a unique tabulation cell. The tabulation cells are the fundamental reporting unit for tabular economic statistics—that is, statistics such as 'total retail sales' are reported at the cell level by aggregating 'retail sales' data over all business units satisfying the geography and SIC classifiers that define the cell. The tabulation structure is two-dimensional because at each level of classification except the lowest levels a unique two-way statistical table is created. This can be seen as follows. Consider a county by three-digit SIC level tabulation cell. The county may be disaggregated into its constituent cities (including balance-of-county, if necessary). The three-digit SIC can be disaggregated into its constituent four-digit SICs. A two-way table is thereby created: a grand total equal to the value of the statistic for the original county by three-digit SIC tabulation cell, row totals equal to the values of the statistic for the county by four-digit SIC tabulation cells, column totals equal to the values of the statistic for the city (and balance-of-county) by the original three-digit SIC tabulation cells, and internal entries presenting the value of the statistic by the corresponding city (and balance-of-county) by four-digit SIC tabulation cells. Two-dimensional tabulation structures can be modeled in a compact and computationally efficient manner, known as mathematical networks (Nemhauser and Wolsey 1988). Networks have been used for disclosure limitation in economic statistics (Cox 1987b, 1995, 1996) and other applications (Causey, Cox, and Ernst 1985; Cox 1987a).

In many U.S. states, the city-county-MSA-state classification is hierarchical and the tabulation structure exhibits the kind of two-dimensional structure described above. Some economic statistics, such as those for retail sales, are further classified by categorical variables such as number of employees, grouped, for example, as 0 employees, 1–10 employees, 11–50 employees, and so on. This results in a three-dimensional tabulation structure. Some of the approaches to disclosure limitation discussed in Section 4 are motivated by analogous uniformities in tabular structure.

Statistical agencies such as national statistical offices (NSOs) collect data pertaining to individual business units (establishments) such as local retail outlets. Aggregates of establishment data over establishments affiliated with the same business enterprise, such as a national retail corporation, naturally create enterprise-level data. Additional enterprise-level data may be collected directly. Often, these data are confidential. In return for receiving confidential data, statistical agencies promise the enterprise that they will keep the data confidential—that is, they will not release identifiable enterprise or establishment data directly and will not release statistics or data summaries from which confidential enterprise or establishment data can be inferred. This promise is referred to as the *confidentiality contract* between the agency and the enterprise. It obligates the agency to identify and limit risk of disclosure to each enterprise and establishment. In order to carry this out on a large, automated scale, disclosure risk must be quantified. Quantification of disclosure risk is the subject of this section and the next.

The tabulation structure introduces disclosure risk in three ways. The first is as follows. Data collected from individual establishments in a tabulation cell aggregate to the value of the statistic of interest for the entire cell—the cell value. If establishments associated with the same enterprise are grouped and their data pooled, then the cell value can be viewed as an aggregation over enterprise data. Each enterprise aggregate is referred to as the *contribution* of the enterprise to the cell value. We assume henceforth that all contributions, and therefore all cell values, are non-negative. This is the typical case, and mathematically the most challenging one.

Release of the cell value compromises confidentiality if the cell value is a close approximation of the contribution of any enterprise contributing to the cell, the largest enterprise in particular. Moreover, if any enterprise in the cell (typically, the second largest enterprise) subtracts its contribution from the cell value, it obtains an upper estimate for the contribution of the largest respondent. According to the technical definition of disclosure attributed to Dalenius (U.S. Department of Commerce 1978), any upper estimate constitutes some disclosure. In practice, the NSO considers that disclosure has occurred if this estimate is 'too close'. How close is "too close" is determined by individual agency policy and often differs from agency to agency and country to country (Federal Committee on Statistical Methodology 1994). The problem of quantifying the notion of "too close" and measuring disclosure and disclosure protection is discussed in Section 3.

As illustrated above, the term *primary disclosure* is used to describe the first way in which the tabulation structure introduces disclosure risk within individual tabulation cells. Primary disclosure occurs when release of the value of a tabulation cell or cell combination constitutes disclosure. Primary disclosure is easily dealt with for individual cells: All primary disclosure cells are suppressed from publication.

The tabulation structure, being additive, also allows for a second form of disclosure risk, known as secondary (or *complementary*) disclosure, as follows. Replace each primary disclosure by a disclosure variable x, and all other cells by their value. This results in a system of linear equations in the disclosure variables, denoted $AX = B$. Maximizing and minimizing each of these variables with respect to this system produces *exact* upper and lower estimates of the values of primary disclosure cells. If any of these estimates is 'too close', then disclosure remains. This process is referred to as *disclosure audit*. If the tabulation structure is two-dimensional, disclosure audit can be accomplished using mathematical networks that ensure efficient computation and integer outputs based on integer inputs (Cox 1995). Nonhierarchical and multidimensional tabulation structures are more complicated theoretically and computationally and do not ensure integer results (Cox 2000).

Secondary disclosure is limited by suppressing additional, nondisclosure cells until all such estimates are sufficiently broad. This process, referred to as complementary cell *suppression*, is the subject of Section 4. Complementary cell suppression has been the focus of considerable research in mathematics, statistics, and operations research since the 1970s (Cox 1980, 1995; Federal Committee on Statistical Methodology 1994; Fischetti and Salazar 1999, 2000; Willenborg and de Waal 2000).

The third way in which the tabulation structure introduces disclosure risk is disclosure within cell combinations, known as *multicell disclosure*. When two cells, such as total retail sales in one county and total retail sales in a second county, are combined, enterprises contributing to both cells contribute the total of their original contributions to the cell union. Such a union may be formed, for example, if both cells are suppressed but their total value can be determined precisely from

$AX = B$. In principle, all such cell unions must be treated as individual cells and examined for (primary) disclosure. Because the number of potential cell unions grows exponentially, this is an unmanageable problem if approached directly.

3. Quantifying Disclosure and Disclosure Protection in Tabular Economic Data

This section is concerned with the following questions. If the statistical agency sets out to thwart the efforts of an unauthorized third party to closely estimate enterprise or establishment data, then how close is 'too close'? How can the notion of 'too close' be quantified? How can this quantification be translated into operational computer software for disclosure limitation and disclosure audit driven by mathematical and statistical methods?

Perhaps the most intuitive notion of 'too close' is to say that the intruder has come too close if the intruder can estimate the contribution e of an enterprise or establishment to a cell (call it cell $\boldsymbol{x}$) to within less than a small percentage (p-percent) of the cell value $v(x)$, where $p \le 100$. Let x denote the disclosure variable corresponding to cell $\boldsymbol{x}$. Write

$$v(x) = \sum_{i=1}^{N} e_i,$$

where N denotes the number of enterprises contributing to $\boldsymbol{x}$, and $e_i \ge 0$ denote the individual enterprise contributions, ordered from largest ($i = 1$) to smallest ($i = N$). Because risk increases as the percentage of the estimate decreases, the greatest risk in releasing $v(x)$ is that of the largest respondent. Therefore, it suffices to assume that $e = e_1$, the largest contribution. Then the release of the cell value $v(x)$ results in disclosure if

$$v(x) < (1 + p/100)e_1 \tag{3.1}$$

or, equivalently, if

$$(p/100)e_1 - \sum_{i=2}^{N} e_i > 0 \tag{3.2}$$

This is an illustration of outsider disclosure because only publicly available information is used to achieve disclosure. As discussed in Section 2, insider disclosure occurs if another contributor to the cell subtracts its contribution e' from the cell value, and uses the result as an upper estimate of e_1. Because the greatest risk––measured by smallest percentage estimate—occurs when the intruder is the second largest respondent, then it suffices to assume that $e' = e_2$, the second largest contribution. Under insider disclosure, the release of the cell value $v(x)$ results in disclosure if

$$v(x) - e_2 < (1 + p/100)e_1 \tag{3.3}$$

$$S_p(x) = (p/100)e_1 - \sum_{i=3}^{N} e_i > 0 \tag{3.4}$$

This notation is convenient, if only to remind us that outsider disclosure implies insider disclosure, and it forms the basis of the remainder of this section.

The p-percent rule can be generalized to incorporate the case of a coalition of enterprises pooling its data in order to closely estimate the contribution of a competitor, as follows. Assume coalitions of size $(n - 1)$ exist, for $n \geq 2$. Then the greatest risk occurs when contributors number 2, ..., n form a coalition against the largest contributor. This results in the disclosure rule:

$$S_{p(n)}(x) = (p/100)e_1 - \sum_{i=n+1}^{N} e_i > 0 \tag{3.5}$$

Again, the formulation is convenient because, for example, it reveals that larger coalitions result in increased disclosure risk.

The U.S. Bureau of the Census uses a p-percent rule to define disclosure in economic tabular data. Before the 1992 economic censuses, it used the n-respondent, k-percent rule, under which disclosure occurs if the sum of the contributions of the n largest respondents is greater than k-percent of the cell value, $k \leq 100$. Namely, if

$$S_{n,k}(x) = \sum_{i=1}^{n} ((100 - k)/k)e_i - \sum_{i=n+1}^{N} e_i > 0 \tag{3.6}$$

This rule is based on coalitions of $(n-1)$ contributors. It is long-standing Census Bureau policy to keep confidential the parameters (n, k, p) used in its disclosure rules for economic tabular data.

Use of the (n, k)-rule at the Census Bureau dates back to the 1940s, when disclosure limitation was done by hand. Institutional history has it that the rule was selected to provide $p = (100 - k)$-percent protection against coalitions, namely, it was intended to have the effect of equation (3.5). Before computers, it was simpler to compute the total contribution of the n largest contributors and compare this to k-percent of the cell value; hence the adoption of the (n, k)-rule.

In the early 1990s, the author undertook a comparative study of the p-percent and (n, k)-rules, for $p = 100 - k$. Direct comparison of the rules (3.5) and (3.6) yields

$$S_{p(n)}(x) - S_{n,k}(x) = (p/100)e_1 - \sum_{i=1}^{n} ((100-k)/k)e_i \tag{3.7}$$

$$= -(p/100)(p/(100-p))e_1 - \sum_{i=2}^{n} (p/(100-p))e_i < 0$$

Thus, $S_{p(n)}(x) \le S_{n,k}(x)$, and consequently the (n, k) rule is stricter than the p-percent rule—that is, it declares disclosure in a larger set of cases.

The two rules were also compared in terms of the minimum percentage protection provided to the largest respondent by the cell value against an insider coalition in a published (nondisclosure) cell. Minimum protection occurs when the corresponding disclosure rule equals zero. The numeric value of the protection provided under both rules equals

$$\sum_{i=n+1}^{N} e_i .$$

When the p-percent rule is zero, the numeric protection of course equates to p-percent minimum protection. When the (n, k)-rule equals zero, the numeric protection equals

$$\sum_{i=1}^{n} (p/(100-p))e_i .$$

As a percentage of e_1, this can range anywhere from a lower bound of $100p/(100-p)$-percent to n times that amount, depending on the relative sizes of the contributions of the largest n contributors. In other words, the (n, k)-rule provides more than p-percent minimum protection but provides widely different percentage protection to different enterprises. If minimum p-percent protection to all enterprises is the goal, then the (n, k)-rule provides *overprotection*; it demands more disclosure limitation, which, as described in the next section, translates to more distortion or abbreviation of the data than is absolutely necessary. Moreover, if as a matter of policy all enterprises are to be treated equally (in terms of limiting disclosure risk, measured by minimum percentage protection), then the (n, k)-rule is inappropriate. In consideration of these and other factors, the U.S. Census Bureau replaced the (n, k)-rule by the p-percent rule during the 1990s.

The mathematical structure of these rules enabled their comparison. There are evident advantages to formulating disclosure rules in this manner. Rules of this sort, based on linear combinations of enterprise contributions, are called *linear sensitivity* measures, studied by Cox (1981). Another linear sensitivity measure is:

$$S_{p;q}(x) = (p/q)e_1 - \sum_{i=3}^{N} e_i > 0 \quad (3.8)$$

This is the p-percent, q-percent ambiguity rule developed and used by Statistics Canada. This rule incorporates both insider knowledge and public knowledge, as follows.

Assume that, from public knowledge, an outsider can estimate the contribution of any enterprise to within q-percent of its value, $q \leq 100$. How then should a disclosure rule be formulated in order to protect the largest enterprise from p-percent disclosure, $p < q$, from the second largest? The rule is provided by (3.8), and is easily generalized to coalitions. Comparison of (3.3) and (3.8) reveals that an ambiguity rule is stricter than a corresponding percentage rule. Indeed, the p-percent rule is precisely the p-percent, (q = 100)-percent ambiguity rule.

The linear sensitivity measures presented above measure the closeness of upper estimates of the contribution of the largest contributor. Technically, they are *upper linear sensitivity measures.* An upper linear sensitivity measure

$$S(x) = \sum_{i=1}^{N} w_i e_i$$

provides a computational algorithm for checking for primary disclosure:

Primary Disclosure Rule: A cell $\boldsymbol{x}$ is a primary disclosure cell if and only if $S(x) > 0$.

Primary disclosure cells are deleted (suppressed) from publication. Analysis based on lower estimation results in an analogous concept, lower linear sensitivity measures. The theory of lower measures, covered in Cox (1981), is analogous and will not be pursued in this chapter.

A disclosure rule, whether expressed by a sensitivity measure or not, is derived from policy and data considerations and must above all be sensible. For example, a disclosure rule given by a measure S for which all $w_i = 0$ would not be sensible because no cell could be a primary disclosure. A rule with some $w_i > 0$ but no $w_j < 0$ is not sensible because all cells would be primary disclosures. Consequently, henceforth some of the enterprise weights in the rule are assumed positive and others negative. Also assume that a smallest negative weight exists.

There are other notions of sensibility for a primary disclosure rule. Given a disclosure rule, consider any two tabulation cells, such as retail sales in a city and retail sales in the remainder of the county containing the city. Assume that neither cell is a primary disclosure cell with respect to the rule. Then it would be unreasonable, and not sensible, if the rule designated the cell union—for example, retail sales within the county—as a disclosure cell. Moreover, it would be operationally infeasible to check for disclosure, because all cell combinations would have to be created and subjected to the rule. This illustrates the concept of subadditivity: An upper linear sensitivity measure S is subadditive if and only if for any two disjoint tabulation cells $\boldsymbol{x}$ and $\boldsymbol{y}$ and the cell union $\boldsymbol{x} \cup \boldsymbol{y}$:

$$S(x \cup y) \leq S(x) + S(y) \tag{3.9}$$

Subadditivity can be verified directly, as follows.

Theorem (Cox 1981): An upper linear sensitivity measure

$$S(x) = \sum_{i=1}^{N} w_i e_i$$

is *subadditive* if and only if $w_i \geq w_j, i \leq j$.

It is easily verified that the p-percent, (n, k), and p-percent, q-percent ambiguity rules are all subadditive. Appropriate combinations of subadditive rules, such as the sum or maximum of two subadditive rules, are also subadditive. Maximum rules are useful as they express the notion: $\boldsymbol{x}$ is a primary disclosure if either Condition A or Condition B holds.

Linear sensitivity measures are then a powerful tool: They provide a compact, algorithmic mechanism for presenting a disclosure rule, they enable comparison of rules, and they can be selected to assure subadditivity. In addition, they enjoy another important property: Given a primary disclosure cell $\boldsymbol{x}$ and its contributing enterprises, the upper linear sensitivity measure $S(x)$ encodes the minimum possible value of any nonprimary disclosure cell x' containing x, as follows (Cox 1981).

The disclosure rule depends only on the positivity or not of the upper linear sensitivity measure S. A nondisclosure cell x' of minimal value containing primary disclosure cell x will satisfy $S(x') = 0$. Let w_r denote the last negative weight of S, and r the first index at which this weight is achieved. The method for computing the minimum possible value of $v(y)$ for which x' is a nondisclosure cell is illustrated below for the simplest case, namely $N > r - 1$. For other cases, the value computed (in 3.10) provides a lower bound on the minimum value.

To determine the minimum possible value of a nonprimary disclosure cell containing cell $\boldsymbol{x}$, it suffices to note that the largest enterprise in $\boldsymbol{x}$ must make the same contribution in the larger cell $\boldsymbol{x'}$ as it does in $\boldsymbol{x}$, and similarly for any coalition of respondents, for otherwise the value of the minimal cell would be inflated beyond the minimum value. To achieve the minimum, it suffices to add as many new contributors as necessary, each contributing at most the current smallest contribution e_N, until $S(x') = 0$ is reached. The weight in S of each of these new contributions equals w_N. Hence, the minimum value for $\boldsymbol{x'}$ is:

$$u(x) = v(x) + S(x)/|w_r| \tag{3.10}$$

$v(y)$ is referred to as the *upper protection limit* for x; it is a lower bound on acceptable upper estimates of $v(x)$. Had a lower sensitivity measure also been applied, a corresponding lower protection limit—an upper bound on acceptable lower estimates of $v(x)$—would be determined. Often in practice, and for simplicity of exposition here, a lower measure is not used and the lower protection limit is assigned in a symmetric manner, thereby equal to:

$$l(x) = \max\{0, v(x) - S(x)/|w_r|\} \tag{3.11}$$

This completes the mathematical theory of primary disclosure. The next section deals with extending these results to complementary disclosure, which is more complicated and based in part on (3.10).

4. Complementary Cell Suppression

As described in Section 3, given a disclosure rule represented as a subadditive upper linear sensitivity measure *S*, primary disclosure cells can be identified directly through cell-by-cell application of the rule. Primary disclosures are suppressed from publication, but, as discussed in Section 2, this alone is not likely to ensure adequate disclosure protection, namely, inference within a system of linear equations A*X* = B in the disclosure variables, within which exact linear estimates of each disclosure variable may be obtained, and, second, the possibility that multi-cell disclosure remains after complementary cell suppression. Each of these issues is examined in this section.

If exact upper and lower estimates of the value of suppressed primary disclosure cells are not sufficiently broad, they may be broadened by suppressing additional, nondisclosure cells from publication, by replacing some of the constants in $AX = B$ by additional variables. The suppressed nondisclosure cells are called complementary suppressions, and the process is complementary cell suppression. Complementary cell suppression is a theoretically and computationally difficult problem. In particular, in the parlance of computational complexity, it has been shown to be an *NP-hard* problem, even for one-dimensional tables (Kelly, Golden, and Assad 1992), rendering unlikely the existence of a polynomial time algorithm for its solution.

The technical difficulty of the complementary cell suppression problem is illustrated in Table 1. Despite what appears to be an adequate pattern of suppressed cells, the value (10) of the suppressed cell in the upper left-hand corner can be computed precisely (subtract the sum of the second and third columns from the

Table 1. Apparently Adequate Cell Suppression

D	**D**	**D**	0	**20**
0	**D**	**D**	0	**20**
D	0	0	**D**	**30**
D	0	0	**D**	**30**
30	**10**	**20**	**40**	**100**

sum of the first two rows). We illustrate the complementary cell suppression problem in the most familiar case, that of a single two-dimensional table, illustrated for concreteness by Table 2.

Table 2: Cell Suppression in a Two-Dimensional Table

20(D)	20	10	20	**70**
10	**20(D)**	5	15	**50**
40	10	**20(D)**	10	**80**
5	15	**10(D)**	5	**35**
75	**65**	**45**	**50**	**235**

Assume that each cell in Table 2 has been subjected to primary disclosure analysis using an upper linear sensitivity measure *S*, and that the four cells in bold and marked *D* have been identified as primary disclosures. Assume that the corresponding protection limits are +/–50 percent of the original values. The subject of this section is: What constitutes an acceptable complementary cell suppression pattern for Table 2, and how can acceptable patterns be computed?

It is instructive to begin with an ideal complementary cell suppression pattern, illustrated in Table 3. It can be verified that the exact linear upper and lower estimates of the original suppressed cells equal their protection limits, so that the pattern is acceptable. Why is it ideal?

Table 3: An Ideal Complementary Cell Suppression Pattern

D	20	**D**	20	**70**
D	**D**	5	15	**50**
40	**D**	**D**	10	**80**
D	15	**D**	5	**35**
75	**65**	**45**	**50**	**235**

There are four rows and three columns of Table 2 that contain primary disclosures. All of these rows and two of these columns contain only one primary disclosure. Consequently, at least four complementary suppressions must be made (one in each of the rows). Table 3 contains precisely four complementary suppressions, and is therefore economical (and optimal) from the standpoint of suppressing the fewest cells. Why is this important? As disclosure limitation is focused on reducing disclosure risk below acceptable levels, it must do so with as little distortion to the quality and completeness of the data as possible. One measure of data distortion is total number of cells suppressed. Under that measure, Table 3 is ideal. Least total number of cells suppressed was the primary measure used by the U.S. Census Bureau for the 1977 and 1982 U.S. economic censuses. If we assign a suppression variable z_{ij} to each table cell (suppress the cell value a_{ij} if $z_{ij} = 1$, publish the cell value if $z_{ij} = 0$), then least number of cells suppressed is achieved by minimizing the objective function:

$$\sum_{i,j} z_{i,j} \tag{4.1}$$

If we treat the z_{ij} as continuous variables on the interval [0, 1], then (4.1) is a linear objective function.

Table 3 also involves the least total value of cells suppressed (35). To see this, note that at least 10 units must be suppressed in each of the first three rows, while at least five units must be suppressed in row four. Another measure of data distortion is total value suppressed. Least total value suppressed is achieved by minimizing the linear objective function:

$$\sum_{i,j} a_{i,j} z_{i,j} \tag{4.2}$$

The measure of data distortion used in the 1987 and subsequent U.S. economic censuses is total value suppressed. It was the secondary measure in 1977 and 1982.

A third measure of data distortion, used by Statistics Canada, is Berg entropy (Sande 1984), which seeks to strike a balance between number suppressed and value suppressed. This measure is represented by the linear objective function:

$$\sum_{i,j} \log\,(1 + a_{i,j}) z_{i,j} \tag{4.3}$$

A mathematical model for the complementary cell suppression problem (in general, as well as for two-way tables) is constructed as follows. Let $Ax = 0$ denote the tabulation structure wherein every cell value v(x) is replaced by a variable x. For primary disclosure $k = 1, \ldots, K$, the vectors $\boldsymbol{x}_k +$ and $\boldsymbol{x}_k^-$ are two copies of the x-variables, and $\boldsymbol{x}_{k,k} +$ and $\boldsymbol{x}_{k,k}^-$ denote the specific variables corresponding to primary disclosure cell k. **a** denotes the vector of true cell values, so that a_k is the value of primary disclosure cell k, which must be protected. Vectors $\boldsymbol{L}$ and $\boldsymbol{U}$ represent, respectively, deviations below and above the cell values that an outsider could reasonably infer from public information (standard defaults are, *e.g.*, 100 percent of the cell value); $\boldsymbol{l}$ and $\boldsymbol{u}$ denote minimum acceptable deviations: For primary disclosure cell k, l_k, u_k denote the minimum acceptable lower and upper protection limits away from the cell value, with $l_k \le a_k$, $l_k + u_k > 0$. Let z be a {0, 1}-vector denoting the decision to publish (0) or suppress (1) the corresponding cell.

A mathematical model for complementary cell suppression is:

$$
\begin{aligned}
\boldsymbol{A}\boldsymbol{x}_k^- &= \boldsymbol{A}\boldsymbol{x}_k^+ = \boldsymbol{0} \\
\max\{\boldsymbol{0}, \boldsymbol{a} - \boldsymbol{L} \cdot z\} &\le \boldsymbol{x}_k^-, \boldsymbol{x}_k^+ \le \boldsymbol{a} + \boldsymbol{U} \cdot z \\
x_{k,k}^- &\le a_k - l_k \\
x_{k,k}^+ &\ge a_k + u_k
\end{aligned}
\tag{4.4}
$$

Feasible solutions to (4.4) correspond to acceptable cell suppression patterns. Note that $z_k = 1$, $k = 1,..., K$. If protection limits are symmetric, exact lower and upper bounds on the values a_k of suppressed primary disclosures computable from (4.4) may be nearly symmetric, with the result that their midpoint closely approximates the true value. This can be addressed by substituting for the last two protection constraints of (4.4) less stringent constraints called range constraints:

$$
\begin{aligned}
x_{k,k}^- &\le a_k \le x_{k,k}^+ \\
x_{k,k}^+ - x_{k,k}^- &\ge l_k + u_k
\end{aligned}
\tag{4.5}
$$

Incorporation of a linear objective function such as (4.1) – (4.3) with the model (4.4), (4.5) results in an integer linear programming problem (ILP). Complementary cell suppression at the cell level therefore can be expressed as an ILP. Building on earlier work of Kelly, Golden, and Assad (1992), Fischetti and Salazar (1999, 2000) have proposed precisely such formulations and, in addition, are demonstrating that even though these programs are massive and involve integer variables, they can be solved in practical applications.

The Fischetti-Salazar model exhibits several attractive features theoretically. The extent to which these features scale to larger and larger tabulation structures is likely to be the focus of much cell suppression research in the near future. Cox (2001a) explores these features in detail.

The first feature is that complementary suppression is performed *globally*, that is, the entire tabulation structure is taken into account as suppressions are performed. This is desirable because, if other and separate parts of the tabulation system are examined separately, it is possible that disclosure can arise when the parts are rejoined. Typically ways of dealing with this problem are to check and resolve disclosure between the parts. Unfortunately, such approaches are ad hoc and two possibilities remain: Either not all residual disclosure is detected and resolved, or the measures taken to resolve the problem suppress more information than would have been suppressed under a global method (oversuppression).

One existing global method is CONFID of Statistics Canada, based on linear programming (Robertson 1993). CONFID was developed by Gordon Sande, who has developed an extension called ACS (Automated Complementary Suppression). Complete details are not available, but it is likely that ACS is also global. The U.S. Census Bureau has used two complementary cell suppression methodologies since 1977—a combinatorial method (Cox 1980) and a method based on mathematical networks. Both of these methods are for the most part based on a single two-way table and are not global. Fischetti-Salazar methods are global.

The second feature is that complementary cell suppression is *simultaneous*, that is, all suppressions are made based on the solution of a single (albeit massive) mathematical program. The advantage of simultaneous suppression is that it takes into account the contribution each suppressed cell makes to reducing the risk of each primary disclosure, thus reducing the likelihood of oversuppression. CONFID and the U.S. Census network models protect one primary disclosure at a time and are not simultaneous. The U.S. Census Bureau combinatorial method was simultaneous, but only for a single table, and additional analysis was required to avoid disclosure as illustrated in Table 1. Fischetti-Salazar methods are simultaneous.

The third feature is that the cell suppression method is self-auditing; a secondary mathematical programming analysis of the tabulation structure to verify disclosure protection is not needed. Cell suppression methods based on mathematical programming are self-auditing, because the protection constraints are an explicit part of the mathematical programming model.

The fourth feature is that the cell suppression method is optimal in that it is capable of producing optimal suppression patterns for a given measure of information loss. Only the Fischetti-Salazar methods are optimal. Strictly speaking, if a method is neither global nor simultaneous, it is not optimal.

Based on (mixed) integer linear programming, Fischetti-Salazar methods can be computationally intensive and sensitive to the size of the problem. If the binary integer suppression variables (z) are relaxed to the form of continuous variables restricted to the interval [0, 1], a method known as partial suppression results, in which a cell is suppressed if its z-variable is non-zero in the final solution. Partial suppression methods are not optimal and are prone to oversuppression. Based on linear programming, CONFID is a partial suppression method.

Other nonsuppressing approaches to complementary cell suppression include releasing exact interval estimates for suppressed cells. Controlled estimates, an approach suggested by Ramesh Dandekar, U.S. Energy Information Administration, releases estimates of suppressed cells taken from within exact intervals and constructed to preserve tabular structure. Cox (2001b) offers an integer linear programming model for controlled estimates.

The Fischetti-Salazar methods emerge as the most promising for complementary cell suppression, provided that good computational performance continues to be demonstrated over a range of problem types and sizes. However, all current methods, including Fischetti-Salazar, do not protect against multicell disclosure. For example, if one primary disclosure cell contains only one enterprise and a second primary disclosure cell also contains only one enterprise, and if the value of the cell combination can be deduced or narrowly estimated from AX = B, then either outsider disclosure occurs (if the enterprises are the same) or insider disclosure occurs (if they differ). Some systems, including CONFID and the U.S. Census Bureau system, check directly for multicell disclosure along single tabulation equations. CONFID enables the analyst to hard-code constraints addressing multicell disclosure in prespecified cell combinations. However, no current method deals systematically with the multicell problem. The multicell problem is important because it is at this level that actual respondents, and not artifactual cell totals, are being protected for unauthorized disclosure.

The method proposed by Cox (2001b), introduced in Cox (1999), incorporates enterprise variables and constraints and protects against multicell disclosure. The method is based on an integer linear programming model amenable to the computational strategies of Fischetti-Salazar. Like Fischetti-Salazar, it can be relaxed to a partial suppression problem. The extent to which this model, which itself is considerably larger than the Fischetti-Salazar model, is computationally efficient and the cases for which it is computationally feasible offer appealing topics for near-term future research.

5. Concluding Comments

Disclosure risk for tabular economic data is based on risk to establishments and enterprises contributing to tabulation cells. However, because of computational limitations and lack of theoretical foundations, early approaches to complementary cell suppression focused on disclosure limitation at the cell level only, meaning that effort was focused exclusively on satisfying protection constraints as defined by the sensitivity rule. Cell-level protection is a necessary but not a sufficient condition for respondent-level protection—ensuring against insider disclosure in all tabulation cells and cell combinations.

Mathematical models for cell-level protection have matured rapidly (*e.g.*, Fischetti and Salazar 1999, 2000) and are ready to be examined for computational efficiency. Models for respondent-level protection (*e.g.*, Cox 2001a) are emerging. Such models provide a theoretical framework for the problem, and they next need to be examined and improved based on computational considerations.

References

Causey, B., L.H. Cox, and L. Ernst (1985) 'Applications of Transportation Theory to Statistical Problems', *Journal of the American Statistical Association,* 80, pp.903–9.

Cox, L.H. (1980) 'Suppression Methodology and Statistical Disclosure Control', *Journal of the American Statistical Association,* 75, pp.377–85.

—— (1981) 'Linear Sensitivity Measures in Statistical Disclosure Control', *Journal of Statistical Planning and Inference,* 5, pp.153–64.

—— (1987a) 'New Results in Disclosure Avoidance for Tabulations', Proceedings of the 45th Session, International Statistical Institute, Voorburg. *Bulletin of the International Statistical Institute,* pp.83–4.

—— (1987b) 'A Constructive Procedure for Unbiased Controlled Rounding', *Journal of the American Statistical Association,* 82, pp.520–4.

—— (1995) 'Network Models for Complementary Cell Suppression', *Journal of the American Statistical Association,* 90, pp.1453–62.

—— (1996) 'Addendum'. *Journal of the American Statistical Association,* 91, p.1757.

—— (1999) 'Invited Talk: Some Remarks on Research Directions in Statistical Data Protection', Statistical Data Protection, Proceedings of the Conference, Lisbon, March 25-27, 1998, Luxembourg: Eurostat, pp.163–76.

—— (2000) 'On Properties of Multi-Dimensional Statistical Tables', submitted.

—— (2001a) 'Complementary Cell Suppression for Tabular Data: An Examination', submitted.

—— (2001b) 'Discussion (on Session 49: Statistical Disclosure Control for Establishment Data),' *ICES II: The Second International Conference on Establishment Surveys—Survey Methods for Businesses, Farms and Institutions, Invited Papers,* Alexandria, VA: American Statistical Association, pp.904-907.

Federal Committee on Statistical Methodology (1994) *Statistical Policy Working Paper 22:Report on Statistical Disclosure Limitation Methodology,* Washington, D.C.: U.S. Office of Management and Budget, Statistical Policy Office (NTIS Document Sales, PB94-165305).

Fischetti, M., and J.J. Salazar (1999) 'Models and Algorithms for the 2-Dimensional Cell Suppression Problem in Statistical Disclosure Control', *Mathematical Programming,* pp.84, pp.283–12.

—— (2000) 'Models and Algorithms for Optimizing Cell Suppression in Tabular Data with Linear Constraints', *Journal of the American Statistical Association,* 95, pp.916–28.

Kelly, J., B. Golden, and A. Assad (1992) "Cell Suppression: Disclosure Protection for Sensitive Tabular Data', *NETWORKS,* 22, pp.397–417.

Nemhauser, G., and L. Wolsey (1988) *Integer and Combinatorial Optimization,* New York: John Wiley & Sons.

Robertson, D. (1993) 'Cell Suppression at Statistics Canada,' *Proceedings of the Annual Research Conference*, Washington, D.C.: U.S. Bureau of the Census, Washington, D.C., pp.107–31.

Sande, G. (1984) 'Automated Cell Suppression to Preserve Confidentiality of Business Statistics', *Statistical Journal of the United Nations ECE,* 2, pp.33–41.

Willenborg, L., and T. de Waal (2000) *Elements of Statistical Disclosure Control*, Lecture Notes in Statistics 155, New York: Springer Verlag.

U.S. Department of Commerce (1978) *Report on Statistical Disclosure and Disclosure Limitation,* Statistical Policy Working Paper 2, Washington, D.C.: Office of Statistical Policy and Standards.

Confidentiality, Disclosure, and Data Access: Theory and Practical Application for Statistical Agencies
Pat Doyle, Julia I. Lane, Jules J.M. Theeuwes and Laura M. Zayatz (Eds)

Chapter 9

Nonperturbative Disclosure Control Methods for Tabular Data*

Sarah Giessing
Federal Statistical Office of Germany

1. Introduction

Tabular data are basically aggregate data. As such, they are not supposed to contain information that can reveal the identity of particular respondents. In many cases, however, table cells indeed contain information on a single or very few respondents, which implies a disclosure risk for the data of those respondents. In such cases, measures for protection of those data have to be put in place. The choice is between suppressing part of the information or perturbing the data—for example, by adding random noise or rounding.

Although perturbation methods may have certain advantages in minimizing information loss, there are compelling objections to their application to tabular data, in particular to tabular data on businesses. These objections come mainly from government statistical agencies, because the reliability of the data they publish is such a fundamental issue. Particularly when data are released not to a few sophisticated users (as is often the case with, for example, scientific use microdata files) but in the form of tables released to the general public, it is hard to make sure that the users will take the perturbation into appropriate account. Because of this objection, the most common disclosure control measure for tabular data is cell suppression, which is the focus of this chapter.

Cell suppression as a statistical disclosure control (SDC) technique comprises two steps. First, the disclosure risk connected to each cell of a table is assessed.

* For the most part, this chapter does not present the author's own concepts, but rather provides a summary and illustration of issues raised in the literature (as cited in the References). Much of the author's insight comes from the work of Cox (1981), Robertson (1993), and Jewett (1993). The author is also grateful to B. Veldhues of the Business Statistics Division of the German Federal Statistical Office and R.D. Repsilber of Landesamt für Datenverarbeitung und Statistik Nordrhein-Westfalen, who have done pioneer work in the field of automated tabular data protection in Germany. Thanks go to these two for numerous interesting discussions. Section 5 is based in part on their ideas and concepts, some but not all of which are illustrated in Veldhues (1999) and Repsilber (1999). In particular, the idea for example 9 is owed to Repsilber.

Section 2 explains what is involved in disclosure risk assessment at the cell level. This assessment tells the data disseminator which cells reveal too much information, and those cells are then suppressed. These so-called primary suppressions are not enough to prevent exact disclosure or close estimation of the sensitive cells, however, because of the additive relationship between the cells of the table. Additional cells must also be suppressed (so-called secondary or 'complementary' suppressions). This is the second step, called secondary cell suppression.

Section 3 presents a mathematical statement of the secondary cell suppression problem and discusses its practical complexity. Section 4 then discusses heuristics for its solution. Section 5 focuses on procedures for coordinating suppression patterns in multiple (linked) tables.

This chapter, which focuses on cell suppression in tabular data on businesses, is addressed to practitioners—in particular, to official statisticians charged with the protection of tabular data—to users of software tools for secondary data suppression, and in particular to new users, who would like to know something about the methodological underpinnings of those tools. Therefore, the discussion focuses primarily on practice. References are made throughout the discussion to Lawrence Cox's discussion in Chapter 8 of this volume, which presents the mathematical properties and bases for many of the points made more heuristically here.

2. Disclosure Risk

This section explains the concepts involved in the assessment and control of disclosure risk for tabular data on business statistics. It first investigates the kind of disclosure that is possible when disclosure control at the cell level is lacking. Then it discusses what is referred to as table-level disclosure risk. This is defined as the risk that intruders may be able to use the linear relations between published and suppressed cell values (revealed by the totals and subtotals in the table) to derive bounds for the suppressed cell entries and, thus, estimates of respondent data for respondents in the 'confidential' cells.

Disclosure Risk at the Cell Level

Statistical agencies charged with protecting the confidentiality of survey respondents typically apply certain measures (called sensitivity measures or sensitivity rules) to assess the disclosure risk connected with release of a certain aggregate (or cell) within a table. Choice of a particular sensitivity rule is usually based on particular attacker scenarios (which involve assumptions about additional knowledge available in public or to particular users of the data) and on some (intuitive) notion about the sensitivity of the variable involved.

- *Attacker scenarios:* With business data, it is usually assumed that those who might be interested in disclosing individual respondent data are competitors of

the respondent or other parties who are generally well informed on the situation in that sector of the economy to which the particular cell relates. It is assumed, specifically, that the attackers are able to identify the largest contributors to a cell. The commonly applied sensitivity rules differ in the particular kind and precision of the additional knowledge assumed.
- *Notion on the sensitivity of the variable:* Some sensitivity rules protect against exact disclosure of individual data only; others go further, and protect the data from being too closely estimable (see Section 2).

The process of applying a sensitivity rule to the cells of a table is called check for primary confidentially, or simply primary confidentiality. A table cell that is indeed sensitive according to the sensitivity rule used is subject to what is called primary suppression.

Sensitivity Rules to Prevent Exact Disclosure. To prevent exact disclosure of respondent data, it is common to use a 'rule of n'. Such a rule defines an aggregate as sensitive when the number of respondents contributing to this aggregate is below a certain threshold. Let n be the number of respondents contributing to the aggregate. We distinguish the following cases:
- $n = 1$: there is a risk of disclosure, because there is only a single contributor.
- $n = 2$: there is a risk of disclosure, because either of the two contributors would be able to calculate the contribution of the other by subtracting his or her own contribution from the aggregate.
- $n = 3$: there is no risk of (exact) disclosure as long as attackers know at most one of the contributions (*i.e.,* their own).
- When coalitions of respondents are assumed to exist, that is, attackers together know $n-1$ of the contributions, there is a disclosure risk unless the number of contributors exceeds n.

Sensitivity Rules to Prevent Confidential Data From Being Closely Estimable. When a particular variable is deemed strongly confidential, preventing only *exact* disclosure may be judged inadequate. The data disseminator may also wish to prevent an attacker from deducing too precise an estimate. This is a risk whenever the aggregate is dominated by a very small number of contributions. Common sensitivity rules to prevent this kind of disclosure are concentration rules—for instance, the 'n respondent, k percent' dominance rules, generally referred to as (n,k) rules. The following discussion clarifies the concept of sensitivity rules by using the $(1,k)$ dominance rule and the p-percent rule for illustration. In a more general way, the sensitivity rules given below can be considered 'upper linear sensitivity measures'. (For the definition and mathematical properties of linear sensitivity measures generally, see Chapter 8 or Cox (1981).)

(1,*k*) rule

The simplest concentration rule is the $(1,k)$ rule.

According to the $(1,k)$ rule, an aggregate is identified as sensitive whenever the largest contribution x_1 is greater than k percent of the total aggregate value X, that is, when

$$x_1 > \frac{k}{100}X.$$

This will ensure that the largest contribution x_1 in any non-sensitive aggregate will be k percent of the table cell value X at most. More precisely, an attacker, estimating the largest contribution to be $\hat{x}_1 = X$, will overestimate x_1 by at least $(100-k)$ % of the aggregate value X, and by at least $100 \cdot \frac{100-k}{k}$ % of x_1 of itself, as can be proven as follows:

$$x_1 \leq \frac{k}{100}X \Leftrightarrow \frac{\hat{x}_1}{x_1} \geq \frac{100}{k} \Leftrightarrow \frac{\hat{x}_1 - x_1}{x_1} \cdot 100 \geq \left(\frac{100}{k} - 1\right) \cdot 100$$

Example 1

Application of the $(1, k)$ rule for $k = 90$, that is, the $(1, 90)$ rule.

Let the total value of a table cell be $X = \$100{,}000$.

Let the largest contribution be $x_1 = \$90{,}000$.

Because $\$90{,}000 \leq 90/100 * \$100{,}000$, according to the (1,90)-rule the cell is safe—no risk of disclosure.

The upper estimate for the largest contribution $\hat{x}_1 = \$100{,}000$ will overestimate x_1 by 11.1% of x_1:

$$\frac{\hat{x}_1 - x_1}{x_1} \cdot 100 = \left(\frac{100}{k} - 1\right) \cdot 100 = 11{,}1\,.$$

Often, however, the second largest contributor is able to derive a much more precise upper estimate of the largest contribution by subtracting his own contribution x_2 from the aggregate total $X(\hat{x}_1 = X - x_2)$. See Example 2.

Example 2

Application of the (1,90) rule.

Let the total value of a table cell be	X= \$100,000.
Let the largest contribution be	x_1 = \$50,000.
Let the second largest contribution be	x_2 = \$49,000.

Because $\$50{,}000 \leq (90/100) * \$100{,}000$, according to the (1,90) rule the cell is safe—no risk of disclosure. But $\hat{x}_1 = \$100{,}000 - \$49{,}000 = \$51{,}000$, hence

$$100 \cdot \frac{\hat{x}_1 - x_1}{x_1} = 100 \cdot \frac{\$51{,}000 - \$50{,}000}{\$50{,}000} = 2 .$$

The second largest contributor in this case is able to derive an upper estimate for the largest contribution which overestimates the true value by only 2 percent—quite a good estimate!

Example 2 shows that, as with the minimum number of respondents' rules in case of a 'rule of 2' (see previous subsection, case $n = 2$), there is indeed a risk of disclosure, because either of the two largest contributors would be able to derive a close upper estimate for the contribution of the other one simply by subtracting his or her own contribution from the aggregate total. One option to prevent this kind of disclosure risk is to use the $(2,k)$ dominance rule instead of the $(1,k)$ dominance rule. The $(2,k)$ dominance rule is based on the percentage of the two largest contributions in the cell instead of only the largest contribution. Because the $(2,k)$ dominance rule has a tendency toward oversuppression (for illustration, see Chapter 8), however, the use of the p-percent rule is preferred.

p-percent rule

According to the p-percent rule, an aggregate is sensitive when the second largest respondent can estimate the largest contribution x_1 to within p percent of x_1, that is, when

$$\frac{(X - x_2) - x_1}{x_1} \cdot 100 < p$$

This rule can be illustrated as follows: On the assumption that there are no respondent coalitions (*i.e.*, that no attackers know more than one of the contributions), the best upper estimate of any other contribution x_2 can be obtained by the second largest contributor by subtracting his own contribution from the cell value X to estimate the largest contribution $(\hat{x}_1 = X - x_2)$. Application of the p-percent rule yields this upper estimate to overestimate the true value by at least p percent for any non-sensitive cell.

The disseminator has to fix threshold p in advance according to the disclosure control requirements in place. (For alternative primary suppression rules and a more analytical discussion of sensitivity rules, see Chapter 8.)

It should be emphasized here that the sensitivity rules introduced so far are rules specifically to ensure that an *upper estimate* of a contribution will overestimate the true value by a certain percentage. We have not considered sensitivity measures for a risk of what might be called enterprise *interval disclosure*, a situation when the attacker can deduce a narrow interval containing the true value. To make them operable, measures for enterprise interval disclosure risk must incorporate assumptions about user knowledge that are additional to those mentioned earlier (*e.g.,* on the assumption that attackers can estimate the contribution of any enterprise to within a certain percentage of its value). Opinions differ as to whether such additional assumptions are justified. In any case, when suitably specified additional user knowledge is assumed, the rules presented above (or similar ones) can also be applied to prevent enterprise interval disclosure.

Table-Level Disclosure Risk

As noted, when a table presents totals or subtotals along with its 'inner' cells, there is a linear relationship between the cells of the table. Because of this linear relationship, if it has been established that release of certain cells would present a disclosure risk of a table, other cells must also be suppressed (complementary or secondary suppressions) to prevent a risk of disclosure at the table level. The risk resulting from the fact that the primary suppressions can be recalculated or closely estimated from the table context by simple linear operations is referred to as table-level disclosure risk.

This section explains and illustrates several strategies that help reduce table-level disclosure risk. These include, for example, the specification of a particular pattern of suppressed cells to avoid exact disclosure in a two-dimensional table or the setting of an interval if a tight lower bound for any cell is assumed known to a potential intruder. Sliding protection ranges and rounding technologies are discussed briefly.

Example 3 shows turnover in a hypothetical food production sector as the basis for subexamples showing the correspondence between primary cell-level sensitivity measures and table-level disclosure control. Example 3.1 begins the discussion by showing the potential problem of table-level disclosure risk when the rule of 3 (see above) is used to identify cell-level sensitivity.

Example 3

Food production sector	Turnover	Number of respondents
Total	T	N_T
bakers	15,000	122
butchers	25,000	95
millers	X	N_X
brewers	Y	N_Y
others	15,000	51

T denotes the overall turnover in the food production sector, X and Y the turnover of the millers and brewers, respectively. N_T, N_X, and N_Y denote the corresponding numbers of respondents.

Example 3.1

Assume that a rule of 3 (see above) is used for primary confidentiality.

Let the number of millers in the table of Example 3 be $N_X = 1$.

Let the number of brewers in the table of Example 3 be $N_Y = 3$.

Then, the turnover of the millers X risks exact disclosure. But if no other cell is suppressed, X can easily be recalculated through subtraction: $T - 15{,}000 - 15{,}000 - 25{,}000 - Y - 0$ ($= X$). A second cell also has to be suppressed (Y for instance) to avoid disclosure. What factors should be considered when making this choice of a complementary suppression?

Minimum Size Requirement for Complementary Suppressions. In a table with totals, it is easy to use subtraction to recalculate the value of a cell union of suppressed cells, such as the union of two primary suppressions or of a primary suppression together with the complementary suppression.

Example 3.2

Assume a rule of 3 is used for primary confidentiality.

Let the overall number of respondents in the table of Example 3 be $N_T = 268$.

Let the number of millers in the table of Example 3 be $N_X = 1$.

Let the number of brewers in the table of Example 3 be $N_Y = 1$.

The table now contains two confidential cells: *X* and *Y*.

If we apply the rule of 3 to the union of these suppressed cells, whose value can be obtained by subtraction from the column total *T*: $X + Y = T - 15{,}000 - 15{,}000 - 25{,}000$, however, we find that yet another cell must be suppressed, because the number of respondents to the cell union is $N_{x+y} = N_x + N_y = 2$. Without an additional suppression, the miller might identify the contribution of the brewer by subtracting his own contribution *X* from the value of the cell union *X* + *Y*, and vice versa.

Example 3.2 illustrates how unions of suppressed cells must also be regarded as knowable by an attacker. The sensitivity rule employed should therefore also be applied to any (linear) combination of suppressed cells within a row or column relation of a table in order to avoid disclosure risk for any particular single contribution to the combination.

Example 3.3

Assume that a (1,85) rule (see earlier discussion of the (1,*k*) rule) is used for primary confidentiality.

Let the total turnover in the food production sector in example 3 be T=55,345.

Let the number of millers in the table of Example 3 be N_X=3.

Let the number of brewers in the table of Example 3 be N_Y=3.

Let the sequence of contributions of distinct respondents to X (turnover of millers) be x_1=300, $x_2 = 20$, $x_3 = 10$.

Let the contribution sequence of Y (turnover of brewers) be y_1=5, $y_2 = 5$, $y_3 = 5$.

It can then be easily verified that *X* is sensitive according to the (1,85) rule, while *Y* is not. However, it is not sufficient to suppress the turnover of brewers along with the turnover of millers to prevent the largest miller from disclosure. This is because the total turnover of brewers is too small to protect the contribution of the largest miller from being closely estimable.

More precisely, the turnover of the largest miller still dominates the combined turnover of millers and brewers $Z = X + Y$ (contribution sequence

$z_1 = 300 > z_2 = 20 > z_3 = 10 > z_4 = 5 = z_5 = 5 = z_6 = 5$) because

$$z_1 = 300 > \frac{85}{100}\left(\sum_{i=1}^{6} z_i\right) = \frac{85}{100} \cdot (300 + 20 + 10 + 5 + 5 + 5) = \frac{85}{100} \cdot 345 = 293{,}25$$

Thus, sufficient protection of the primary suppression requires that one either suppress a different cell (instead of the turnover of brewers) or an additional cell as well.

Which cells would give sufficient protection? It has been proven in general (Cox 1981) that a necessary condition for the union of a sensitive cell with an arbitrary secondary suppressed cell to be non-sensitive is that the value of the secondary suppression Y exceed a given minimum size. If the minimum size condition is not fulfilled for a complementary suppression Y to a sensitive cell X, then the combination of X and Y (*i.e.*, the 'combined' cell) will still be sensitive.

This minimum size depends on the particular degree of sensitivity of the sensitive cell according to the specific sensitivity rule used. The formulas in Table 1 give calculations of the minimum size for each of the sensitivity rules discussed earlier. They specify the minimum size for an adequate non-sensitive secondary suppression to protect a sensitive cell with a total cell value X and N distinct respondents with contributions $x_1 \geq x_2 \geq \ldots \geq x_N$.

Table 1. Minimum Size Condition for Feasible Complements

Sensitivity rule	Minimum size for a feasible complement
(1,*k*) rule	$\frac{100}{k}x_1 - X$
(2,*k*) rule	$\frac{100}{k}(x_1 + x_2) - X$
(*n*,*k*) rule	$\frac{100}{k}(x_1 + x_2 + \ldots + x_n) - X$
p-percent rule	$\frac{p}{100}x_1 - (X - x_1 - x_2)$

(See Chapter 8 or Cox (1981) for a more general formulation of the minimum size requirement.)

Note two important cases, however, when the minimum size requirement is not a sufficient criterion for the combined cell to be safe: (1) when the complementary suppression is a sensitive cell itself and (2) when a single respondent can contribute to more than one component cell in a combined cell. In these two cases, combinations of a sensitive cell and its complement may still be sensitive, even when the minimum size requirement is fulfilled for the complement. (For further discussion of this problem, often referred to as 'multicell disclosure', see Chapter 8.)

Suppression Patterns in Two-Dimensional Tables. Size is not the only issue to be considered in choosing a complementary suppression. A second important issue influencing its suitability for protecting primary suppression is the 'pattern' of the suppressions within the table.

In the case of simple two-dimensional tables with row and column totals, people often presume that it is not possible to recalculate any suppressed entries if there are either none or at least two suppressions in any row and column of the table. This is, of course, a necessary criterion, but it is not sufficient in certain cases. (For an instance, see Table 1 of Chapter 8.)

Indeed, it has been proven (Geurts 1992, p.10) that a suppressed cell in a two-dimensional table is safe from exact disclosure if, and only if, the cell is contained in a 'cycle' of suppressed cells.

A cycle is defined to be a sequence on non-zero cells with indices,

$\{(i_0, j_0), (i_1, j_0), (i_1, j_1)(i_2, j_1), \ldots, (i_n, j_n), (i_0, j_n)\}$, where all i_k and j_1 for $k = 0, 1, \ldots, n$, and $l = 0, 1, \ldots., n$, respectively (n: length of cycle), are different, *e.g.*, $i_{k_1} \neq i_{k_2}$ unless $k_1 = k_2$, and $j_{l_1} \neq j_{l_2}$ unless $l_1 = l_2$. Example 4 (taken from Geurts 1992, Table 5, p.10) illustrates the case of a table containing a cycle of suppressed cells:

Example 4

	1	**2**	**3**	**4**	**Total**
1	X_{11}	X_{12}	2	7	13
2	2	X_{22}	X_{23}	7	13
3	3	2	8	1	14
4	X_{41}	2	X_{43}	3	18
Total	11	10	19	18	58

Note that a pattern of cells forming the corner points of a rectangle is a cycle.

Unfortunately, the cycle criterion is not sufficient in the case of higher- (more than three-) dimensional tables. This means in a higher-dimensional table it may still be possible to recalculate the true value of a suppressed cell, even if it is contained in a cycle of suppressed cells. For higher-dimensional tables, no such simple criterion that is both necessary and sufficient has yet been found. What we do know is that a suppressed cell in a higher-dimensional table is protected from exact disclosure if it is contained in a pattern of suppressed, non-zero cells that form the corner points of a hypercube. This 'hypercube criterion' is a sufficient but not a necessary criterion for a 'safe' suppression pattern.

Suppression Interval. Even though a suppression pattern may be 'safe' in the sense that no suppressed cell can be disclosed exactly, it is always possible to derive upper and lower bounds for the true value of any particular suppressed cell entry of a table by using the linear relations between published and suppressed cell values in the table. This holds both for tables with non-negative values and for tables containing negative values when, instead of zero, some other (possibly tight) lower bound for any cell is assumed available to data users. The interval given by these bounds is called the 'suppression interval'. Example 5 illustrates the calculation of the suppression interval in the case of a simple two-dimensional table where all cells may assume only non-negative values:

Example 5

	1	**2**	**Total**
1	X_{11}	X_{12}	7
2	X_{21}	X_{22}	3
3	3	3	6
Total	9	7	16

For this table the following linear relations hold:

$$X_{11} + X_{12} = 7$$
$$X_{21} + X_{22} = 3$$
$$X_{11} + X_{21} = 6$$
$$X_{12} + X_{22} = 4$$
with $X_{ij} \geq 0$ for all (i, j)

Solving this linear problem for a particular suppressed cell, for X_{11} for instance, subject to the constraints $X_{ij} \geq 0$ for all (i,j), yields $3 \leq X_{11} \leq 6$, and the suppression interval for X_{11} is [3;6].

Linear programming methodology makes it possible to derive an upper bound (X_{ij}^{max}) and a lower bound (X_{ij}^{min}) for the set of feasible values for any suppressed cell (i, j) in a table. Feasible here means with respect to the linear relations between published and unpublished cell values given by the table, and also with respect to some *a priori* constraints for the suppressed cell values, such as the assumption that they may only assume non-negative values. For cell (1,1) in Example 5 these bounds are $(X_{11}^{\min}) = 3$ and $(X_{11}^{\max}) = 6$.

If X_{ij} can also assume negative values, the value 0 in the constraint $X_{ij} \geq 0$ should be replaced by some other general or cell-related lower bound l or l_{ij} for X_{ij} if such a bound is assumed known to data users. The same thing should be done for positive values if a bound greater than 0 is assumed known. If it is assumed that some general or cell-related upper bound(s) u or u_{ij} be also known to data users, then constraints $X_{ij} \leq u_{ij}$ should be added to the set of constraints as well.

The Protection Interval. A more serious disclosure control problem has arisen with the availability of freeware software able to do the computations required to calculate the suppression intervals of suppressed cells. Proper selection of complementary suppression thus requires the disseminator to determine upper and lower safety bounds for any *primary* suppression. We call the interval between these upper and lower bounds the 'protection interval'. A proper suppression procedure should ensure that no suppression pattern is considered feasible that yields a suppression interval that does not enclose the protection interval for any sensitive cell.

To be more specific: According to the common sensitivity rules discussed earlier, there is an unacceptable risk of disclosure if the upper bound of the suppression interval does not sufficiently exceed the true value of the sensitive cell. If the distance between upper bound and true value is below the minimum size for a feasible complement calculated according to the formulas of Table 1, then this upper bound can be used to derive estimates for single contributions to the sensitive cell that are too close, as defined by the sensitivity criterion used to judge primary confidentiality. This is illustrated in Example 6.

Example 6

> The turnover of millers in Example 3.3 above ($X = 330$ with a sequence of contributions of distinct respondents $x_1 = 300, x_2 = 20, x_3 = 10$) is confidential according to the (1,85) rule. But if the upper bound $X^{\max}$ for this confidential value is below $\frac{100}{85}x_1 = 352{,}94$, it will be dominated by the largest single contribution x_1.

This illustrates that the bounds of the protection interval should be specified according to the rules applied for primary confidentiality. In particular, the distance between upper bound and true cell value should not be below the minimum size for a feasible (sufficiently large) complement (see Table 1). It is therefore recommended that the upper bound of the protection interval be determined by adding the cell value X to this minimum size. Symmetry considerations might then suggest determining its lower bound by subtracting this minimum size from the cell value. The protection interval given by these bounds would normally, according to the primary confidentiality rule, be sufficient to protect the sensitive cell. (One exception is the case of multicell disclosure, which cannot be avoided just by proper definition of the protection interval. For further discussion of this issue, see Chapter 8.)

But wider protection intervals also tend to require larger or more suppressions—and protection intervals that are larger than necessary tend to yield oversuppression. Thus, the protection interval for any sensitive cell should be defined just wide enough to protect the relevant respondent's data against approximate disclosure.

When tables have both positive and negative values and no *a priori* lower bounds are assumed known, users (or abusers) are unable to obtain finite suppression intervals. In this case, it is sufficient to protect only against exact disclosure, and protection intervals are not necessary.

Sliding Protection Ranges and Rounding Techniques. So far we have discussed how to determine an upper bound for the protection interval. We have also seen which kind of disclosure risk is involved in releasing data from which the user can estimate the true value of a sensitive cell below this upper bound limit. The literature (Kelly *et al.* 1992) has suggested using sliding protection as a way to reduce the size and number of secondary suppressions. This would involve, for example, controlling only for the width of the protection interval, without setting any conditions regarding its location (except that it contain the true value of the primary suppression). Notwithstanding its advantages, sliding protection does not meet the requirements implied by the commonly used definition of cell sensitivity referred to earlier in the subsection on sensitivity rules. For evidence, remember the point made at the end of the sensitivity rules subsection that (without some particular assumption as to users' knowledge) the common sensitivity criteria are, in fact, criteria to prevent a close upper estimate from being derivable for confidential data. With sliding protection, for example, a suppression pattern could potentially occur in which the upper bound value of an estimate of the primary suppression is identical to its true value—causing a risk of disclosure.

It was also mentioned at the end of that subsection, however, that under some additional specific assumptions about user knowledge, the sensitivity rules could be interpreted as rules to prevent enterprise interval disclosure. In that case, sliding protection might be considered sufficient: Even in the unfortunate case when the upper bound for the value of the primary suppression is identical to its true value, sliding protection can ensure that the lower bound is so small that the interval an attacker could derive for single contributions to the sensitive cell will be sufficiently wide to avoid close estimation. This is also relevant to the following remarks on rounding.

The kind of protection that rounding, as opposed to cell suppression, can give to business data is also a form of sliding protection. The common rounding techniques do not prevent a user from deriving an upper estimate for the value of a protected (rounded) sensitive cell that is arbitrarily close or even identical to its true value. Therefore, rounding techniques should be used as the sole means of disclosure protection only when the disseminator either feels it is sufficient to prevent exact disclosure of sensitive information or aims at protecting enterprise interval disclosure.

Release of Additional Information. As demonstrated in the subsection on the suppression interval, users of a published table can derive upper and lower bounds for the true value of any particular suppressed cell entry by making use of the linear relations between published and suppressed cells of the table. A suppression procedure should therefore ensure that no suppression pattern is considered unless disclosure of all these bounds causes no risk of disclosure for individual respondent data. Accordingly, when tabular data have been protected in this manner, a data disseminator might as well publish these bounds along with the protected data, thereby improving the publicly available information content greatly. Along the same lines, disseminators could support their data analysis by publishing perturbed values to replace suppressed original cell entries. Such perturbed values should be located between the upper and lower bounds, matching subtotals and totals of the protected table. This will ensure that the original additive relations between the cells of the table will be maintained.

3. The Secondary Cell Suppression Problem in Theory and Practice

The secondary cell suppression problem is how to apply complementary suppressions to a set of sensitive cells in such a way as to ensure that the complementary suppressions create the required uncertainty about the true values of the sensitive cells, while still preserving as much information in the table as possible.

A suppression pattern is assumed here to create the required uncertainty about the true values of the sensitive cells if, for any sensitive cell, it contains the protection interval as defined by the disseminator.

To find a good balance between protection of individual response data and provision of information—in other words, to take control of the information loss that obviously cannot be avoided completely because of the requirements of disclosure control—it is necessary to somehow rate the information content of data. Equating a minimum loss of information with the smallest number of suppressions is probably the most natural rating concept, yet experience has shown that this concept often yields suppression patterns in which many larger cells are suppressed, removing considerable amounts of information. (Chapter 8 gives the most common measures of information loss.) Note that several criteria other than the numeric value may influence a user's perception of a particular cell's importance, such as its situation within the table (totals and subtotals are often rated as highly important) or its category (certain categories of variables are often considered to be of secondary importance).

Mathematical Statement of the Secondary Cell Suppression Problem

Chapter 8 gives a mathematical statement of secondary cell suppression as an integer linear programming problem. According to that statement, the task of an algorithm for secondary cell suppression is to select from the set of feasible suppression patterns (*i.e.,* the set of feasible solutions to (4.4) in Chapter 8) that pattern for which an objective function (*cf.* (4.1), (4.2), or (4.3) in Chapter 8) assumes the smallest value. If possible, one defines the objective function as the information loss function selected. This ensures that the information loss due to suppression will be minimized.

How now to solve the secondary cell suppression problem? Imagine a very simple, straightforward algorithm: First, find any possible set of cells within the table containing the set of sensitive cells. Check its feasibility as a suppression pattern. Then calculate the information loss for any feasible set according to (4.1), (4.2), or (4.3) of Chapter 8 and select the set for which the value of the objective function is smallest.

In practice, this algorithm cannot work. First, as a result of combinatorial explosion, for any real-life sized statistical table the number of possible sets of cells (*e.g.* sets of cell combinations) would be huge. Second, checking for the feasibility of a particular suppression pattern requires a considerable number of calculations. These two factors together would result in a gigantic computational burden. In fact, it has been proven (*cf.* Geurts 1992) that the problem is 'np-hard'. That is, the number of required computations grows exponentially (non-polynomially) with the size of the table. So one either has to find a much more sophisticated algorithm to solve this problem or look for a good heuristic approach that can at least find a pattern with a relatively small value of the objective function. Because the objective function is determined heuristically anyway, finding an approximate solution to the minimization problem is a reasonable goal.

Secondary Cell Suppression in Complex Tables

As mentioned in Section 2, all the linear relations between published and unpublished values of the table have to be considered in setting up the secondary cell suppression problem for a table. This leads us to a crucial question: What is a table anyway? In the absence of confidentiality concerns, a statistician creates a table in order to show certain properties of a dataset or to enhance comparison between different variables. So a single table might mix apples and oranges. In addition, statisticians wishing to present a large number of those 'properties' may release multiple tables from a particular dataset. Where one table ends and the next begins is a matter of choice. The ideal table might even be one that fits nicely on a standard-size sheet of paper. With respect to the secondary cell suppression problem, however, we have to think of tables in a different and very precise way.

Definition of Tables in the Context of Secondary Cell Suppression. The first thing to consider in defining tables in the context of secondary cell suppression is the data basis for the table. The microdata file we think of in this context is a file containing one record per respondent, with each record containing a number of key codes and a number of entries giving respondent data on a response variable. The key codes may be regarded as respondent data on some categorical 'explanatory' variables. They can be used to group the respondents according to certain criteria—such as their economic activity, region, size class of turnover, or legal form. In the set of response variables, an additive relation may often be present, for instance, if one of the response variables is actually the sum of the others in the set. As an example, consider the following relation: total investment = investment in building + investment in ground + investment in technical equipment + other investment. Any entry (cell) within a table will then present the total of a particular response variable for a specific group of respondents, such as the total investment of the group of millers with a turnover of between $250,000 and $500,000 in community XXX.

When we talk about the number of dimensions in a table, we usually mean the number of explanatory variables used to specify the groups. A cell in a table exhibits the value of the response variable for the group of respondents falling into the same category for each explanatory variable. When the table presents data on more than one response variable, and the response variables are linearly related as in the investment example above, we consider the table to have one more dimension. Thus, the table is defined by the set of explanatory and response variables. The cell described above might be contained in a table presenting the investment of businesses by their economic activity, size class of turnover, and region, which has three dimensions, or four, if it also presents data for the other investment categories.

Several important principles should be noted in connection with this definition of a table.

- We consider a table as more than one table if it presents data on more than one response variable, unless there is a linear relationship between these response variables.
- We assume that any table also presents data for the overall category of any of its dimensions. That means we assume that the table contains totals in any dimension. Otherwise the so-called 'dimension' would not be considered a dimension of that table but a separate table.
- We assume a table to be 'complete'. That is, any respondent in the microdata set will belong to at least two categories (one being the overall category) of each explanatory variable used to make up the table.
- If we plan to publish two tables constructed exactly the same way using identical sets of response and explanatory variables but with the second table based only

on a subset of the microdata set used to calculate the first table, we do not consider them separate tables. Instead, we create a third table based on the remainder of the dataset and consider these three tables as a single table for the purpose of data protection. Note that this new 'single' table has an additional dimension, with the explanatory variable corresponding to this additional dimension having three categories: 'overall', 'respondent belongs to subset', and 'respondent does not belong to subset'. This extends analogously to the case of more than one subset, as long as the subsets do not overlap.

It may also extend to the case of two tables constructed from the same set of explanatory variables and differing only in the response variable. If, for example, one of the response variables describes a certain fraction of the magnitudes given by the other (such as 'investment in building' as a fraction of 'total investment'), then these two tables should be protected as a single table as well, together with a corresponding 'remainder' table ('non-building investment').

A common approach used in cases such as this is to go through the table protection procedure only for the first table and then simply suppress those entries in the second and third tables that correspond to the suppressed entries in the first table. Though this approach is certainly valuable in reducing the effort for data protection, it should not be used when the information given by the second variable has to be considered both sensitive and identifying. If, in other words, it has to be assumed that attackers are able not only to identify those respondents with extraordinarily large responses to the 'overall' variable (*e.g.*, the investment) but also to guess large respondents to the second variable, then this second variable has to be considered sensitive as well. For our building investment example, this might be the case. Assume, for example, that one miller in a small area has built an expensive new mill. Assume further that the other millers did not have major building costs. Then many of those other millers can be assumed to know that this particular firm has a very large share in their common published value for building investment, and be able to closely estimate this single contribution. Thus, this cell needs to be suppressed even if the 'overall investment' may turn out to be safe from disclosure.

Hierarchical Tables. Data collected within government statistical systems must usually meet the requirements of many users, who differ widely in the particular interest they take in the data. Some may need community-level data, while others may need detailed data on a particular branch of the national economy but no regional detail. Statisticians try to cope with this range of user interests by using elaborate classification schemes to categorize respondents at various levels of detail. Thus, a respondent will often belong to various categories of the same classification scheme—for instance, a particular community, within a particular county, within a particular state—and may thus fall into four categories of the regional classification depicted here:

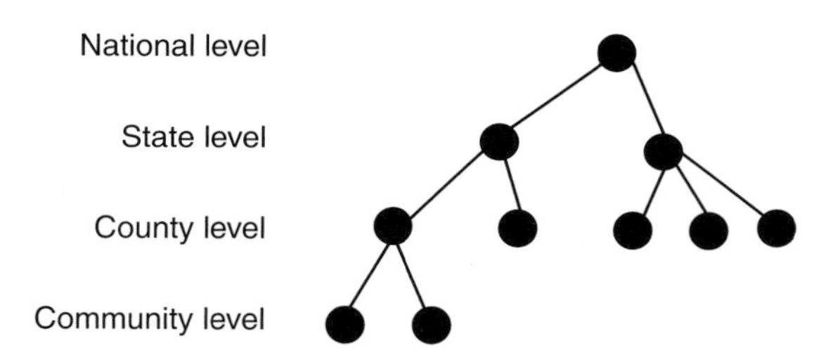

Such a table is a hierarchical table, defined as a table in which at least one of the explanatory variables has a hierarchical structure. The structure between the categories of hierarchical variables also implies a substructure for the table. From this table with substructure, we can create subtables without substructure as follows:

For any explanatory variable we pick one particular non-bottom-level category (the food production sector, for instance). Then we construct a subvariable. This subvariable consists only of the category picked in the first step and those categories of the level below belonging to this category (bakers, butchers, *etc.*). After that is done for each explanatory variable, the table specified through a set of these subvariables is free from substructure and is a subtable of the original one.

Any cell within the subtable also belongs to the original table, and many cells of the original table appear in more than one subtable.

Instead of constructing subtables without any substructure, we may simply wish to construct tables with a less complex substructure. In that case we would not construct the subvariables described above for each explanatory variable, but only for a part of them. Then the table specified through original (hierarchical) variables in some dimensions and subvariables in the other dimensions will have a less complex structure than the original table.

Once again, because these subtables of the same table have cells in common, we should not protect them separately. Otherwise the same cell that is suppressed in one subtable as a secondary suppression remains unsuppressed within another table. In such a case, a user comparing the two subtables would be able to disclose confidential cells in the first table. A common alternative approach is to protect subtables separately but note any complementary suppression belonging also to one of the other subtables, suppress it in this subtable as well, and repeat the cell suppression procedure for this table. This approach is sometimes called a backtracking procedure. Although a backtracking process will usually require the cell-suppression procedure to be repeated several times for each subtable, the number of computations required for the process will still be much smaller than when the entire table is protected all at once.

It must be stressed, however, that a backtracking procedure is not global, as denoted in Chapter 8. (See that chapter for discussion of problems related to non-global methods for secondary cell suppression.)

It should be noted in this context that we can easily turn the secondary cell suppression problem for a hierarchical table into the equivalent problem of protecting a corresponding single table without substructure. To do so, we only have to create one new explanatory variable for each level of each hierarchical explanatory variable. The set of categories for these new variables must equal the set of categories of the corresponding hierarchical variable at the particular level plus one (only one!) 'overall' category. The corresponding table will be specified through the new set of non-hierarchical variables together with those of the 'old' original variables that had no substructure. See Example 7.

Example 7

Consider a one-dimensional table. Let the table present, say, the turnover in computer production by geography. Let geography be a three-level hierarchy. The table will present on the first level the value T as overall total, on the second level the values U (USA) and E (EU), and on the third level values U_1 (California), U_2 (New York), U_3 (other US), E_1 (UK), E_2 (Germany), and E_3 (other EU). These values are interrelated in the following way:

$T = U + E$; $U = U_1 + U_2 + U_3$; $E = E_1 + E_1 + E_1$, as reflected in the following one-dimensional hierarchical table:

USA/EU	**USA**				**EU**			
Total	**Total**	**California**	**New York**	**Other USA**	**Total**	**UK**	**Germany**	**Other EU**
T	U	U_1	U_2	U_3	E	E_1	E_2	E_3

The secondary cell suppression problem for this table is then actually equal to the secondary cell suppression problem for the following simple (non-hierarchical) two-dimensional table:

	Total	**California**	**New York**	**Other USA**	**UK**	**Germany**	**Other EU**
Total USA/EU	T	U_1	U_2	U_3	E_1	E_2	E_3
USA	U	U_1	U_2	U_3	0	0	0
EU	E	0	0	0	E_1	E_2	E_3

This equivalency results from the fact that the linear relations between the cell values, which can be derived from the second table, are either the same as or equal to those of the first table or they are identity equations, such as $U_1 = U_1 + 0$.

4. Solving the Secondary Cell Suppression Problem

This section first describes several heuristic approaches for solving the secondary cell suppression problem that are already in regular practical use and for which software packages have been developed. It then discusses certain issues concerning the applicability of these packages.

Heuristic Approaches

The first two approaches presented here reduce the computational burden through two important properties they have in common. First, instead of considering as a possible suppression pattern every combination of cells in the table that includes the primary suppressions, they concentrate on particular patterns known to be 'safe' from theory (*i.e.,* known to protect the sensitive cells at least against exact disclosure). Second, both approaches handle complex tables by partitioning them into subtables that are protected successively within a backtracking procedure.

Hypercube Approach. This method has been described in Repsilber (1994 and 1999). It is based on the fact, mentioned earlier, that a suppressed cell in an n-dimensional table cannot be disclosed exactly if that cell is contained in a pattern of suppressed, non-zero cells forming the corner points of a hypercube.

The method subdivides n-dimensional tables with hierarchical structure into a set of n-dimensional subtables without substructure (as described in Section 3). These subtables are then protected successively in an iterative procedure that starts from the highest level. Successively for each primary suppression in the current subtable, all possible hypercubes with this cell as one of the corner points are constructed.

For each hypercube, a lower bound is calculated for the width of the suppression interval for the primary suppression that would result from the suppression of all corner points of the particular hypercube. To compute that bound, it is not necessary to implement the time-consuming solution to the linear programming problem (see discussion below). If it turns out that the bound is sufficiently large, the hypercube becomes a feasible solution. For any of the feasible hypercubes, the loss of information associated with the suppression of its corner points is calculated. The particular hypercube that leads to minimum information loss is selected, and all its corner points are suppressed.

After all subtables have been protected once, the procedure is repeated in an iterative fashion. Within this procedure, when cells belonging to more than one subtable are chosen as secondary suppressions in one of these subtables, in further processing they will be treated like sensitive cells in the other subtables they belong to.

This method has been developed at the Landesamt für Datenverarbeitung und Statistik in Nordrhein-Westfalen, Germany, and implemented in the GHQUAR software. It should be noted that the technique of partitioning a complex table into separately treated subtables is not essential for the hypercube method, and a new version of the method is in development that will be able to handle tables with hierarchical substructure as a single problem.

Another issue that bears repeating concerning the hypercube method is that the 'hypercube criterion' is a sufficient but not a necessary criterion for a 'safe' suppression pattern. Thus, for particular subtables, the 'best' suppression pattern may not be a set of hypercubes—in which case, of course, the hypercube method will miss the best solution and lead to some oversuppression. How much difference this tendency toward oversuppression makes in practice is another question.

Network-Flow Approach. For application to two-dimensional tables, the U.S. Census Bureau uses a network-flow approach, as suggested, for example, in Cox (1992). This method makes use of the fact, mentioned earlier, that in two-dimensional tables, a suppressed cell is safe from exact disclosure if, and only if, the cell is contained in a cycle of suppressed cells. The approach of the Census Bureau subdivides a table into subtables with hierarchical tree substructure allowed in one of the two dimensions. Such a structure can be represented as a network, and any closed path in such a network forms a cycle in the sense of the criterion presented in Section 2. For each primary suppression in the subtable, all cycles within the network containing it are constructed and a feasible cycle selected. The appropriate criterion for a cycle to be feasible reflects the following condition: Suppressing all 'arcs' of the cycle yields suppression intervals that give sufficient protection to the sensitive cells or cell combinations within the cycle.

A feasible cycle is selected by applying a general-purpose network-flow subroutine. In a way similar to the approach described in the previous subsection, the algorithm is applied to all subtables successively and repeated within an iterative procedure. Unlike the subtable partitioning for the hypercube method, this partitioning is an essential component of the network approach, because it is impossible to represent the entire complex table as a single network.

Linear Programming Approach. A linear programming approach is used in the CONFID program package of Statistics Canada and in the commercial package ACS (which is an improved version of CONFID). The algorithm and its improved version have been described in Robertson (1993, 1994, 2000) and Sande (1999). The U.S. Census Bureau uses a similar approach for protection of three- and four-dimensional tables. Unlike the two methods explained above, the linear programming approach does not concentrate on assessing suppression patterns of a particular shape, such as hypercubes or closed cycles. So even if the 'ideal' pattern is not simply a set of hypercubes or cycles, there is still a chance for this method to find it. On the other hand, the method is computationally extremely burdensome, because the only simplification used to make the secondary cell suppression problem tractable is the standard linear programming simplification (LP-relaxation).

Ideally, all cells and their interrelationships from an entire set of multiple interrelated tables disseminated from the same survey can be treated together with this method. In practice, however, the computational burden and enormous computer resource requirements may prevent its use for large applications and force the user to split problems into subproblems.

Integer Linear Programming Approach. The integer linear programming (LP) approach described in Fischetti and Salazar (1998) has been implemented in the second version of the τ-ARGUS program and yields the exact solution of the integer linear programming problem. Integer linear programming has a higher computational burden than the (continuous variable) linear programming described in the previous subsection, however. The advantage of the integer LP approach is that using integer variables in the objective function (specifically binary variables) allows one to express exactly the most natural objective functions for the suppression problem (*e.g.,* the number of suppressed cells, or the total value of the suppressed cells). With the continuous variables used for expressing the constraints when applying continuous variable linear programming, these objective functions can be expressed only approximately. (For further discussion of these methods, see Chapter 8.) Because the approach has such a huge computational burden, further improvement is required so as to make it applicable to real-life complex tables. Refinement work is being carried out from 2001 to 2003, subsidized by a grant of the European Commission (see also Section 6 below).

Applicability of Software for Secondary Cell Suppression

Traditionally, complementary suppressions had to be assigned manually. Staff from statistical institutes would check any additive relations between table cells and assign complementary suppressions to make sure the value of sensitive cells could not be recalculated by subtracting values of published cells in a row or column from the published row or column total. This procedure had to be repeated over and over again, because a newly assigned complement in, for example, a row might require an additional suppression in a column already checked, and so forth.

Apart from being extremely laborious, such a procedure is not fail-safe (*cf.* Section 2). Therefore, although several statistical institutes have developed similar procedures into software tools, we do not identify them in this chapter.

The following systems for automated secondary cell suppression are at least able to consider all the relations between the cells of simple two-dimensional tables at once: CONFID/ACS, GHQUAR, τ–ARGUS, and the software used at the U.S. Census Bureau. The methodology of those systems has been described briefly in this section and in Chapter 8. (For a detailed comparison of them, see Giessing 1998, 1999a, or 1999b.). Here it may be useful to mention that a system's usability and performance are influenced not only by the particular heuristic approach for solution of the secondary cell suppression problem used, but also by two key qualities that have an impact on size and complexity of structure of tables that can be handled efficiently:

- *Computing time and computer source requirements:* Because computing time is such a critical issue for cell suppression algorithms, at some point in the process, all of the existing systems store any information relevant for the selection of secondary suppressions for any cell of the table, or at least a large part of it, in the computer main memory. This imposes a serious restriction on the maximum size of problems to which an algorithm can be applied.
- *Data structure and software implementation:* In principle, a backtracking procedure (*cf.* Section 3) makes it possible to deal with tables of arbitrary size and complexity. Thus it is basically not a question of the cell suppression algorithm itself, but merely a matter of 1) the particular software implementation of an algorithm; 2) the data structure used by the system, and the data management; 3) whether large, complex tables can be handled at all; and 4) how much user interaction is required to control the backtracking procedure—for example, how comfortably the system can be applied to those tables. The GHQUAR software, for instance, can be applied effectively, with no requirement for user interaction, within a few CPU-minutes on an IBM mainframe, to tables with up to 1 million cells in up to seven dimensions, any of which may be hierarchically substructured into up to eight levels of depth.

5. Multiple Tables

Technological advance has made it much easier for users of statistical data to compare and analyze suppression patterns in different tables. For overlapping (linked) tables, this increases the risk of disclosure. Therefore, using proper procedures for coordination of suppression patterns in multiple (linked) tables is becoming an issue of growing importance. Most of the concepts suggested below will be of practical relevance only when software for automated protection of linked tables is available. (For the methodological concept of suitable software, see Giessing 2001a.)

Linked Tables

Usually, some of the tables in a set of multiple tables published from the same source (*e.g.*, response data from a survey) will be overlapping. For instance, let a table T1 present turnover by two enterprise employee size classes (column A of Example 8), a table T1.1 present turnover by these two enterprise employee size classes and two sections of the economy (columns A, D, and G of Example 8), and a table T1.2 present turnover by the two enterprise employee size classes and by two enterprise legal forms (columns A, B, and C of Example 8). Thus, T1 is a one-dimensional table defined by the variable SIZE, T1.1 is a two-dimensional table defined by the variables SIZE and SECTION, and T1.2 is a two-dimensional table defined by the variables SIZE and LEGAL FORM.

Example 8

		Total			**Section 1**			**Section 2**		
		Total	**Legal Form 1**	**Legal Form 2**	**Total**	**Legal Form 1**	**Legal Form 2**	**Total**	**Legal Form 1**	**Legal Form 2**
		A	*B*	*C*	*D*	*E*	*F*	*G*	*H*	*I*
Total	1									
Size Class 1	2			s						
Size Class 2	3									

Then T1 is a subtable of T1.1, as well as of T1.2, at least if T1.1 and T1.2 are based on the same dataset, and the two categories of 'employee size class' are identical for both tables T1.1 and T1.2.

A cell of the overlap table T1 will be a sensitive cell of T1.1 if, and only if, it is also a sensitive cell of T1.2. In Example 8 we assume there is only one sensitive cell, the turnover for legal form 2 and employee size class 1 in table T1.1.

When secondary cell suppression is carried out for T1.1 and T1.2 individually, it is not unlikely that there will be T1 cells unsuppressed in T1.2 that are complementary suppressions in T1.1, and *vice versa*. Assume, for instance, that for protection of the sensitive cell in T1.1 column *C* the two cells in rows 2 and 3 of column A were selected in addition to the cell in row 3. Because there are no sensitive cells in T1.2, the column A cells will remain unsuppressed in T1.2 when secondary cell suppression is carried out for T1.1 and T1.2 individually. In this case, any user given access to both tables will be able to discover these values and recalculate the sensitive cell in T1.1.

There are ways of preventing this situation. One alternative would be to protect the 'full' table T1.3: 'turnover by enterprise employee size class, section of the economy, and enterprise legal form' (columns A to I in Example 8) and suppress in T1.1 and T1.2 any cells that also were suppressed in T1.3. Another alternative would be to apply a table-to-table protection procedure. Within a similar type of backtracking procedure as that described in Section 3, one would first apply secondary cell suppression to, for example, table T1.1, and then to table T1.2, keeping track of any secondary suppressions in the overlap table T1. Secondary suppressions in T1, as resulting from protecting T1.1, will be treated like primary suppressions when protecting T1.2, and *vice versa*. The procedure will be repeated over and over again, until a step of the iteration is reached where no new secondary suppressions have been selected from T1. After the table-to-table protection procedure is finished, any cell of the overlap table T1 is either suppressed in both T1.1 and T1.2, or unsuppressed in both T1.1 and T1.2. Moreover, none of the suppressions can be disclosed by making use of the additive relationship between suppressed and unsuppressed cells in either T1.1 or T1.2.

Table Design for Multiple Tables

From the tabular data protection point of view, ideally statisticians should create one big, multidimensional table from the microdata file resulting from some survey (ideally even resulting from several surveys if they share common respondents and common queries). In principle, this table should contain any result ever to be published on the basis of that surveys. Then, if there were no limits on computing time and computer resources, we could protect this table in a single run of the cell suppression software. For Example 8, this would mean protecting the 'full' table T1.3, instead of T1.1 and T1.2. The method should also be capable of dealing with this table as a single problem, instead of protecting several subtables within a backtracking procedure.

In the real world, of course, this perfect approach will rarely be an option. We should be clear, however, that any deviation from it may cause risk of disclosure. And the greater the deviation, the higher the resulting risk. So, if the avoidance of disclosure risk is the most important aim of the disclosure control procedure, we need to make our tables as large as our disclosure control software allows.

We should also be aware that, unless we use a software definitively able to avoid the suppression of higher-level cells except when really 'necessary', this approach will result in some oversuppression—more or less serious depending on the efficiency of our software in avoiding subtotals.

Consider the following instance: a county containing (for simplicity) two communities and a sector of the economy split into two subsectors. We want to publish the data—on turnover, for instance—at the community level for the entire sector only, and *vice versa,* at the subsector level for the entire county only. In the table below, in other words, we do not intend to publish the cells shaded gray.

Example 9

	subsector 1	subsector 2	sector
community 1	0	10	10
community 2	22	2	24
county	22	12	34

Imagine now that the cell (community 2, subsector 2) with a cell value of 2 is sensitive, while all the other cells are non-sensitive. If, in that case, we publish all the white cells, any party who knows that there is no sector 1 enterprise in community 1 will be able to disclose the value of the sensitive cell. This danger could not have been detected if data had been arranged in the following two tables and disclosure control for each table had been conducted separately:

	subsection 1	**subsection 2**	**section**
county	22	12	34

and

	section
community 1	10
community 2	24
county	34

A backtracking procedure could not have helped either, of course, because there are no sensitive cells at all in the tables taken separately.

Assume now that there are three subsectors and three communities in our example:

	subsector 1	**subsector 2**	**subsector 3**	**section**
community 1	0	10	10	20
community 2	22	2	0	24
community 3	10	0	10	20
county	32	12	20	64

Then, publishing the set of white (unshaded) cells only might not cause any disclosure risk to the respondents of the confidential cell (community 2, subsector 2), because the suppression pattern given by the light gray cells is a cycle of non-zero cells—which is a feasible suppression pattern in a two-dimensional table. Had we tried to protect this table using a software that selects secondary suppressions according to the hypercube method, however, this suppression pattern, not having the shape of corner points of a rectangle, would not have been suggested, and some of the white cells could have been suppressed—a clear case of oversuppression.

Cell Suppression Within the Process of Statistical Data Production. Ideally, as noted, a table-to-table protection procedure should be applied to the full set of tables potentially releasable from a data source. This option seems less and less realistic, however, as technological advances progressively change the process of statistical data production and release. Formerly, the set of cells/tables published from a particular survey would have been largely fixed in advance. Now, however, the process of releasing data is becoming more and more user-demand-driven and less preplanned, even to the extent of providing public use statistical database query systems. Inability to forward-plan causes serious trouble for cell suppression.

The situation can be improved to some extent if data are 'pooled' to keep track of all suppressions in tables already released. This 'data pool' should contain one and only one record for each cell of any table already protected, and this record should contain an entry on the suppression status of the cell. When a new table is to be protected, the protection procedure can investigate the data pool for any cell of the new table. If any cell has already been used as a secondary suppression in one of the tables previously released, a backtracking procedure can treat it as a primary suppression. But what if the cell is 'unsuppressed' according to the data pool entry but selected for secondary cell suppression in the new table? Because it is impossible to 'undo' previous release of a cell, the inconsistency could only be avoided if the new table, or at least the affected part of it, were abandoned.

Preferences. Particularly in the context of table-to-table protection, facilities that allow the data preparer to specify preferences for certain cells or sets of cells to remain unsuppressed or, on the contrary, to be preferred as complementary suppressions, may improve the performance of the protection procedure considerably. GHQUAR, for instance, has an option that makes the program avoid selecting (sub-)totals as complementary suppression. This proves to be quite useful, as tables are normally linked by those cells. In Example 8 above, for instance, it is table T1 that 'links' tables T1.1 and T1.2. But table T1 is, in fact, a marginal table in both T1.1 (formed by summing over the variable SECTION) and T1.2 (formed by summing over the variable LEGAL FORM).

Another extraordinarily useful option in this context is the option to mark cells 'frozen', that is, ineligible for suppression—as offered, for instance, by the software of the U.S. Census Bureau. This option makes it possible, in principle, to deal with the problem of linked tables when some of the tables are already released by marking all the previously published cells in the overlap section of the 'new' tables as frozen. In fact, this may cause the program to break down in situations where no suppression pattern of 'unfrozen' cells can be found that protects the confidential cells sufficiently. Instead of 'freezing' these cells completely, a weaker variant would be to give them a low probability to be selected as secondary suppression—for example, by increasing the 'costs' assigned to suppression of such cells.

Alternatively, it is sometimes desirable to mark certain cells as preferable for suppression. Sometimes, for instance, cells have to be included because secondary cell suppression requires the table to be complete (*cf.* Section 3), even though one does not intend to release these cells. Or a cell that is part of an overlap table of a set of linked tables may be well suited for secondary suppression in one of the other tables to be processed later. Preferred status is implemented by decreasing the regular 'costs' assigned to the cell.

As another application for these 'preferences', imagine the situation of a table published periodically (*e.g.*, monthly, quarterly, or annually). Some of the sensitive cells will then be sensitive in every period. As a simple illustration, assume a table without substructure that contains some cells that are 'forever' sensitive. Assume, further, that there is more than one feasible suppression pattern, and that the costs for each pattern differ only slightly. If nothing is done, the suppression pattern is very likely to change from period to period. This might be undesirable on general principles; worse, it could cause a risk of disclosure if the variation across periods in the cell values of the secondary suppressions is only small. In this example, the problem could be solved by preferential suppression of cells suppressed in the previous period.

6. Summary

As this chapter has explained, suitable methodology and software tools are available for the protection of tabular data by cell suppression. However, research and development is still needed and is ongoing in this field. In 2001, for example, the CASC (Computational Aspects of Statistical Confidentiality) research project began, subsidized by a grant from the European Union. The general objective of this project is to integrate best praxis tools and methods concerning tabular data protection into the software τ-ARGUS, thus facilitating the transfer of these technologies and of the associated methodological know-how. For details on CASC project objectives, see Giessing (2001) or Giessing and Hundepool (2001). For research and development plans concerning tabular data protection software at the U.S. Census Bureau, see Massell (2001).

References

Cox, L.H. (1981) 'Linear Sensitivity Measures in Statistical Disclosure Control', *Journal of Planning and Inference*, 5, pp.153-64.

—— (1992) 'Solving Confidentiality Problems in Tabulations Using Network Optimization: A Network Model for Cell Suppression in the U.S. Economic Censuses', *Proceedings of the International Seminar on Statistical Confidentiality*, Dublin.

Fischetti, M., and J.J. Salazar (1998) 'Modelling and Solving the Cell Suppression Problem for Linearly-Constrained Tabular Data', in J. Domingo-Ferrer (ed), *Statistical Data Protection '98, Conference Proceedings,* March 25-27, Lisbon.

Geurts, J. (1992) *Heuristics for Cell Suppression in Tables,* Working Paper, Voorburg: Netherlands Central Bureau of Statistics.

Giessing, S. (1998) 'Looking for Efficient Automated Secondary Cell Suppression Systems: A Software Comparison', *Research in Official Statistics,* 1:2, pp.69–86.

Giessing, S. (1999a) 'A Survey on Packages for Automated Secondary Cell Suppression', *Proceedings of the Eurostat/UN-ECE Work Session on Statistical Data Confidentiality* 1999, March, Thessaloniki.

—— (1999b) 'Vergleich der Software zur Maschinellen Durchführung der Sekundären Geheimhaltung', in *Forum der Bundesstatistik, Band 31/1999: Methoden zur Sicherung der Statistischen Geheimhaltung* (in German).

—— (2001) 'New Tools for Cell Suppression in τ-ARGUS: One Piece of the CASC Project Work Draft', in *Pre-Proceedings of the Eurostat/UN-ECE Work Session on Statistical Data Confidentiality 2001* (Conference held in Hersonissos, June).

Giessing, J., and A. Hundepool (2001) 'The CASC Project: Integrating Best Practice Methods for Statistical Confidentiality', in *Pre-Proceedings of the NTTS & ETK 2001 Conference* (Hersonissos, June).

Jewett, R. (1993) 'Disclosure Analysis for the 1992 Economic Census', unpublished manuscript, Washington, D.C.: U.S. Bureau of the Census, Economic Statistical Methods and Programming Division.

Kelly, J.P., B.L. Golden, and A.A. Assad (1992) 'Cell Suppression: Disclosure Protection for Sensitive Tabular Data', *Networks,* 22, pp.397–417.

Massell, P.B. (2001) *Cell Suppression and Audit Programs Used for Economic Magnitude Data*, SRD Research Report Series No. RR2001/01, Washington, D.C.: U.S. Bureau of the Census.

Repsilber, R.D. (1994) 'Preservation of Confidentiality in Aggregated Data', paper presented at the Second International Seminar on Statistical Confidentiality, November, Luxemburg.

— (1999) 'Wahrung der Geheimhaltung in Aggregierten Daten Quaderverfahren mit Intervallschutz für Vollständige Tabellen', in *Forum der Bundesstatistik, Bd. 31/1999, Methoden zur Sicherung der Statistischen Geheimhaltung.*

Robertson, D. (1993) 'Cell Suppression at Statistics Canada', *Proceedings of the Annual Research Conference,* Washington, D.C.: U.S. Bureau of the Census.

Robertson, D. (1994) 'Automated Disclosure Control at Statistics Canada', paper presented at the Second International Seminar on Statistical Confidentiality, November, Luxemburg.

Robertson, D. (2000) 'Improving Statistics Canada's Cell Suppression Software (CONFID)', in J.G. Bethlehem and P.G.M. van der Hejden (eds) *Proceedings in Computational Statistics 2000,* Heidelberg: Physica-Verlag.

Sande, G., 'Structure of the ACS Automated Cell Suppresion System', *Proceedings of the Joint Eurostat/UN-ECE Work Session on Statistical Data Confidentiality 1999,* March, Thessaloniki.

Veldhues, B. (1999) 'Tabellierung und Geheimhaltung der Ergebnisse der Handwerkszählung 1995', in *Forum der Bundesstatistik, Bd. 31/1999, Methoden zur Sicherung der Statistischen Geheimhaltung.*

Confidentiality, Disclosure, and Data Access: Theory and Practical Application for Statistical Agencies
Pat Doyle, Julia I. Lane, Jules J.M. Theeuwes and Laura M. Zayatz (Eds)

Chapter 10

Disclosure Limitation in Longitudinal Linked Data*

John M. Abowd
Cornell University, U.S. Census Bureau, CREST, and NBER

Simon D. Woodcock
Cornell University

1. Introduction

We consider longitudinal linked data, defined as microdata that contain observations from two or more related sampling frames with measurements for multiple time periods from all units of observation. Our prototypical longitudinal linked dataset contains observations from work histories and data on the individuals and employers observed in those work histories. We are primarily interested in the problem of confidentiality protection when data from all three sampling frames are combined for statistical analysis. Our goal is to develop and illustrate techniques that are appropriate for a variety of statistical analyses in widespread use in government agencies, such as INSEE and the U.S. Census Bureau, and in academic research in the social sciences.

Current measures for confidentiality protection in linked datasets pose a number of problems for analysts. In particular, because the datasets that are linked are frequently constructed by different statistical agencies, the set of disclosure limitation requirements for the linked data are generally the union of disclosure limitation requirements of the several agencies. In practice, this can severely limit the useful-

* The research reported in this paper was partially sponsored by the U.S. Census Bureau, the National Science Foundation (SES-9978093), and the French Institut National de la Statistique et des Etudes Economiques (INSEE) in association with the Cornell Restricted Access Data Center. The views expressed in the paper are those of the authors and not of any of the sponsoring agencies. The data used in this paper are confidential, but the authors' access is not exclusive. No public use datasets were released as a part of this research. Restricted access to the French data was provided to Abowd by INSEE through an agreement with Cornell University. The authors thank Benoit Dostie, Sam Hawala, Janet Heslop, Paul Massell, Carol Murphree, Philip Steel, Lars Vilhuber, Marty Wells, Bill Winkler, and Laura Zayatz for helpful comments on earlier versions of this research.

ness of the linked data. These limitations on the usefulness of the resulting data motivate a unified approach to confidentiality protection in linked data.

In analyses of longitudinal linked data, analysts generally choose one of the underlying sampling frames as the reference population for the statistical modeling. Thus, the data matrix consists of rows that have been sampled from a specific population (*e.g.*, individuals, jobs, or employers) and the columns consist of functions of the linked data appropriate for that analysis (*e.g.*, sales/worker or the identity of the employing firm in an analysis of individuals; characteristics of the distribution of employees at several points in time for an analysis of employers). Confidentiality of any of the contributing data files can, thus, be compromised by elements of either the rows or columns of the resulting data matrix. For example, linked individual information such as birth date or education can compromise the confidentiality of the individual data; linked information from the work history such as wage rates can compromise the confidentiality of both the individual and employer; linked information from the employer such as annual sales can compromise the confidentiality of the employer data. Our goal is to study methods that mask data from each of the source files in a manner that statistically preserves as much of the complicated relationships among the variables as possible.

Statistical Concepts

We assume that the analyst is interested in the results of a statistical analysis of the form:

$$Y = f(X, \beta, \varepsilon) \tag{1}$$

where $[Y \ \ X]$ is the matrix of all available data (confidential and disclosable); $f(\cdot)$ is a (possibly) nonlinear function of X; β is a set of statistical parameters; and ε is a statistical error term distributed according to $p_{Y|X}(\varepsilon \mid X, \Omega)$. For completeness, note that X follows the joint distribution function $p_X(x|\Theta)$. We will consider methods for protecting the confidential data matrix $[Y \ \ X]$, primarily the use of multivariate multiple imputation techniques where $[Y \ \ X]$ is drawn from the predictive density, based on $p_{Y|X}(\varepsilon \mid X, \Omega) p_X(x|\Theta)$ and appropriate prior distributions on $(\Omega \ \ \Theta)$, to protect confidentiality for an entire analysis.

Background

Most statistical agencies assert that preserving confidentiality in longitudinal linked data and creating a statistically useful longitudinal public use product are incompatible goals[1]. Perhaps for this reason, other researchers have not addressed the issue of disclosure limitation in longitudinal linked data. This is evident from

1 See, *e.g.*, Nadeau, Gagnon, and Latouche (1999) for a discussion of issues surrounding the creation of public use files for Statistics Canada's longitudinal linked Survey of Labour and Income Dynamics.

the material presented in Appendix A, which contains a comprehensive, annotated bibliography of recent research on disclosure limitation. However, a number of authors have proposed methods for disclosure limitation for general microdata, some of which are directly relevant to our proposed method. Because these are discussed in detail in Appendix A, we only briefly summarize these works here.

Kim and Winkler (1997) describe a two-stage masking technique applied to matched CPS-IRS microdata. The first stage of their technique is to mask variables with additive noise from a multivariate normal distribution with mean zero and the same correlation structure as the unmasked data. In the second stage, the authors randomly swap quantitative data within collapsed (age $\times$ race $\times$ sex) cells for records that pose an unacceptable disclosure risk. This approach preserves means and correlations in the subdomains on which the swap was done and in unions of these subdomains. However, the swapping algorithm may severely distort means and correlations on arbitrary subdomains. Subsequent analysis of the masked data (*e.g.*, Moore 1996a; Winkler 1998) indicates that the Kim and Winkler (1997) approach adequately preserves confidentiality and generates data that yield valid results for some analyses.

Our proposed approach draws heavily on the related suggestions of Rubin (1993), Fienberg (1994), and Fienberg, Makov, and Steele (1998). These authors suggest releasing multiple datasets consisting of synthetic data; Rubin (1993) suggests generating these data using multiple-imputation techniques similar to those applied to missing data problems; Fienberg (1994) suggests generating these data by bootstrap methods. There are numerous advantages to masking data via such methods. For example, valid statistical analyses of microdata masked by other methods generally require 'not only knowledge of which masking techniques were used, but also special-purpose statistical software tuned to those masking techniques' (Rubin 1993, p. 461). In contrast, analysis of multiply-imputed synthetic data can be validly undertaken using standard statistical software simply by repeated application of complete-data methods. Furthermore, an estimate of the degree to which the disclosure proofing technique influences estimated model parameters can be inferred from between-imputation variability. Finally, since the released data are synthetic, in other words, contain no data on actual units, they pose no disclosure risk. Fienberg, Makov, and Steele (1998) have presented an application of such methods to categorical data; Fienberg and Makov (1998) apply these ideas to develop a measure of disclosure risk.

In a series of related articles, Kennickell (1991, 1997, 1998, 2000) describes the Federal Reserve Imputation Technique Zeta (FRITZ) algorithm, which is based on Rubin's (1993) suggestion and has been applied to disclosure limitation in cross-sectional survey data (the Survey of Consumer Finances, SCF). FRITZ is a sequential, iterative algorithm for imputing missing data and masking confidential data using a sequence of regression models (see Appendix A). In line with the above-mentioned proposals, the algorithm generates multiply-imputed, masked data. Unlike the suggestions of Rubin (1993) and Fienberg (1994), the released data are not synthetic. Rather, only a subset of cases and variables are masked, and the remaining data are left unmasked. The FRITZ algorithm has proven quite successful in application to the SCF, and for this reason we suggest its extension to longitudinal linked data.

Organization of the Paper

The organization of the paper is as follows. Section 2 presents the details of data masking and data simulation techniques applied to longitudinal linked data files. Section 3 summarizes the use of conventional complete-data methods for analyzing multiply-masked or simulated data. Section 4 applies our methods to confidential longitudinal linked data from INSEE, the French national statistical institute. Section 5 provides a brief summary and conclusions. We include an extensive appendix that relates our methods to those already in the disclosure limitation literature.

2. Masking Confidential Data by Multiple Imputation

Consider a database with confidential elements Y and disclosable elements X. Both Y and X may contain missing data. Borrowing notation from Rubin (1987), let the subscript *mis* denote missing data and the subscript *obs* denote observed data, so that $Y = (Y_{mis}, Y_{obs})$ and $X = (X_{mis}, X_{obs})$. We assume throughout that the missing data mechanism is ignorable.

The database in question is represented by the joint density $p(Y,X,\theta)$, where θ are unknown parameters. Following the related suggestions of Rubin (1993) and Fienberg (1994), the basic idea behind our disclosure limitation method is to draw masked data $\tilde{Y}$ from the posterior predictive density

$$p(\tilde{Y}|Y_{obs}, X_{obs}) = \int p(\tilde{Y}|X_{obs},\theta)p(\theta|Y_{obs}, X_{obs})d\theta \qquad (2)$$

to produce M multiply-imputed masked data files $(\tilde{Y}^m, X^m)$, where $m = 1, ..., M$. In practice, it is simpler to first complete the missing data using standard multiple-imputation methods and then generate the masked data as draws from the posterior predictive distribution of the confidential data given the completed data. For example, first generate M imputations of the missing data (Y^m_{mis}, X^m_{mis}), where each implicate m is a draw from the posterior predictive density

$$p(Y_{mis}, X_{mis}|Y_{obs}, X_{obs}) = \int p(Y_{mis}, X_{mis}|Y_{obs}, X_{obs}, \theta)p(\theta|Y_{obs}, X_{obs})d\theta\,. \qquad (3)$$

With completed data $Y^m = (Y^m_{mis}, Y_{obs})$ and $X^m = (X^m_{mis}, X_{obs})$ in hand, draw the masked data implicate $\tilde{Y}^m$ from the predictive density

$$p(\tilde{Y}|Y^m, X^m) = \int p(\tilde{Y}|X^m,\theta)p(\theta|Y^m, X^m)d\theta \tag{4}$$

for each imputation m.

The longitudinal linked databases that we consider in this paper are very large and contain a variety of continuous and discrete variables. Furthermore, they are characterized by complex dynamic relationships between confidential and disclosable elements. For these reasons, specifying the joint probability distribution of all data, as in (3) and (4), is unrealistic. Instead, we approximate these joint densities using a sequence of conditional densities defined by generalized linear models. Doing so provides a simple way to model the complex interdependencies between variables that is computationally and analytically tractable. This method also provides a simple means of accommodating both continuous and categorical data by choice of an appropriate generalized linear model. We impute missing data in an iterative fashion using a generalization of Sequential Regression Multivariate Imputation (SRMI) developed by Raghunathan, Lepkowski, Van Hoewyk, and Solenberger (1998). The SRMI approach and its generalization to the case of longitudinal linked data are described in the following section. Given the multiply-imputed completed data, we produce masked data on a variable-by-variable basis as draws from the posterior predictive distribution defined by an appropriate generalized model under an uninformative prior. Hence, if we let y_k denote a single variable among the confidential elements of our database, masked values $\tilde{y}_k$ are draws from

$$p(\tilde{y}_k|Y^m, X^m) = \int p(\tilde{y}_k|Y^m_{\sim k}, X^m,\theta)p(\theta|Y^m, X^m)d\theta \tag{5}$$

where $Y^m_{\sim k}$ are completed data on confidential variables other than y_k.

The SRMI Approach to Missing Data Imputation

For simplicity, consider a dataset consisting of N observations on $K + P$ variables, ignoring for the moment the potential complications of longitudinal linked data. Let X be an $N \times P$ design or predictor matrix of variables with no missing values. Let Y be an $N \times K$ matrix of variables with missing values, and denote a particular variable in Y by y_k. Without loss of generality, assume they are ordered by their number of missing values, so that y_1 has fewer missing values than y_2, and so on, though the missing data pattern need not be monotone. Model-based imputations can use the density for Y given by

$$\begin{aligned} &p(y_1,y_2, \ldots,y_K|X,\theta_1,\theta_2, \ldots,\theta_K) \\ &= p_1(y_1|X,\theta_1)p_2(y_2|X,y_1,\theta_2)\ldots p_K(y_K|X,y_1,y_2, \ldots,y_{K-1},\theta_K) \end{aligned} \tag{6}$$

where p_k are conditional densities and θ_k is a vector of parameters in the conditional density of y_k, $k = 1, \ldots, K$. The SRMI approach is to model each of these conditional densities using an appropriate generalized linear model with unknown parameters θ_k, then impute missing values by drawing from the corresponding predictive density of the missing data given the observed data. Again for simplicity, assume a diffuse prior on the parameters, in other words, $\pi(\theta) \propto 1$.

The SRMI imputation procedure consists of L rounds. Denote the completed data in round $l + 1$ on some variable y_k by $y_k^{(l+1)}$. In round $l+1$, missing values of y_k are drawn from the predictive density corresponding to the conditional density

$$f_k(y_k | y_1^{(l+1)}, y_2^{(l+1)}, \ldots, y_{k-1}^{(l+1)}, y_{k+1}^{(l)}, \ldots, y_k^{(l)}, X, \theta_k) \tag{7}$$

where the conditional density f_k is specified by an appropriate generalized linear model and θ_k are the parameters of that model. Hence, under SRMI, at each round l, the variable under imputation is regressed on all non-missing data and the most recently imputed values of missing data. The imputation procedure stops after a predetermined number of rounds or when the imputed values are stable. Repeating the procedure M times yields M multiply-imputed datasets.

Note that if the missing data pattern is monotone (see Rubin 1987), the imputations obtained in round 1 are approximate draws from the joint posterior predictive density of the missing data given the observed data. Furthermore, in certain cases, the SRMI approach is equivalent to drawing from the posterior predictive distribution under a fully parametric model. For example, if all elements of Y are continuous and each conditional regression model is a normal linear regression with constant variance, the SRMI algorithm converges to the joint posterior predictive distribution under a multivariate normal distribution with an improper prior for the mean and covariance matrix (Raghunathan, Lepkowski, Van Hoewyk, and Solenberger 1998, p. 11).

The SRMI method can be considered an approximation of Gibbs sampling. A Gibbs sampling approach to estimating (6) proceeds as follows. Conditional on the values $\theta_2^{(l)}, \ldots, \theta_K^{(l)}$ and $Y_1^{(l)}, \ldots, Y_K^{(l)}$ drawn in round l, draw $\theta_1^{(l+1)}$ from its conditional posterior density, which is based on (6). Next, draw the missing values of y_1 conditional on the new value $\theta_1^{(l+1)}$, the completed data $X, y_2^{(l)}, \ldots, y_K^{(l+1)}$, and round l parameter estimates $\theta_2^{(l)}, \ldots, \theta_K^{(l)}$. That is, in round $l+1$ the missing values in y_k are drawn from

$$p_k^* (y_k | X, y_1^{(l+1)}, \ldots, y_{k-1}^{(l+1)}, y_{k+1}^{(l)}, \ldots, y_K^{(l)}, \theta_1^{(l+1)}, \ldots, \theta_k^{(l+1)}, \theta_{k+1}^{(l)}, \ldots, \theta_k^{(l)}), \tag{8}$$

which is computed based on (6). Though such an approach is conceptually feasible, it is often difficult to implement in practice, especially when Y consists of a mix of continuous and discrete variables. SRMI approximates the Gibbs sampler to the extent that (7) approximates (8).

A Prototypical Longitudinal Linked Dataset

Before discussing the details of imputing missing values and masking longitudinal linked data, we must first introduce some basic notation. The prototypical longitudinal linked dataset that we consider contains observations about individuals and their employers linked by means of a work history that contains information on the jobs each individual held with each employer. The data are longitudinal because complete work history records exist for each individual during the sample period and because longitudinal data exist for the employer over the same period. Suppose we have linked data on I workers and J firms with the following file structure. There are three data files. The first file contains data on workers, U, with elements denoted u_i, $i=1, ..., I$. In the application below these data are time-invariant, but in other applications they need not be. We refer to U as the individual characteristics. The second data file contains longitudinal data on firms, Z, with elements z_{jt}, $j=1, ..., J$ and $t=1, ..., T_j$. We refer to Z as the employer characteristics. The third data file contains work histories, W, with elements w_{it}, $i=1, ..., I$ and $t=1, ..., T_i$. The data U and W are linked by a person identifier. The data Z and W are linked by a firm identifier; we conceptualize this by the link function $j=J(i,t)$, which indicates the firm j at which worker i was employed at date t. For clarity of exposition, we assume throughout that all work histories in W can be linked to individuals in U and firms in Z and that the employer link $J(i,t)$ is unique for each (i,t)[2].

Applying SRMI to Missing Data Imputation in Longitudinal Linked Data

With notation in hand, we now discuss applying SRMI to longitudinal linked data. The methods described in this section and the next are applied to a particular linked longitudinal database in Section 4.

When imputing missing data in each of the three files, we should condition the imputation on as much available information as possible. For example, when imputing missing data in the worker file U, we should condition not only on the non-missing data in U (individual characteristics) but also on characteristics of the jobs held by the individual (data in W) and the firms at which the individual was employed (data in Z). Similarly, when conditioning the imputation of missing data in W and Z, we should condition on non-missing data from all three files. This necessitates some data reduction. To understand the data reduction, consider imputing missing data in the individual characteristics file U. Because individuals have work histories with different dynamic configurations of employers, explicitly conditioning the missing data imputation of individual characteristics on every variable corresponding to each job held by each worker is impractical—there are a different

---

[2] The notation to indicate a one-to-one relation between work histories and individuals when there are multiple employers is cumbersome. Our application properly handles the case of multiple employers for a given individual during a particular sample period.

number of such variables for each observation to be imputed. A sensible alternative is to condition on some function of the available data, which is well defined for each observation. For example, one could compute the person-specific means of time-varying work history and firm variables and condition the missing data imputation of variables in U on these[3]. Similar functions of person- and job-specific variables can be used to condition missing data imputation in the firm file Z. In what follows, we use the functions g, h, m, and n to represent these data reductions.

It is also appropriate to condition the imputation of time-varying variables not only on contemporaneous data, but also on leads and lags of available data (including the variable under imputation). Because the dynamic configuration of work histories varies from worker to worker and the pattern of firm 'births' and 'deaths' varies from firm to firm, not every observation with missing data will have the same number of leads and lags available as conditioning variables. In some cases, there will be no leads and lags available at all. We suggest grouping observations by the availability of dynamic conditioning data (*i.e.*, the number of leads and lags available to condition missing data imputations) and separately imputing missing data for each group. This maximizes the set of conditioning variables used to impute each missing value. Again, some data reduction is generally necessary to keep the number of groups reasonable. For example, one might only condition on a maximum of s leads and lags, with $s=1$ or $s=2$. We parameterize the set of dynamic conditioning data available for a particular observation by κ_{it} in the work history file and γ_{jt} in the firm file.

It may also be desirable to split the observations into separate groups on the basis of some observable characteristics, for example, gender, full-time/part-time employment status, or industry. We parameterize these groups by λ_i in the individual file, μ_{it} in the work history file, and ν_{jt} in the firm file.

Given an appropriate set of conditioning data, applying SRMI to missing data imputation in longitudinal linked data is straightforward. The key aspects of the algorithm remain unchanged—one proceeds sequentially and iteratively through variables with missing data from all three files, at each stage imputing missing data conditional on all non-missing data and the most recently imputed values of missing data. As in the general case, the optimal imputation sequence is in increasing degree of missingness. As each variable in the sequence comes up for imputation, observations are split into groups based on the value of κ_{it}, γ_{jt}, λ_i, μ_{it}, and/or ν_{jt}. The imputes are drawn from a separate predictive density for each group. After the imputes are drawn, the source file for the variable under imputation is reassembled from each of the group files. Before proceeding to the next variable, all three files must be updated with the most recent imputations, because the next variable to be imputed may reside in another file (U, W, or Z). At the same time, the functions of

[3] Because the individual characteristics in our application are time-invariant, we use this approach, but it is easy to generalize to the case where the individual characteristics (as distinct from the job characteristics) vary over time.

conditioning data (including leads and lags) described above generally need to be recomputed. As in the general case, the procedure continues for a prespecified number of rounds or until the imputed values are stable.

Explicitly specifying the posterior predictive densities from which the imputations are drawn is notationally cumbersome. For completeness, we give these in (9), (10), and (11). For a particular variable under imputation, subscripted by k, we denote by $U_{<k}$ the set of variables in U with less missing data than variable k; $W_{<k}$ and $Z_{<k}$ are defined analogously. We denote by $U_{>k}$ the set of variables in U with more missing data than variable k, and define $W_{>k}$ and $Z_{>k}$ similarly. As in Section 2, we use the subscript *obs* to denote variables with no missing data. We also subscript conditioning variables by i, j, and t as appropriate to make clear the relationships between variables in the three data files. The predictive densities from which the round $l+1$ imputations are drawn are

$$\int f_{u_k}\left(\begin{matrix} u_k \middle| U^{(l+1)}_{<k,i}, U^{(l)}_{>k,i}, U_{obs,i}, g_k(\{Z^{(l+1)}_{<k,J(i,t)}, Z^{(l)}_{>k,J(i,t)}, Z_{obs,J(i,t)}\}_{t=1}^{t=T_i}), \\ h_k(\{W^{(l+1)}_{<k,it}, W^{(l)}_{>k,it}, W_{obs,it}\}_{t=1}^{t=T_i}), \lambda_i, \theta_k \end{matrix}\right) p_k(\theta_k|.)d\theta_k \quad (9)$$

$$\int f_{w_k}\left(\begin{matrix} w_k \middle| U^{(l+1)}_{<k,i}, U^{(l)}_{>k,i}, U_{obs,i}, \{Z^{(l+1)}_{<k,J(i,\tau)}, Z^{(l)}_{>k,J(i,\tau)}, Z_{obs,J(i,\tau)}\}_{\tau=t-s}^{\tau=t+s}, \\ \{w^{(l)}_{k,i\tau}\}_{\tau=t-s}^{\tau=t+s}, \{W^{(l+1)}_{<k,i\tau}, W^{(l)}_{>k,i\tau}, W_{obs,i\tau}\}_{\tau=t-s}^{\tau=t+s}, \kappa_{it}, \mu_{it}, \theta_k \end{matrix}\right) p_k(\theta_k|.)d\theta_k \quad (10)$$

$$\int f_{z_k}\left(\begin{matrix} z_k \middle| m_k(U^{(l+1)}_{<k,J^{-1}(i,t)}, U^{(l)}_{>k,J^{-1}(i,t)}, U_{obs,J^{-1}(i,t)}), \\ \{z^{(l)}_{k,j\tau}\}_{\tau=t-s,\tau\neq t}^{\tau=t+s}, \{Z^{(l+1)}_{<k,j\tau}, z^{(l)}_{>k,j\tau}, Z_{obs,j\tau}\}_{\tau=t-s}^{\tau=t+s}, \\ n_k(\{w^{(l+1)}_{<k,J^{-1}(i,\tau)\tau}, W^{(l)}_{>k,J^{-1}(i,\tau)\tau}, W_{obs,J^{-1}(i,\tau)\tau}\}_{\tau=t-s}^{\tau=t+s}), \gamma_{jt}, \nu_{jt}, \theta_k \end{matrix}\right) p_k(\theta_k|.)d\theta_k \quad (11)$$

where the posterior densities $p_k(\theta_k|.)$ are conditioned on the same information as the probability model for the kth variable.

Masking the Completed Data

Repeating the missing data imputation method of the previous section M times yields M sets of completed data files (U^m, W^m, Z^m) which we shall call the completed data implicates $m = 1, ..., M$. The implicates are masked independently by drawing masked values of confidential data from an appropriate predictive distribution such as (5). We call the resulting M masked data files the masked data implicates $m = 1, ..., M$. Although in many ways the masking procedure is similar to the missing data imputation method described above, an important difference is that masking is not iterative. Masked data are drawn only once per observation-confidential variable-implicate triple.

As in the missing data imputation and for the same reasons, some data reduction is required when specifying the conditioning set for each confidential variable. Similarly, dynamic conditioning data (leads and lags) available for masking a particular variable will vary from observation to observation. Hence, as in the missing data imputation, it is useful to group observations by the set of such data available to condition the masking regressions. It is also useful to group observations on the basis of some key variables, such as gender, full-time/part-time employment status, and industry, for which we would expect parameters of the predictive distribution to differ. We retain the same notation for parameterizing these groups as defined above.

The masking algorithm for a single implicate is as follows. First, split each of the three files into groups as described above. Then, for each confidential variable, estimate an appropriate generalized linear model on each group, conditioning on a well-chosen subset of the available data from all three files. Given the posterior distribution of the parameters of this generalized linear model, compute a draw from the predictive distribution for each confidential variable in each group. The masked data are these draws from the predictive distributions. The final step is to reassemble the masked data files from the various group files. Repeating this procedure on each completed data implicate yields multiply-imputed masked data. As before, the predictive densities are defined by an appropriate regression model and prior. For a given variable k from one of the source files, we draw its masked implicate from the posterior predictive density corresponding to an appropriate generalized linear model and an uninformative prior. For a particular implicate, these predictive densities are

$$\int f_{u_k}(u_k | U^m_{\sim k,i}, g_k(\{Z^m_{J(i,t)}\}_{t=1}^{t=T_i}), h_k(\{W^m_{it}\}_{t=1}^{t=T_i}), \lambda_i, \theta_k)\ p_k(\theta_k|.)d\theta_k \tag{12}$$

$$\int f_{w_k}\begin{pmatrix} w_k | U^m_i, \{Z^m_{J(i,\tau)}\}_{\tau=t-s}^{\tau=t+s}, \\ \{w^m_{k,i\tau}\}_{\tau=t-s,\tau\neq t}^{\tau=t+s}, \{W^m_{\sim k,i\tau}\}_{\tau=t-s}^{\tau=t+s}, \kappa_{it}, \mu_{it}, \theta_k \end{pmatrix} p_k(\theta_k|.)d\theta_k \tag{13}$$

$$\int f_{z_k}\begin{pmatrix} z_k | m_k(\{U^m_i\}_{i\in\{i|j=J(i,t)\}}), \\ \{z^m_{k,j\tau}\}_{\tau=t-s,\tau\neq t}^{\tau=t+s}, \{Z^m_{\sim k,j\tau}\}_{\tau=t-s}^{\tau=t+s}, \\ n_k(\{W^m_{i\tau}\}_{\tau=t-s,i\in\{i|j=J(i,t)\}}^{\tau=t+s}), \gamma_{jt}, \upsilon_{jt}, \theta_k \end{pmatrix} p_k(\theta_k|.)d\theta_k \tag{14}$$

where the posterior density of the parameters, $p_k(\theta_k|.)$, is conditioned on the same information as the conditional density of the variable being masked, and the subscript $\sim k$ refers to all other variables in the same source file. As always, there is a tradeoff between the analytic usefulness of the masked data file and the degree of confidentiality it affords. Below, we discuss various means of understanding the choices involved in these conflicting objectives.

Improving Confidentiality Protection. Our masking procedure preserves the configuration of the longitudinal histories in the three data files. That is, although all cases of confidential variables are masked, links between records in the three files are not perturbed. This preserves particular dynamic aspects of the database, such as individual work histories and firm births and deaths, as well as the history of worker-firm matches. In principle, the assumption of disclosable history configurations could be relaxed—for example, by perturbing some links between files, censoring some job records, or suppressing data on particular individuals or firms. We do not explore these issues in detail here, but note that perturbing the configuration of histories in the masked or completed data implicates may lead to substantial increases in confidentiality protection.

A final step before releasing the masked data is to remove unique person and firm identifiers in the various data files. These can be replaced with randomly generated ones. Note that the identifiers used in the released data need not be the same in each implicate. In fact, using different identifiers in each implicate will serve to increase confidentiality protection, because this prevents an intruder from easily combining information about firms or individuals across implicates. To do so, records in each implicate would first need to be statistically matched to records from the other implicates.

Improving Analytic Usefulness. In most applications, substantial improvements in the analytic quality of the masked data can be achieved by imposing a priori restrictions on the masked values. In general, such restrictions will reduce between-implicate variability, and hence reduce the level of confidentiality protection. Restricting the masked values can be done in a variety of ways. Sampling from a posterior predictive density proceeds in two steps—first, sampling from the posterior density of model parameters and, second, sampling from the predictive density conditional on the parameter draw. Importance sampling of the parameters and/or masked values is one way to improve the analytic quality of the masked data. In cases where the data are highly collinear, the usual case in the type of data we are considering, estimates of parameter covariances are likely to be imprecise. In such cases, restricting parameter draws to central regions of the posterior density can dramatically improve the quality of the masked data. Specifying parsimonious masking models will also be useful in such situations. We demonstrate an application of these methods in Section 4.

For a given parameter draw, restricting the draws of the masked values themselves will also serve to improve the analytic quality of the masked data. One approach is to restrict draws to central regions of the predictive density using standard methods. Another approach applicable to continuous variables is to individually restrict the masked values to lie inside an interval around true values. The interval can be specified in absolute or percentage terms. For example, one could restrict the masked values to lie within p percent of the completed values. To do so for a particular observation, sample from the predictive density until the restriction is satisfied or until a prespecified number of draws have been taken, at which time the masked value is set equal to one of the endpoints of the interval. An application of this method is described in Section 4.

Outliers provide two conflicting types of information. First, they may indicate that the original data have measurement errors (*e.g.,* earnings data that have been miscoded). Second, they may indicate that the underlying population is quite heterogeneous (*e.g.,* sales data within broadly defined industry groups). Either case has the potential to severely distort estimates of the masking equations and hence the masked data. We suggest treating outliers of the first type during the missing data imputation stage. Data values determined to be outliers of the first type can be set to missing and imputed along with other missing data. This procedure reduces the influence of these observations on both the missing data imputation and the data masking. It may substantially improve the analytic quality of the masked data. An important feature of our masking procedure, when combined with restricting the masked values to a range around the completed values, is that it is robust to outliers of the second type—the outlying values are perturbed in a manner consistent with the underlying data, without exerting undue influence on the masking of other observations.

We have not yet been explicit about the set of cases to be masked. In principle, not all observations need to be masked, though our method easily accommodates masking any number of cases in the three data files. Masking only a subset of cases will obviously improve the analytic usefulness of the data, though it does so at the expense of confidentiality protection. An example of such an application is Kennickell (1997), who masks sensitive data on a small subset of cases in a cross-sectional file of individuals using methods related to those presented here. Details of the Kennickell (1997) application can be found in the Appendix.

Traditional disclosure limitation methods may prove useful in preserving information when used in conjunction with our regression masking method. For example, some variables in the database may not pose a disclosure risk in aggregated analyses (*e.g.*, occupation or industry) but at a disaggregated level provide an easy means of compromising the confidentiality of records. For such variables, the overall analytic usefulness of the database may be better preserved by collapsing some cells or using data coarsening methods other than outright masking. We provide some examples below.

Simulated Data Based on Disclosable Summary Statistics

Disclosable summary statistics are defined as cross tabulations of discrete variables, conditional moments of continuous variables, generalized linear model coefficients, estimated covariance matrices, and estimated residual variances from such models. We construct disclosable summary statistics using an automated set of checks for conditions that are often associated with confidentiality preservation in tabulated data. Such checks normally include cell size and composition restrictions that generate primary suppressions as well as complementary suppressions generated by transformation tests that prevent the recovery of a suppressed cell from the released cells, conditional moments, or estimated model statistics. We build a data simulator that uses this disclosable statistical information to produce simulated draws from the predictive densities summarized by (12), (13), and (14). For comparability with our analyses of completed and masked data, we assume that there are also some variables X^m, possibly multiply-imputed for missing data, that can be released in microdata form. We note that if the variables in X^m are all discrete, then such a release is equivalent to releasing the full cross tabulation of all of the columns of X^m.

We provide M simulated draws from equation (4) using an approximation for the solution to (2). For each simulated implicate of Y, we compute our approximation based on

$$p(\tilde{Y}|D(Y^m), X^m) = \int p(\tilde{Y}|D(Y^m), X^m, \theta) p(\theta|D(Y^m), X^m) d\theta \tag{15}$$

where the relation $D(Y^m)$ means that we have replaced the values of the original Y^m with aggregated values based on disclosable moments. To put the simulation procedure in context, we summarize the relation between the completed, masked, and simulated data as follows. The original observed data are (Y_{obs}, X_{obs}). The completed data are multiple imputations based on an approximation to (3). The masked data are multiple imputations conditional on the completed data and based on an approximation to equation (4). The simulated data are multiple imputations conditional on traditionally disclosable functions of the completed data and based on an approximation to equation (15).

The procedure we use to estimate (15) is analogous to the masking procedure described in Section 2. The data in each file (U, W, and Z), are grouped according to the same conditions that are used to form λ_i, κ_{it}, μ_{it} and γ_{jt}, ν_{jt} in the data masking with the following exception. Each data configuration implied by these conditioning sets is subjected to an automatic traditional disclosure analysis that confirms that, within each file, the configurations are mutually exclusive, the cell sizes meet a minimum criterion, and there is no dominant unit. When cells fail such a test, they are collapsed. If no collapse is possible, the offending cell is suppressed. No marginal cells are used; hence, the margins constitute the complementary suppression where needed. To avoid notational clutter, we use the same symbols for these data configurations as in Section 2. For each data file and each data configuration within the file, we compute the conditional means of Y^m from completed data implicate m. We form $D(Y^m)$ by replacing, for each observation in each data file, the value of Y^m with the appropriate conditional mean[4].

The exact simulation equations are given by

$$\int f_{u_k}(u_k|D(U^m_{\sim k,i}), D(g_k(\{Z^m_{J(i,t)}\}_{t=1}^{t=T_i})), D(h_k(\{W^m_{it}\}_{t=1}^{t=T_i})), \lambda_i, \theta_k)\, p_k(\theta_k|.)d\theta_k \tag{16}$$

which is based on equation (12),

$$\int f_{w_k}\begin{pmatrix} w_k|D(U^m_i), D(\{Z^m_{J(i,\tau)}\}_{\tau=t-s}^{\tau=t+s}), \\ D(\{w^m_{k,i\tau}\}_{\tau=t-s,\tau\neq t}^{\tau=t+s}, \{W^m_{\sim k,i\tau}\}_{\tau=t-s}^{\tau=t+s}), \kappa_{it}, \mu_{it}, \theta_k \end{pmatrix} p_k(\theta_k|.)d\theta_k \tag{17}$$

---

4 Additional moments can be used, but we have not implemented this feature in our simulator.

which is based on equation (13),

$$\int f_{z_k}\begin{pmatrix} z_k \big| D(m_k(\{U_i^m\}_{i \in \{i|j = J(i,t)\}})), \\ D(\{z_{k,j\tau}^m\}_{\tau = t-s, \tau \neq t}^{\tau = t+s}, \{Z_{\sim k,j\tau}^m\}_{\tau = t-s}^{\tau = t+s}), \\ D(n_k(\{W_{i\tau}^m\}_{\tau = t-s, i \in \{i|j = J(i,t)\}}^{\tau = t+s})), \gamma_{jt}, \nu_{jt}, \theta_k \end{pmatrix} p_k(\theta_k|.)d\theta_k, \tag{18}$$

which is based on equation (14), and where $p_k(\theta_k|.)$ is conditioned on $D(Y^m)$, X^m. For each posterior distribution $p_k(\theta_k|.)$, we simulate a draw using M implicates based on the generalized linear model statistics estimated for the appropriate masking equation. These statistics are also collapsed and suppressed to conform to the disclosure criteria used to form $D(Y^m)$.

3. Using the Completed, Masked, and Simulated Data for Statistical Analysis

One of the principal advantages of multiply-imputed data is that valid statistical inferences can be obtained using standard complete-data methods. We illustrate these formulas for a generic statistic of interest, $\hat{Q}$, to be computed on multiple data implicates. The multiple implicates can be the result of missing data imputation (to produce completed data), masking (to produce masked data), or simulation (to produce simulated data). The formulas relating the complete-data methods and the multiple-imputation methods use standard relations derived in Rubin (1987). For convenience, we reproduce these formulas here.

The quantity of interest, Q, may be either a scalar or a k-dimensional column vector. Assume that, with access to complete confidential data, inferences for Q would be based on

$$(Q - \hat{Q}) \sim N(0, V)$$

where $\hat{Q}$ is a statistic estimating Q, $N(0, V)$ is the normal distribution of appropriate dimension, and V is the covariance of $(Q - \hat{Q})$. Valid inferences can be obtained using the statistics $\hat{Q}$ and V computed on each of the data implicates. Denote the values obtained on each of the implicates by $\hat{Q}_1, ..., \hat{Q}_M$ and $V_1, ..., V_M$. The M complete-data statistics are combined as follows. Let

$$\bar{Q}_M = \frac{1}{M}\sum_{m=1}^{M} \hat{Q}_m$$

denote the average of the complete-data estimates and

$$\bar{V}_M = \frac{1}{M} \sum_{m=1}^{M} V_m$$

be the average of the complete-data variances. The between-implicate variance of the statistics $\hat{Q}_1, ..., \hat{Q}_M$ is

$$B_M = \frac{1}{M-1} \sum_{m=1}^{M} (\hat{Q}_m - \bar{Q}_M)(\hat{Q}_m - \bar{Q}_M)^T,$$

and the total variance of $(Q - \bar{Q}_M)$ is

$$T_M = \bar{V}_M + \frac{M+1}{M} B_M.$$

The standard error of a particular element of $\bar{Q}_M$ is the square root of the appropriate diagonal element of T_M. Examples of statistical analyses based on multiply-imputed masked data are given in the next section.

4. An Illustration Using French Longitudinal Linked Data

To illustrate the missing data imputation, masking, and simulation procedure described above, we apply these methods to a French longitudinal linked database on individuals and their employers. The data consist of both survey and administrative records collected by INSEE (Institut National de la Statistique et des Etudes Economiques). The data structure is the same as the prototypical longitudinal linked dataset described in Section 2. These data are described in detail in Abowd, Kramarz, and Margolis (1999).

Individual and Work History Data

Individual characteristics and work history data are derived from the 'Déclarations Annuelles des Données Sociales' (DADS), a large-scale administrative database of matched employer-employee information collected by INSEE. The work history data are based on mandatory employer reports of the gross earnings of each employee subject to French payroll taxes. These taxes apply to all 'declared' employees and to all self-employed individuals—essentially all employed individuals in the economy.

The Division des Revenus prepares an extract of the DADS for scientific analysis that consists of all individuals employed in French enterprises who were born in October of even-numbered years, excluding civil servants. Our data span the years 1976 through 1996, with 1981, 1983, and 1990 excluded because the underlying administrative data were not collected in those years. Each record corresponds to a unique individual-year-establishment combination. Observations in the DADS file include an identifier that corresponds to the employee (ID), an identifier that corresponds to the establishment (SIRET), and an identifier that corresponds to the economic enterprise of the establishment (SIREN). Because the employer data are reported at the enterprise level, we are concerned primarily with the enterprise identifier.

Because our purposes are mainly illustrative, we select a 20 percent random subsample of individuals in the DADS for the example application. A strict 10 percent subsample of the DADS, known as the Echantillon Démographique Permanent (EDP), includes detailed demographic information on variables such as education. Our 20 percent subsample consists of the 10 percent of individuals in the EDP, plus an additional 10 percent random subsample of the other individuals in the DADS. The resulting subsample consists of 3,213,374 work history records on 362,913 individuals.

Time-invariant individual variables selected from the DADS for this illustration are gender, year of birth (range 1912 to 1980), and years of education. Time-varying job characteristics included are real annual compensation (annualized wage), occupation, geographic location of employment, full-time/other status, and number of days paid in the year (range 1 to 360)[5]. Of these individual and work history variables, year of birth, years of education, real annual compensation, and days paid are selected for masking. Occupation is collapsed to five categories, and geography is collapsed to two: employed in Ile-de-France (metropolitan Paris) and otherwise.

Firm Data

The primary source for our firm-level data is the 'Enquête Annuelle d'Entreprises' (EAE) collected by INSEE and organized by SIREN. This survey collects detailed information from economically related establishments with the same owner (called enterprises) and annual year-end employment greater than 20 employees. Variables selected from the EAE for this illustration are type of industry, annual sales, average employment over the year, and capital stock. Of these, annual sales, average employment, and capital stock are masked. Type of industry is collapsed to 40 categories prior to 1992 and 40 (different) categories thereafter[6].

---

[5] Days paid is an administrative variable that indicates the part of the year for which an employee received payments. Thus, by law, 360 days paid is a full year of work. Days paid from 1 to 359 represent payments for less than a full year of work. All individuals in the sample are permitted paid days of leave in accordance with French law and collective bargaining agreements, which cover more than 90 percent of all jobs.

[6] These categories correspond to standard French industrial classifications. In 1993, French industrial classifications changed from the Nomenclature d'Activités Productives (NAP) system to the Nomenclature d'Activités Francaises (NAF) system. There is no one-to-one mapping between these classification systems.

The sample of firms used for the example consists of firms in the EAE matched to work history records in our 20 percent subsample of the DADS. The firm sample is not representative of the French economy as a whole, which precludes certain analyses at the firm level. However, it serves to demonstrate the masking methods presented above. Our firm sample consists of 470,812 annual records on 105,813 enterprises.

We note that not all job records in the DADS can be linked to enterprises in the EAE. Non-matches arise when individuals are employed at enterprises with fewer than 20 employees and/or nonrespondent enterprises. This complicates the missing data imputation and masking because not all worker and job history records have firm data available as conditioning variables.

Missing Data Imputation Stage

Four of the variables selected for the illustration have missing data. These are education (missing for approximately half of individuals—those not in the EDP subsample), annual sales (missing in 47,796 matched EAE records), average employment (missing in 150,833 matched EAE records), and capital stock (missing in 35,989 matched EAE records). Hence the imputation sequence is capital stock, sales, employment, then education. This admits some computational efficiencies because the work history and person files need to be updated with imputed firm data only once per round (after the imputation of all three firm variables). Before estimating imputation models, observations are grouped as described in Section 2. For the firm variable imputations, these groups are defined purely by the availability of leads and lags (four groups)[7]. For the wage outlier imputations, groups are defined by the availability of firm data (due to non-matches), by the availability of leads and lags, by gender, and by full-time/part-time status (20 groups). For the education imputations, we define groups on the basis of gender and the availability of firm data (four groups).

The firm variables with missing data are all continuous, so we use linear regression models for the imputation. Because all three are positive and highly skewed, these regression models are estimated in logarithms. Education is recorded in eight categories, so the appropriate imputation model is multinomial logistic regression.

The annualized wage variable was the only one with problematic outliers. Using the method described in Section 2, we set outlying values to missing and impute these along with other missing data. We detect outliers at the end of the first round of imputation on the other variables with missing data—the first point at which we have complete data on which to condition an outlier detection model. Outliers are detected via a simple log wage regression. Wage values more than five standard deviations from their predicted value are considered outliers. After outliers are set to

[7] The number of groups given in this section correspond to rounds 2 through 10 of the imputation procedure. There are more groups in round 1 due to missing data on the variables that define the groups.

missing, the annualized wage variable joins the imputation sequence as the first variable imputed in round 2.

Initial experiments with imputing missing data in our database demonstrated the importance of specifying a parsimonious imputation model. For the logistic regressions, model selection is done manually. For the linear regressions, we automate the model selection procedure. For each linear model estimated, we specify a set of candidate conditioning variables. The model is first estimated on all candidate variables. Only variables that meet the Schwarz (1978) criterion are retained. The imputation model is then re-estimated on the reduced set of conditioning variables, and imputed values are drawn from the corresponding predictive distribution. The set of candidate variables are selected along the lines described in Section 2. For the firm variables with missing data, candidate variables include up to one lead and lag of the variable under imputation (where available), contemporaneous values and up to one lead and lag of the other firm variables, firm-specific means of contemporaneous values of work history variables for employees, and mean characteristics of the workers employed at the firm in that period[8]. Candidate variables for imputing wage outliers include up to one lead and lag of the log annualized wage (where available), contemporaneous values and up to one lead and lag of other work history variables, contemporaneous firm variables for the firm at which the worker was employed, and worker characteristics. Conditioning variables for imputing missing education are manually selected from a candidate set of worker characteristics and from worker-specific means of work history and firm variables.

To further improve the quality of the imputed data when drawing from the predictive density, we restrict parameter draws for all estimated posterior distributions to lie within three standard deviations of the posterior mode[9]. Initial experiments demonstrated remarkable improvements in the quality of the imputed data as a result of this restriction because it reduced collinearity of the conditioning data, which increased the precision of posterior parameter covariances.

The imputation procedure consists of 10 rounds. Posterior parameter distributions in the imputation models change little after the sixth round. We repeat the procedure 10 times, yielding 10 completed data implicates.

Masking Stage

Confidential variables in each of the completed data implicates were masked using the methods described in Section 2. Observations were split into the same groups as described in the previous section. We used the same model selection techniques as in the missing data imputation, and we restricted parameter draws in the same way. In addition, we restricted the masked values of continuous variables to lie

[8] For categorical variables in the work history and worker files, proportions in a given category are used in place of means.

[9] For the logistic regressions, parameters are drawn from the normal approximation to the posterior density.

within $p = 20$ percent of the true or imputed value. Masked values were re-drawn until they were within this interval; if after 100 draws the candidate implicate remained outside the interval, the masked value was set equal to the closest endpoint.

The models used to mask variables that had missing data were the same as those described above. The additional masked variables, year of birth and days paid, were both treated as continuous and masked using linear regression. The days paid variable takes values only in the interval between 1 and 360, so we apply a logit-like transformation to this variable for masking[10]. After masking, both year of birth and days paid were rounded to the nearest integer.

Simulation Stage

We simulated the same list of confidential variables as in the masking stage. The automatic disclosure proofing resulted in the suppression of data for 434 enterprise-years and the associated work histories. No individual data were suppressed. Observations were grouped according to the same methods used in the previous section. Parameter draws from the posterior distribution were restricted as in the imputation and masking stages. There is no access to the confidential microdata in the simulation stage, so the simulated values cannot be restricted to lie within an interval around the 'true' value. Instead, the simulated values of days paid and year of birth were restricted to the observed sample range of these variables. The models used to simulate the variables were exactly the same models used to mask these variables, except that some models could not be used because they did not pass the disclosure tests.

Statistical Properties of the Completed, Masked, and Simulated Data

Tables 1, 2, and 3 present basic univariate properties of confidential variables in the completed, masked, and simulated data. Tables 1 and 2 present these for the individual and work history variables by gender. Table 3 presents statistics for the firm data. It is apparent that the masked and simulated data retain the basic univariate properties of the completed data. Biases in the means and variances of masked and simulated variables are generally small. In relative (percentage) terms, the bias is larger though still well with within acceptable limits. These biases are smaller in the masked data than in the simulated data, and smaller for firm variables than for individual and work history variables. The masking and simulation procedures lead to considerable relative increases in the variance of univariate statistics, as we would expect. These increases in variance are much more pronounced in the simulated data than in the masked data. In the masked firm data, the variance of variable means and variances are at times lower than in the completed data. This is likely

[10] This transformation is *logit(days paid)* = $\log\left(\dfrac{days\ paid}{365 - days\ paid}\right)$.

Table 1. Univariate Statistics on Completed, Masked, and Simulated Data: Men

Variable	N	Average Mean or Proportion in Category	Average Variance of Mean	Between-Implicate Variance of Mean	Total Variance of Mean	Relative Bias	Relative Increase in Variance	Average Variance	Average Variance of Variance	Between-Implicate Variance of Variance	Total Variance of Variance	Relative Bias	Relative Increase in Variance
Completed Data													
Year of Birth	201,906	1952	0.001	0	0.001			250.2	0.41	0	0.41		
No Diploma	201,906	0.299	1.03E–06	2.68E–03	2.95E–03			0.207	1.69E–07	3.78E–04	4.16E–04		
Elementary School	201,906	0.180	7.31E–07	7.83E–05	8.68E–05			0.148	2.99E–07	3.26E–05	3.62E–05		
Middle School	201,906	0.112	4.90E–07	4.09E–04	4.51E–04			0.099	2.92E–07	2.42E–04	2.67E–04		
High School	201,906	0.060	2.77E–07	4.67E–05	5.17E–05			0.056	2.14E–07	3.69E–05	4.08E–05		
Basic Vocational School	201,906	0.215	8.32E–07	6.00E–04	6.61E–04			0.168	2.69E–07	1.98E–04	2.18E–04		
Advanced Vocational School	201,906	0.060	2.81E–07	6.13E–05	6.78E–05			0.057	2.16E–07	4.62E–05	5.10E–05		
Technical College or University	201,906	0.036	1.71E–07	4.57E–05	5.04E–05			0.034	1.46E–07	3.85E–05	4.25E–05		
Graduate School	201,906	0.039	1.84E–07	8.16E–06	9.16E–06			0.037	1.57E–07	6.96E–06	7.81E–06		
Log Real Annual Compensation (1980 FF 000)	1,893,555	4.164	4.70E–07	2.37E–09	4.73E–07			0.891	4.19E–06	8.79E–09	4.20E–06		
Days Paid (max 360)	1,893,555	263.0	0.009	0	0.009			17987	180	0	180		
Masked Data													
Year of Birth	201,906	1951	0.001	0.009	0.011	–0.001	7.856	236.4	0.43	1.75	2.36	–0.055	4.734
No Diploma	201,906	0.309	1.03E–06	5.32E–03	5.86E–03	0.033	0.982	0.209	1.61E–07	5.35E–04	5.89E–04	0.007	0.414
Elementary School	201,906	0.181	7.31E–07	8.90E–04	9.80E–04	0.005	10.284	0.148	2.93E–07	3.72E–04	4.09E–04	–0.001	10.319
Middle School	201,906	0.105	4.63E–07	6.16E–04	6.78E–04	–0.061	0.503	0.093	2.83E–07	3.90E–04	4.29E–04	–0.056	0.608
High School	201,906	0.056	2.63E–07	7.66E–05	8.46E–05	–0.054	0.636	0.053	2.06E–07	5.98E–05	6.60E–05	–0.051	0.620
Basic Vocational School	201,906	0.208	8.12E–07	9.62E–04	1.06E–03	–0.031	0.602	0.164	2.73E–07	3.44E–04	3.79E–04	–0.024	0.737
Advanced Vocational School	201,906	0.070	3.22E–07	2.82E–04	3.10E–04	0.163	3.579	0.065	2.35E–07	2.07E–04	2.28E–04	0.147	3.468

Table 1. Univariate Statistics on Completed, Masked, and Simulated Data: Men (Continued)

Variable	N	Average Mean or Proportion in Category	Average Variance of Mean	Between-Implicate Variance of Mean	Total Variance of Mean	Relative Bias	Relative Increase in Variance	Average Variance	Average Variance of Variance	Between-Implicate Variance of Variance	Total Variance of Variance	Relative Bias	Relative Increase in Variance
Technical College or University	201,906	0.032	1.53E–07	6.41E–05	7.07E–05	–0.105	0.402	0.031	1.33E–07	5.54E–05	6.11E–05	–0.103	0.438
Graduate School	201,906	0.038	1.83E–07	1.13E–05	1.26E–05	–0.009	0.376	0.037	1.56E–07	9.66E–06	1.08E–05	–0.008	0.380
Log Real Annual Compensation (1980 FF 000)	1,893,555	4.146	4.58E–07	4.97E–06	5.92E–06	–0.004	11.520	0.867	3.87E–06	4.07E–06	8.34E–06	–0.027	0.987
Days Paid (max 360)	1,893,555	257.4	0.009	0.058	0.073	–0.021	6.672	17039	166	1080	1353	–0.053	6.519
Simulated Data													
Year of Birth	201,906	1951	0.001	0.157	0.173	0.000	138.920	174.8	0.24	1.41	1.79	–0.301	3.359
No Diploma	201,906	0.347	1.08E–06	1.02E–02	1.12E–02	0.160	2.798	0.217	1.32E–07	4.60E–04	5.06E–04	0.049	0.215
Elementary School	201,906	0.142	6.02E–07	6.88E–04	7.57E–04	–0.209	7.722	0.122	3.03E–07	3.47E–04	3.82E–04	–0.177	9.573
Middle School	201,906	0.107	4.72E–07	5.90E–04	6.49E–04	–0.041	0.440	0.095	2.86E–07	3.59E–04	3.96E–04	–0.038	0.483
High School	201,906	0.059	2.72E–07	2.77E–04	3.05E–04	–0.016	4.910	0.055	2.08E–07	2.17E–04	2.39E–04	–0.019	4.866
Basic Vocational School	201,906	0.217	8.26E–07	3.39E–03	3.73E–03	0.012	4.651	0.167	2.56E–07	1.04E–03	1.14E–03	–0.007	4.247
Advanced Vocational School	201,906	0.064	2.95E–07	2.94E–04	3.24E–04	0.056	3.775	0.060	2.21E–07	2.16E–04	2.38E–04	0.049	3.653
Technical College or University	201,906	0.031	1.49E–07	8.54E–05	9.41E–05	–0.133	0.866	0.030	1.29E–07	7.42E–05	8.18E–05	–0.129	0.924
Graduate School	201,906	0.033	1.58E–07	2.87E–05	3.17E–05	–0.144	2.458	0.032	1.38E–07	2.52E–05	2.78E–05	–0.140	2.559
Log Real Annual Compensation (1980 FF 000)	1,893,555	4.172	4.17E–07	2.20E–04	2.43E–04	0.002	512.000	0.789	2.00E–06	5.93E–04	6.54E–04	–0.115	154.885
Days Paid (max 360)	1,893,555	256.5	0.007	2.690	2.967	–0.025	311.300	13990	144	41251	45521	–0.222	251.871

Notes: Education categories are the highest degree attained. Relative bias and variance are computed in comparison to the completed data.
Sources: Authors' calculations are based on the INSEE DADS and EDP data 1976–1996.

Table 2. Univariate Statistics on Completed, Masked, and Simulated Data: Women

Variable	N	Average Mean or Proportion in Category	Average Variance of Mean	Between-Implicate Variance of Mean	Total Variance of Mean	Relative Bias	Relative Increase in Variance	Average Variance	Average Variance of Variance	Between-Implicate Variance of Variance	Total Variance of Variance	Relative Bias	Relative Increase in Variance
Completed Data													
Year of Birth	161,007	1954	0.001	0	0.001			224.7	0.52	0	0.52		
No Diploma	161,007	0.264	1.19E–06	2.74E–03	3.01E–03			0.192	2.66E–07	5.47E–04	6.02E–04		
Elementary School	161,007	0.200	9.93E–07	1.17E–04	1.30E–04			0.160	3.57E–07	4.24E–05	4.70E–05		
Middle School	161,007	0.150	7.90E–07	3.84E–04	4.24E–04			0.127	3.84E–07	1.89E–04	2.08E–04		
High School	161,007	0.082	4.65E–07	8.58E–05	9.49E–05			0.075	3.24E–07	5.92E–05	6.54E–05		
Basic Vocational School	161,007	0.148	7.81E–07	1.90E–04	2.10E–04			0.126	3.86E–07	9.62E–05	1.06E–04		
Advanced Vocational School	161,007	0.074	4.27E–07	8.23E–05	9.10E–05			0.069	3.08E–07	5.85E–05	6.47E–05		
Technical College or University	161,007	0.058	3.37E–07	7.89E–05	8.71E–05			0.054	2.63E–07	6.03E–05	6.66E–05		
Graduate School	161,007	0.025	1.50E–07	3.14E–06	3.60E–06			0.024	1.35E–07	2.83E–06	3.25E–06		
Log Real Annual Compensation (1980 FF 000)	1,319,819	3.777	8.11E–07	4.32E–09	8.16E–07			1.071	6.39E–06	1.65E–08	6.40E–06		
Days Paid (max 360)	1,319,819	260.9	0.014	0	0.014			18122	245	0	245		
Masked Data													
Year of Birth	161,007	1953	0.001	0.016	0.018	0.000	12.245	212.1	0.50	3.36	4.20	–0.056	7.089
No Diploma	161,007	0.239	1.09E–06	6.79E–03	7.47E–03	–0.094	1.482	0.176	3.01E–07	1.05E–03	1.15E–03	–0.084	0.913
Elementary School	161,007	0.193	9.65E–07	6.09E–04	6.70E–04	–0.034	4.147	0.155	3.60E–07	2.33E–04	2.57E–04	–0.029	4.462
Middle School	161,007	0.161	8.31E–07	1.55E–03	1.71E–03	0.074	3.034	0.134	3.71E–07	6.82E–04	7.50E–04	0.052	2.601
High School	161,007	0.082	4.67E–07	1.25E–04	1.38E–04	0.006	0.457	0.075	3.25E–07	8.94E–05	9.87E–05	0.005	0.508
Basic Vocational School	161,007	0.164	8.43E–07	1.26E–03	1.39E–03	0.107	5.616	0.136	3.72E–07	5.89E–04	6.49E–04	0.079	5.109
Advanced Vocational School	161,007	0.073	4.21E–07	2.85E–04	3.14E–04	–0.013	2.449	0.068	3.02E–07	2.03E–04	2.24E–04	–0.015	2.458

Table 2. Univariate Statistics on Completed, Masked, and Simulated Data: Women (Continued)

Variable	N	Average Mean or Proportion in Category	Average Variance of Mean	Between-Implicate Variance of Mean	Total Variance of Mean	Relative Bias	Relative Increase in Variance	Average Variance	Average Variance of Variance	Between-Implicate Variance of Variance	Total Variance of Variance	Relative Bias	Relative Increase in Variance
Technical College or University	161,007	0.063	3.66E–07	1.69E–04	1.86E–04	0.094	1.139	0.059	2.77E–07	1.29E–04	1.42E–04	0.086	1.136
Graduate School	161,007	0.025	1.50E–07	3.16E–06	3.63E–06	0.001	0.008	0.024	1.36E–07	2.87E–06	3.29E–06	0.001	0.013
Log Real Annual Compensation (1980 FF 000)	1,319,819	3.760	7.89E–07	1.38E–06	2.31E–06	–0.005	1.828	1.042	5.83E–06	3.95E–06	1.02E–05	–0.027	0.588
Days Paid (max 360)	1,319,819	254.5	0.013	0.089	0.111	–0.024	7.055	17010	222	1188	1529	–0.061	5.254
Simulated Data													
Year of Birth	161,007	1954	0.001	0.231	0.255	0.000	181.465	160.0	0.28	1.62	2.06	–0.288	2.970
No Diploma	161,007	0.262	1.15E–06	8.66E–03	9.53E–03	–0.007	2.163	0.185	2.69E–07	1.27E–03	1.40E–03	–0.032	1.324
Elementary School	161,007	0.172	8.80E–07	4.92E–04	5.42E–04	–0.143	3.162	0.142	3.76E–07	2.25E–04	2.48E–04	–0.114	4.265
Middle School	161,007	0.162	8.36E–07	1.05E–03	1.16E–03	0.077	1.731	0.135	3.75E–07	4.84E–04	5.33E–04	0.058	1.557
High School	161,007	0.089	4.99E–07	3.39E–04	3.73E–04	0.085	2.936	0.080	3.34E–07	2.28E–04	2.51E–04	0.074	2.840
Basic Vocational School	161,007	0.158	8.22E–07	7.62E–04	8.39E–04	0.069	2.988	0.132	3.78E–07	3.75E–04	4.13E–04	0.052	2.891
Advanced Vocational School	161,007	0.082	4.63E–07	3.42E–04	3.76E–04	0.097	3.134	0.075	3.20E–07	2.38E–04	2.63E–04	0.085	3.058
Technical College or University	161,007	0.053	3.08E–07	1.74E–04	1.92E–04	–0.090	1.201	0.050	2.44E–07	1.37E–04	1.51E–04	–0.087	1.267
Graduate School	161,007	0.024	1.48E–07	1.24E–05	1.38E–05	–0.014	2.833	0.024	1.34E–07	1.13E–05	1.26E–05	–0.014	2.866
Log Real Annual Compensation (1980 FF 000)	1,319,819	3.789	7.41E–07	1.59E–03	1.75E–03	0.003	2140.856	0.978	3.01E–06	3.92E–03	4.32E–03	–0.087	673.020
Days Paid (max 360)	1,319,819	253.3	0.010	5.795	6.384	–0.029	463.974	13327	185	41295	45609	–0.265	185.503

Notes: Education categories are the highest degree attained. Relative bias and variance are computed in comparison to the completed data.
Sources: Authors' calculations are based on the INSEE DADS and EDP data 1976–1996.

Table 3. Univariate Statistics on Completed, Masked, and Simulated Data: Firms

Variable	N	Average Mean or Proportion in Category	Average Variance of Mean	Between-Implicate Variance of Mean	Total Variance of Mean	Relative Bias	Relative Increase in Variance	Average Variance	Average Variance of Variance	Between-Implicate Variance of Variance	Total Variance of Variance	Relative Bias	Relative Increase in Variance
Completed Data													
Log Sales (FF millions)	470,812	10.58	4.39E–06	3.19E–05	3.95E–05			2.07	2.59E–05	4.95E–05	8.04E–05		
Log Capital Stock (FF millions)	470,812	8.50	9.51E–06	6.86E–06	1.71E–05			4.48	1.32E–04	6.28E–05	2.01E–04		
Log Average Employment	470,812	4.33	2.57E–06	9.44E–05	1.06E–04			1.21	1.20E–05	7.45E–05	9.40E–05		
Masked Data													
Log Sales (FF millions)	470,812	10.56	4.37E–06	2.54E–05	3.23E–05	–0.002	–0.183	2.06	2.56E–05	4.32E–05	7.30E–05	–0.012	–0.091
Log Capital Stock (FF millions)	470,812	8.48	9.45E–06	1.47E–05	2.56E–05	–0.002	0.502	4.45	1.28E–04	6.23E–05	1.96E–04	–0.028	–0.021
Log Average Employment	470,812	4.31	2.55E–06	8.93E–05	1.01E–04	–0.004	–0.054	1.20	1.17E–05	7.85E–05	9.81E–05	–0.024	0.044
Simulated Data													
Log Sales (FF millions)	470,378	10.59	1.45E–06	2.30E–04	2.54E–04	0.001	5.439	0.68	2.16E–06	1.22E–04	1.37E–04	0.704	0.704
Log Capital Stock (FF millions)	470,378	8.51	3.79E–06	2.52E–03	2.77E–03	0.002	161.704	1.78	1.93E–05	1.83E–03	2.04E–03	9.154	9.154
Log Average Employment	470,378	4.34	7.26E–07	5.55E–04	6.11E–04	0.002	4.739	0.34	8.70E–07	2.83E–04	3.12E–04	2.318	2.318

Notes: Relative bias and variance are computed in comparison to the completed data. Unweighted statistics.
Sources: Authors' calculations are based on the INSEE EAE data 1978–1996.

a result of our sampling frame for this illustration not being representative at the enterprise level.

Tables 4, 5, and 6 present basic bivariate properties of the confidential variables in the completed, masked, and simulated data. The bivariate statistics are presented as tables of correlation coefficients (below the diagonal) and the between-implicate variance of the correlation coefficient (above the diagonal). These two statistics provide the most summary detail without cluttering the tables excessively. Table 4 presents the correlations among the time-invariant personal characteristics in the individual data file for both genders combined. Table 5 presents the correlations among the time-invariant and time-varying variables linked to the work history data for both genders combined. Table 6 presents the correlations for the firm-level data.

Tables 4 and 5 demonstrate that the masked data fully preserve the bivariate structure of the confidential data at the worker and work history levels. There is almost no bias and relatively little between-implicate variance complicating the inference about a correlation coefficient. The masked data do have substantially more between-implicate variance than the completed data; however, they never have enough to substantially affect inferences about the magnitude of the correlations. The simulated data preserve the correlation structure remarkably well given the limitations inherent in the simulation. There is more bias in the simulated data than in the masked data, and the between-implicate variance is substantially greater than with the completed data but usually not large enough to affect the inference about the correlation coefficient.

Table 6 shows that the masked firm data substantially preserve the correlation structure with between-implicate variation comparable to the completed data. The simulated data display some biases (underestimation of the correlation coefficient) and substantially increased between-implicate variation. Given the sampling frame used to construct the firm-level data for this simulation, neither of these outcomes is surprising. The correlation structure of the simulated data is biased toward zero but not enough to substantially change economically meaningful conclusions about the bivariate relationships.

Given the structure of our imputation, masking and simulation equations, it is perhaps not surprising that the masked and simulated data preserve the first two moments so effectively. Our next analyses are based on models of substantive economic interest. The models predict variables in the work history file based on information in all three linked files. They thus provide a very stringent test of the scientific quality of the masked and imputed data for addressing questions about job-level outcomes.

Table 4. Bivariate Statistics for Individuals: Men and Women Combined

		(1)	(2)	(3)	(4)	(5)	(6)	(7)	(8)	(9)	(10)
Correlations in Completed Data (Between-Implicate Variance Above Diagonal)											
(1)	Male	1	*0.00000*	*0.00320*	*0.00016*	*0.00135*	*0.00071*	*0.00081*	*0.00015*	*0.00082*	*0.00008*
(2)	Year of Birth	–0.07	1	*0.00029*	*0.00008*	*0.00038*	*0.00006*	*0.00003*	*0.00004*	*0.00003*	*0.00001*
(3)	No Diploma	0.04	0.15	1	*0.00071*	*0.00017*	*0.00019*	*0.00036*	*0.00035*	*0.00018*	*0.00011*
(4)	Elementary School	–0.03	–0.34	–0.30	1	*0.00030*	*0.00006*	*0.00033*	*0.00007*	*0.00004*	*0.00002*
(5)	Middle School	–0.06	0.18	–0.24	–0.19	1	*0.00011*	*0.00043*	*0.00010*	*0.00007*	*0.00004*
(6)	High School	–0.04	0.09	–0.17	–0.13	–0.10	1	*0.00012*	*0.00003*	*0.00002*	*0.00001*
(7)	Basic Vocational School	0.08	–0.02	–0.30	–0.23	–0.18	–0.13	1	*0.00011*	*0.00008*	*0.00005*
(8)	Advanced Vocational School	–0.03	–0.02	–0.17	–0.13	–0.10	–0.07	–0.13	1	*0.00002*	*0.00001*
(9)	Technical College or University	–0.05	0.02	–0.14	–0.11	–0.08	–0.06	–0.10	–0.06	1	*0.00001*
(10)	Graduate School	0.04	–0.06	–0.12	–0.09	–0.07	–0.05	–0.09	–0.05	–0.04	1
Correlations in Masked Data (Between-Implicate Variance Above Diagonal)											
(1)	Male	1	*0.00001*	*0.01511*	*0.00340*	*0.00204*	*0.00052*	*0.00470*	*0.00268*	*0.00114*	*0.00011*
(2)	Year of Birth	–0.07	1	*0.00110*	*0.00014*	*0.00074*	*0.00013*	*0.00008*	*0.00004*	*0.00007*	*0.00001*
(3)	No Diploma	0.08	0.13	1	*0.00121*	*0.00030*	*0.00028*	*0.00117*	*0.00052*	*0.00026*	*0.00014*
(4)	Elementary School	–0.02	–0.30	–0.30	1	*0.00064*	*0.00012*	*0.00029*	*0.00027*	*0.00014*	*0.00004*
(5)	Middle School	–0.08	0.17	–0.24	–0.18	1	*0.00032*	*0.00083*	*0.00020*	*0.00022*	*0.00011*
(6)	High School	–0.05	0.08	–0.17	–0.13	–0.10	1	*0.00017*	*0.00005*	*0.00004*	*0.00002*
(7)	Basic Vocational School	0.06	–0.02	–0.30	–0.23	–0.19	–0.13	1	*0.00012*	*0.00012*	*0.00005*
(8)	Advanced Vocational School	–0.01	–0.02	–0.17	–0.13	–0.11	–0.07	–0.13	1	*0.00006*	*0.00002*
(9)	Technical College or University	–0.07	0.02	–0.14	–0.10	–0.08	–0.06	–0.11	–0.06	1	*0.00002*
(10)	Graduate School	0.04	–0.05	–0.11	–0.09	–0.07	–0.05	–0.09	–0.05	–0.04	1

Table 4. Bivariate Statistics for Individuals: Men and Women Combined (Continued)

		(1)	(2)	(3)	(4)	(5)	(6)	(7)	(8)	(9)	(10)
Correlations in Simulated Data (Between-Implicate Variance Above Diagonal)											
(1)	Male	1	*0.00038*	*0.02479*	*0.00319*	*0.00360*	*0.00286*	*0.00578*	*0.00311*	*0.00212*	*0.00041*
(2)	Year of Birth	–0.08	1	*0.00067*	*0.00010*	*0.00006*	*0.00015*	*0.00019*	*0.00015*	*0.00014*	*0.00003*
(3)	No Diploma	0.09	0.02	1	*0.00132*	*0.00061*	*0.00040*	*0.00031*	*0.00106*	*0.00053*	*0.00013*
(4)	Elementary School	–0.04	0.01	–0.29	1	*0.00029*	*0.00013*	*0.00085*	*0.00013*	*0.00006*	*0.00005*
(5)	Middle School	–0.08	0.01	–0.26	–0.17	1	*0.00025*	*0.00116*	*0.00010*	*0.00010*	*0.00007*
(6)	High School	–0.06	–0.02	–0.18	–0.12	–0.11	1	*0.00053*	*0.00007*	*0.00003*	*0.00003*
(7)	Basic Vocational School	0.07	0.01	–0.32	–0.21	–0.19	–0.14	1	*0.00021*	*0.00018*	*0.00020*
(8)	Advanced Vocational School	–0.03	0.00	–0.19	–0.12	–0.11	–0.08	–0.13	1	*0.00003*	*0.00002*
(9)	Technical College or University	–0.05	–0.02	–0.14	–0.09	–0.08	–0.06	–0.10	–0.06	1	*0.00001*
(10)	Graduate School	0.03	–0.05	–0.12	–0.07	–0.07	–0.05	–0.08	–0.05	–0.04	1

Notes: N=362,913.

Sources: Authors' calculations are based on the INSEE DADS and EDP data.

Table 5. Bivariate Statistics Based on the Work History File: Men and Women Combined

	(1)	(2)	(3)	(4)	(5)	(6)	(7)	(8)	(9)	(10)	(11)	(12)	(13)	(14)	(15)	(16)	(17)	(18)	(19)	(20)	(21)	(22)
Correlations in Completed Data (Between Implicate Variance Above Diagonal)																						
(1) Male	1									*0.00330*	*0.00018*	*0.00126*	*0.00079*	*0.00107*	*0.00019*	*0.00109*	*0.00010*		*0.00000*	*0.00000*	*0.00000*	*0.00000*
(2) Full-Time Employee	0.167	1								*0.00008*	*0.00001*	*0.00004*	*0.00002*	*0.00004*	*0.00000*	*0.00003*	*0.00001*		*0.00000*	*0.00000*	*0.00000*	*0.00004*
(3) Engineer, Professional, or Manager	0.097	−0.008	1							*0.00011*	*0.00004*	*0.00004*	*0.00007*	*0.00006*	*0.00006*	*0.00013*	*0.00009*		*0.00000*	*0.00000*	*0.00000*	*0.00000*
(4) Technician or Technical White Collar	0.001	0.031	−0.155	1						*0.00006*	*0.00004*	*0.00003*	*0.00001*	*0.00004*	*0.00004*	*0.00001*	*0.00000*		*0.00000*	*0.00000*	*0.00000*	*0.00000*
(5) Other White Collar	−0.380	−0.088	−0.205	−0.309	1					*0.00037*	*0.00006*	*0.00017*	*0.00011*	*0.00024*	*0.00004*	*0.00011*	*0.00001*		*0.00000*	*0.00000*	*0.00000*	*0.00001*
(6) Skilled Blue Collar	0.314	0.132	−0.176	−0.265	−0.351	1				*0.00049*	*0.00007*	*0.00006*	*0.00002*	*0.00035*	*0.00003*	*0.00003*	*0.00000*		*0.00000*	*0.00000*	*0.00000*	*0.00001*
(7) Unskilled Blue Collar	0.030	−0.063	−0.158	−0.238	−0.316	−0.271	1			*0.00006*	*0.00007*	*0.00005*	*0.00000*	*0.00007*	*0.00001*	*0.00001*	*0.00000*		*0.00000*	*0.00000*	*0.00000*	*0.00000*
(8) Works in Ile–de–France	−0.018	−0.007	0.133	0.066	0.034	−0.092	−0.104	1		*0.00003*	*0.00000*	*0.00001*	*0.00001*	*0.00001*	*0.00001*	*0.00002*	*0.00003*		*0.00000*	*0.00000*	*0.00000*	*0.00000*
(9) Year of Birth	−0.070	−0.106	−0.094	−0.053	0.082	−0.053	0.083	−0.037	1	*0.00019*	*0.00009*	*0.00031*	*0.00005*	*0.00001*	*0.00002*	*0.00003*	*0.00001*		*0.00000*	*0.00000*	*0.00000*	*0.00006*
(10) No Diploma	0.043	−0.046	−0.102	−0.125	−0.039	0.072	0.166	−0.018	0.086	1	*0.00089*	*0.00023*	*0.00024*	*0.00060*	*0.00039*	*0.00024*	*0.00013*		*0.00009*	*0.00021*	*0.00016*	*0.00020*
(11) Elementary School	−0.021	0.028	−0.087	−0.063	0.006	0.073	0.041	−0.033	−0.286	−0.269	1	*0.00030*	*0.00006*	*0.00043*	*0.00010*	*0.00005*	*0.00003*		*0.00002*	*0.00002*	*0.00003*	*0.00003*
(12) Middle School	−0.070	−0.035	0.003	0.042	0.073	−0.088	−0.034	0.015	0.177	−0.197	−0.185	1	*0.00010*	*0.00050*	*0.00010*	*0.00007*	*0.00005*		*0.00007*	*0.00007*	*0.00003*	*0.00007*
(13) High School	−0.047	−0.032	0.088	0.084	0.023	−0.097	−0.071	0.049	0.100	−0.147	−0.137	−0.101	1	*0.00014*	*0.00003*	*0.00002*	*0.00001*		*0.00005*	*0.00002*	*0.00003*	*0.00002*
(14) Basic Vocational School	0.095	0.076	−0.076	−0.021	−0.018	0.112	−0.022	−0.064	0.006	−0.287	−0.269	−0.198	−0.147	1	*0.00017*	*0.00011*	*0.00008*		*0.00003*	*0.00001*	*0.00002*	*0.00002*
(15) Advanced Vocational School	−0.032	0.015	0.025	0.065	0.036	−0.060	−0.059	0.012	−0.010	−0.151	−0.141	−0.104	−0.077	−0.151	1	*0.00004*	*0.00001*		*0.00003*	*0.00009*	*0.00008*	*0.00007*
(16) Technical College or University	−0.064	−0.013	0.111	0.136	−0.025	−0.098	−0.083	0.046	0.030	−0.126	−0.117	−0.086	−0.064	−0.126	−0.066	1	*0.00001*		*0.00006*	*0.00004*	*0.00004*	*0.00004*
(17) Graduate School	0.043	−0.026	0.291	0.021	−0.064	−0.083	−0.073	0.089	−0.060	−0.103	−0.097	−0.071	−0.053	−0.104	−0.054	−0.045	1		*0.00004*	*0.00001*	*0.00001*	*0.00001*
(18) Paid Days	0.008	0.239	0.054	0.077	−0.048	0.047	−0.110	−0.045	−0.297	−0.095	0.094	−0.072	−0.041	0.066	0.026	−0.001	0.010	1	*0.00000*	*0.00000*	*0.00000*	*0.00002*
(19) Log Real Annual Compensation	0.191	0.529	0.207	0.096	−0.120	0.055	−0.168	0.098	−0.165	−0.116	−0.013	−0.026	0.012	0.048	0.049	0.048	0.082	0.110	1	*0.00000*	*0.00000*	*0.00001*
(20) Log Sales	0.060	0.013	0.044	0.068	0.016	−0.042	−0.063	0.030	−0.007	−0.054	−0.038	0.032	0.030	−0.003	0.032	0.025	0.041	0.067	0.131	1	*0.00001*	*0.00001*
(21) Log Capital Stock	0.066	0.126	0.046	0.077	0.006	−0.024	−0.082	0.004	−0.072	−0.074	−0.017	0.018	0.025	0.011	0.031	0.024	0.044	0.157	0.185	0.894	1	*0.00001*
(22) Log Average Employment	0.063	−0.033	−0.004	0.039	−0.019	−0.016	0.003	0.002	−0.059	−0.036	−0.011	0.011	0.008	0.003	0.021	0.010	0.027	0.042	0.073	0.929	0.848	1
Correlations in Masked Data (Between Implicate Variance Above Diagonal)																						
														Masked Variables								
(1) Male	1								*0.00003*	*0.02205*	*0.00458*	*0.00237*	*0.00077*	*0.00748*	*0.00363*	*0.00161*	*0.00014*	*0.00000*	*0.00000*	*0.00000*	*0.00000*	*0.00000*
(2) Full-Time Employee	0.167	1							*0.00000*	*0.00069*	*0.00020*	*0.00010*	*0.00001*	*0.00016*	*0.00006*	*0.00002*	*0.00000*	*0.00000*	*0.00000*	*0.00000*	*0.00000*	*0.00000*
(3) Engineer, Professional, or Manager	0.097	−0.008	1						*0.00000*	*0.00014*	*0.00006*	*0.00010*	*0.00010*	*0.00008*	*0.00017*	*0.00017*	*0.00013*	*0.00000*	*0.00000*	*0.00000*	*0.00000*	*0.00000*
(4) Technician or Technical White Collar	0.001	0.031	−0.155	1					*0.00000*	*0.00011*	*0.00007*	*0.00005*	*0.00001*	*0.00013*	*0.00012*	*0.00008*	*0.00000*	*0.00000*	*0.00000*	*0.00000*	*0.00000*	*0.00000*
(5) Other White Collar	−0.380	−0.088	−0.205	−0.309	1				*0.00000*	*0.00325*	*0.00064*	*0.00042*	*0.00012*	*0.00160*	*0.00085*	*0.00016*	*0.00001*	*0.00000*	*0.00000*	*0.00000*	*0.00000*	*0.00000*
(6) Skilled Blue Collar	0.314	0.132	−0.176	−0.265	−0.351	1			*0.00000*	*0.00321*	*0.00089*	*0.00020*	*0.00004*	*0.00147*	*0.00022*	*0.00010*	*0.00001*	*0.00000*	*0.00000*	*0.00000*	*0.00000*	*0.00000*

Table 5. Bivariate Statistics Based on the Work History File: Men and Women Combined (Continued)

	(1)	(2)	(3)	(4)	(5)	(6)	(7)	(8)	(9)	(10)	(11)	(12)	(13)	(14)	(15)	(16)	(17)	(18)	(19)	(20)	(21)	(22)
(7) Unskilled Blue Collar	0.030	–0.063	–0.158	–0.238	–0.316	–0.271	1		*0.00000*	*0.00022*	*0.00014*	*0.00008*	*0.00002*	*0.00016*	*0.00004*	*0.00004*	*0.00001*	*0.00000*	*0.00000*	*0.00000*	*0.00000*	*0.00000*
(8) Works in Ile–de–France	–0.018	–0.007	0.133	0.066	0.034	–0.092	–0.104	1	*0.00000*	*0.00022*	*0.00002*	*0.00003*	*0.00005*	*0.00004*	*0.00003*	*0.00009*	*0.00006*	*0.00000*	*0.00000*	*0.00000*	*0.00000*	*0.00000*
(9) Year of Birth	–0.075	–0.110	–0.091	–0.050	0.086	–0.057	0.079	–0.035	1	*0.00096*	*0.00013*	*0.00065*	*0.00015*	*0.00007*	*0.00006*	*0.00011*	*0.00002*	*0.00000*	*0.00000*	*0.00000*	*0.00000*	*0.00000*
(10) No Diploma	0.079	–0.037	–0.099	–0.116	–0.040	0.086	0.141	–0.022	0.073	1	*0.00154*	*0.00040*	*0.00043*	*0.00189*	*0.00061*	*0.00039*	*0.00020*	*0.00030*	*0.00072*	*0.00000*	*0.00000*	*0.00001*
(11) Elementary School	–0.013	0.027	–0.078	–0.058	–0.001	0.071	0.041	–0.034	–0.251	–0.264	1	*0.00076*	*0.00015*	*0.00045*	*0.00038*	*0.00023*	*0.00007*	*0.00003*	*0.00023*	*0.00000*	*0.00000*	*0.00000*
(12) Middle School	–0.089	–0.039	0.001	0.037	0.071	–0.087	–0.026	0.015	0.158	–0.197	–0.180	1	*0.00036*	*0.00104*	*0.00028*	*0.00026*	*0.00015*	*0.00026*	*0.00014*	*0.00000*	*0.00000*	*0.00000*
(13) High School	–0.056	–0.031	0.079	0.079	0.026	–0.095	–0.065	0.052	0.091	–0.149	–0.135	–0.102	1	*0.00024*	*0.00007*	*0.00005*	*0.00002*	*0.00006*	*0.00004*	*0.00000*	*0.00000*	*0.00000*
(14) Basic Vocational School	0.063	0.069	–0.072	–0.023	–0.010	0.093	–0.013	–0.060	–0.002	–0.287	–0.258	–0.196	–0.147	1	*0.00024*	*0.00018*	*0.00008*	*0.00009*	*0.00021*	*0.00000*	*0.00000*	*0.00000*
(15) Advanced Vocational School	–0.003	0.019	0.027	0.061	0.021	–0.050	–0.052	0.006	–0.014	–0.158	–0.143	–0.108	–0.081	–0.155	1	*0.00009*	*0.00003*	*0.00001*	*0.00017*	*0.00000*	*0.00000*	*0.00000*
(16) Technical College or University	–0.087	–0.015	0.097	0.124	–0.011	–0.096	–0.078	0.044	0.031	–0.126	–0.114	–0.086	–0.064	–0.123	–0.068	1	*0.00003*	*0.00005*	*0.00008*	*0.00001*	*0.00000*	*0.00001*
(17) Graduate School	0.046	–0.029	0.278	0.023	–0.061	–0.080	–0.071	0.095	–0.050	–0.106	–0.095	–0.072	–0.054	–0.104	–0.057	–0.046	1	*0.00000*	*0.00003*	*0.00000*	*0.00000*	*0.00000*
(18) Paid Days	0.011	0.246	0.055	0.078	–0.051	0.050	–0.111	–0.047	–0.299	–0.092	0.094	–0.071	–0.038	0.067	0.027	–0.003	0.003	1	*0.00000*	*0.00000*	*0.00000*	*0.00000*
(19) Log Real Annual Compensation	0.193	0.535	0.210	0.098	–0.121	0.056	–0.171	0.100	–0.164	–0.102	–0.009	–0.030	0.009	0.042	0.050	0.037	0.077	0.113	1	*0.00000*	*0.00000*	*0.00000*
(20) Log Sales	0.008	–0.005	0.003	–0.006	–0.007	0.006	0.005	0.004	0.013	0.003	–0.005	0.002	0.001	–0.001	–0.001	–0.003	0.002	–0.010	–0.004	1	*0.00000*	*0.00001*
(21) Log Capital Stock	0.008	–0.005	0.008	–0.005	–0.005	0.002	0.001	–0.003	0.017	0.004	–0.007	0.004	0.002	–0.002	0.000	–0.003	0.003	–0.008	–0.002	0.737	1	*0.00000*
(22) Log Average Employment	0.009	–0.001	0.008	–0.003	–0.008	0.005	0.001	0.006	0.018	0.003	–0.008	0.003	0.003	–0.001	0.000	–0.002	0.002	–0.007	0.001	0.773	0.661	1

Correlations in Simulated Data (Between Implicate Variance Above Diagonal)

									Simulated Variables													
(1) Male	1								*0.00052*	*0.03068*	*0.00383*	*0.00423*	*0.00416*	*0.00715*	*0.00402*	*0.00326*	*0.00056*	*0.00006*	*0.00036*	*0.00000*	*0.00000*	*0.00000*
(2) Full-Time Employee	0.167	1							*0.00001*	*0.00071*	*0.00009*	*0.00008*	*0.00008*	*0.00025*	*0.00012*	*0.00009*	*0.00001*	*0.00014*	*0.00064*	*0.00000*	*0.00000*	*0.00000*
(3) Engineer, Professional, or Manager	0.097	–0.008	1						*0.00002*	*0.00029*	*0.00007*	*0.00014*	*0.00042*	*0.00024*	*0.00020*	*0.00055*	*0.00042*	*0.00000*	*0.00001*	*0.00000*	*0.00000*	*0.00000*
(4) Technician or Technical White Collar	0.001	0.031	–0.155	1					*0.00002*	*0.00015*	*0.00009*	*0.00006*	*0.00013*	*0.00021*	*0.00021*	*0.00003*	*0.00000*	*0.00000*	*0.00000*	*0.00000*	*0.00000*	*0.00000*
(5) Other White Collar	–0.380	–0.088	–0.205	–0.309	1				*0.00008*	*0.00391*	*0.00045*	*0.00059*	*0.00056*	*0.00139*	*0.00076*	*0.00032*	*0.00006*	*0.00001*	*0.00005*	*0.00000*	*0.00000*	*0.00000*
(6) Skilled Blue Collar	0.314	0.132	–0.176	–0.265	–0.351	1			*0.00011*	*0.00439*	*0.00070*	*0.00022*	*0.00009*	*0.00232*	*0.00031*	*0.00010*	*0.00002*	*0.00001*	*0.00005*	*0.00000*	*0.00000*	*0.00000*
(7) Unskilled Blue Collar	0.030	–0.063	–0.158	–0.238	–0.316	–0.271	1		*0.00002*	*0.00019*	*0.00018*	*0.00007*	*0.00003*	*0.00019*	*0.00005*	*0.00003*	*0.00001*	*0.00000*	*0.00000*	*0.00000*	*0.00000*	*0.00000*
(8) Works in Ile–de–France	–0.018	–0.007	0.133	0.066	0.034	–0.092	–0.104	1	*0.00001*	*0.00006*	*0.00001*	*0.00007*	*0.00006*	*0.00003*	*0.00002*	*0.00007*	*0.00006*	*0.00000*	*0.00000*	*0.00000*	*0.00000*	*0.00000*
(9) Year of Birth	–0.094	–0.039	–0.099	–0.063	0.079	–0.034	0.081	–0.058	1	*0.00065*	*0.00007*	*0.00007*	*0.00023*	*0.00015*	*0.00016*	*0.00019*	*0.00004*	*0.00001*	*0.00002*	*0.00000*	*0.00000*	*0.00001*
(10) No Diploma	0.083	0.023	–0.102	–0.126	–0.049	0.120	0.127	–0.013	0.018	1	*0.00158*	*0.00096*	*0.00061*	*0.00047*	*0.00152*	*0.00080*	*0.00018*	*0.00004*	*0.00122*	*0.00001*	*0.00000*	*0.00001*
(11) Elementary School	–0.035	–0.004	–0.075	–0.063	0.030	0.033	0.048	–0.035	0.012	–0.266	1	*0.00038*	*0.00018*	*0.00112*	*0.00020*	*0.00011*	*0.00007*	*0.00001*	*0.00014*	*0.00000*	*0.00000*	*0.00001*
(12) Middle School	–0.087	–0.019	0.027	0.062	0.055	–0.089	–0.049	0.020	0.005	–0.255	–0.164	1	*0.00037*	*0.00165*	*0.00013*	*0.00015*	*0.00010*	*0.00000*	*0.00023*	*0.00000*	*0.00001*	*0.00000*
(13) High School	–0.062	–0.023	0.101	0.102	0.011	–0.102	–0.078	0.046	–0.018	–0.183	–0.117	–0.114	1	*0.00074*	*0.00009*	*0.00004*	*0.00005*	*0.00001*	*0.00024*	*0.00000*	*0.00001*	*0.00000*
(14) Basic Vocational School	0.079	0.038	–0.072	–0.028	–0.011	0.087	0.001	–0.058	0.008	–0.316	–0.205	–0.199	–0.142	1	*0.00032*	*0.00029*	*0.00030*	*0.00001*	*0.00020*	*0.00000*	*0.00000*	*0.00000*

Table 5. Bivariate Statistics Based on the Work History File: Men and Women Combined (Continued)

	(1)	(2)	(3)	(4)	(5)	(6)	(7)	(8)	(9)	(10)	(11)	(12)	(13)	(14)	(15)	(16)	(17)	(18)	(19)	(20)	(21)	(22)
(15) Advanced Vocational School	−0.025	−0.011	0.009	0.050	0.037	−0.051	−0.044	−0.001	0.000	−0.186	−0.118	−0.113	−0.081	−0.141	1	*0.00004*	*0.00003*	*0.00001*	*0.00026*	*0.00000*	*0.00001*	*0.00000*
(16) Technical College or University	−0.061	−0.016	0.104	0.122	−0.024	−0.090	−0.073	0.039	−0.023	−0.137	−0.087	−0.084	−0.060	−0.105	−0.060	1	*0.00001*	*0.00002*	*0.00031*	*0.00000*	*0.00000*	*0.00000*
(17) Graduate School	0.024	−0.034	0.244	0.026	−0.053	−0.075	−0.063	0.079	−0.046	−0.114	−0.073	−0.071	−0.051	−0.089	−0.051	−0.037	1	*0.00001*	*0.00009*	*0.00000*	*0.00000*	*0.00000*
(18) Paid Days	0.013	0.258	0.039	0.068	−0.053	0.053	−0.092	−0.040	−0.049	−0.014	−0.006	0.002	0.003	0.012	0.003	0.007	−0.001	1	*0.00007*	*0.00000*	*0.00000*	*0.00000*
(19) Log Real Annual Compensation	0.198	0.549	0.218	0.101	−0.124	0.056	−0.177	0.085	−0.069	−0.027	−0.034	0.001	0.021	0.010	0.005	0.025	0.041	0.199	1	*0.00000*	*0.00000*	*0.00000*
(20) Log Sales	−0.005	−0.003	0.003	0.004	0.004	−0.008	−0.006	0.011	0.000	−0.001	0.000	0.001	0.001	−0.002	0.000	0.001	0.002	−0.001	−0.002	1	*0.00014*	*0.00041*
(21) Log Capital Stock	−0.005	0.006	0.002	0.004	0.003	−0.003	−0.007	0.012	−0.001	−0.001	0.000	0.000	0.001	−0.001	0.000	0.002	0.002	0.001	0.004	0.562	1	*0.00061*
(22) Log Average Employment	0.010	0.017	0.004	−0.006	−0.018	0.016	0.011	0.019	−0.001	0.004	−0.002	−0.002	−0.001	0.001	−0.001	−0.001	0.001	0.005	0.014	0.358	0.382	1

Notes: N=3,213,374.
Sources: Authors' calculations are based on the INSEE DADS and EDP data.

Table 6. Bivariate Statistics in the Firm Data

	(1)	(2)	(3)
CORRELATIONS IN COMPLETED DATA			
(Between-Implicate Variance Above Diagonal)			
(1) Log Sales	1	*0.000001*	*0.000013*
(2) Log Capital Stock	0.733	1	*0.000003*
(3) Log Average Employment	0.767	0.656	1
CORRELATIONS IN MASKED DATA			
(Between-Implicate Variance Above Diagonal)			
(1) Log Sales	1	*0.000001*	*0.000011*
(2) Log Capital Stock	0.737	1	*0.000004*
(3) Log Average Employment	0.773	0.661	1
CORRELATIONS IN SIMULATED DATA			
(Between-Implicate Variance Above Diagonal)			
(1) Log Sales	1	*0.000138*	*0.000410*
(2) Log Capital Stock	0.562	1	*0.000614*
(3) Log Average Employment	0.358	0.382	1

Notes: N=470,812 in completed and masked data; N=470,378 in the simulated data.
Sources: Authors' calculations are based on the INSEE EAE data.

Modeling Wages With Fixed Individual and Employer Effects. Our first substantive model predicts the log wage rate (real, full-time, full-year compensation) based on individual characteristics, employer characteristics, and unobservable individual and employer effects. We chose this example for two related reasons. First, this model can only be estimated using linked longitudinal employer-employee data (see Abowd, Kramarz, and Margolis 1999). Second, the dependent variable is among the most studied job-level outcomes in economics; hence, we can use substantial prior information to interpret the reasonableness of the estimated statistical models. We include only one observation per individual per time period, selecting the dominant job (based on days paid) for that time period.

The statistical model is a two-factor analysis of covariance with main effects only for the two factors. The covariates consist of time-varying characteristics of the individual, job, and employer. The first factor is an individual effect that is decomposed into a function of time-invariant personal characteristics and unobservable personal heterogeneity. The second factor is a firm effect that consists of unobservable employer heterogeneity. All components of the full design matrix of the model are non-orthogonal. We compute the full least squares estimator for all the effects by direct solution of the model normal equations using the conjugate gradient algorithm specialized to the sparse representation of our normal equations[11]. We calculate the identifiable person and firm effects, for subsequent use in statistical analysis, using a graph-theoretic analysis of the group structure of the underlying design matrix (see Abowd and Kramarz 2000).

Table 7 presents coefficient estimates for the time-varying and time-invariant regressors in the model. All estimated regression coefficients are shown in the table. The completed data results are essentially the same as other analyses of these data (*e.g.*, Abowd, Kramarz, and Margolis 1999). Wage rates increase with labor force experience at a decreasing rate. Profiles for women are flatter than those for men. There is a premium for working in Ile-de-France. Wage rates increase as the size of the firm increases (using log sales as the size measure). The quadratic term in the log sales relationship is often negative, but it is essentially zero in these data. As regards time-invariant personal characteristics, for men, each schooling diploma increases earnings relative to the no-diploma reference group, and the usual elementary school, middle school, high school, college/university, graduate school progression holds. Similar results hold for women, except that elementary school completion is less valuable than no-diploma[12].

To consider the reliability of the estimates from the masked and simulated data, we first examine the experience profiles. Figure 1 shows the comparison of the completed, masked, and simulated experience profiles for men and women. The horizontal axis is years of potential labor force experience (years since finishing school). The vertical axis is the log wage rate. The slope of the log wage profile, called the return to experience, is interpreted as the percentage change in the wage rate associated with a small increase in experience given the current level of experience. Although the masked coefficients differ from the completed data coefficients for both genders, the estimated profiles are essentially identical. The additional variation associated with the between-implicate component of the variance of the profile would not affect inferences about the experience profiles in any meaningful way. On the other hand, the profiles estimated from the simulated data are substantially flatter than those estimated from the completed or masked data, and the profile for men is slightly flatter than the one for women. These are meaningful differences that would materially affect conclusions drawn from the simulated data. One might reasonably ask if there were indications in the simulated data analysis that the conclusions for this variable would be sensitive to the simulation. While the standard errors of the model coefficients are somewhat larger for the analysis of the simulated data, they are not enough larger to provide the necessary signal about the discrepancies shown in Figure 1.

[11] Robert Creecy of the U.S. Census Bureau programmed the sparse conjugate gradient implementation as well as the graph-theoretic identification algorithm. Both programs are used by the Bureau's Longitudinal Employer-Household Dynamics program.

[12] In the EDP 'no-diploma' means that the respondent to the French census declared that he or she did not complete elementary school.

Table 7. Summary of Coefficient Estimates for the Analysis of a Linear Model Predicting Log Real Annual Wage Rates with Fixed Person and Firm Effects

	Completed Data					Masked Data					Simulated Data				
	Average Coefficient	Average Variance of Coefficient	Between-Implicate Variance of Coefficient	Total Variance of Coefficient	Standard Error of Average Coefficient	Average Coefficient	Average Variance of Coefficient	Between-Implicate Variance of Coefficient	Total Variance of Coefficient	Standard Error of Average Coefficient	Average Coefficient	Average Variance of Coefficient	Between-Implicate Variance of Coefficient	Total Variance of Coefficient	Standard Error of Average Coefficient
Time-Varying Observables															
Male x Experience	0.0757	6.88E–07	3.86E–06	4.93E–06	0.0022	0.0470	5.93E–07	4.19E–07	1.05E–06	0.0010	0.0128	6.85E–07	2.31E–05	2.61E–05	0.0051
Male x (Experience^2)/100	–0.2676	4.24E–05	1.54E–04	2.12E–04	0.0146	–0.0858	4.27E–05	4.47E–05	9.19E–05	0.0096	–0.0257	5.63E–05	8.29E–05	1.47E–04	0.0121
Male x (Experience^3)/1,000	0.0519	4.17E–06	1.06E–05	1.58E–05	0.0040	0.0055	4.76E–06	7.63E–06	1.32E–05	0.0036	0.0066	6.62E–06	6.09E–06	1.33E–05	0.0036
Male x (Experience^4)/10,000	–0.0041	4.76E–08	8.96E–08	1.46E–07	0.0004	0.0001	6.02E–08	1.28E–07	2.01E–07	0.0004	–0.0007	8.43E–08	7.51E–08	1.67E–07	0.0004
Female x Experience	0.0386	1.46E–06	8.82E–07	2.44E–06	0.0016	0.0329	1.31E–06	5.36E–06	7.20E–06	0.0027	0.0167	1.64E–06	7.36E–06	9.74E–06	0.0031
Female x (Experience^2)/100	–0.0709	9.66E–05	4.39E–05	1.45E–04	0.0120	–0.0405	1.01E–04	4.20E–04	5.63E–04	0.0237	–0.0425	1.50E–04	2.69E–04	4.45E–04	0.0211
Female x (Experience^3)/1,000	0.0075	9.91E–06	3.95E–06	1.42E–05	0.0038	0.0006	1.18E–05	4.48E–05	6.11E–05	0.0078	0.0121	1.92E–05	3.17E–05	5.41E–05	0.0074
Female x (Experience^4)/10,000	–0.0002	1.16E–07	4.08E–08	1.61E–07	0.0004	0.0004	1.53E–07	5.27E–07	7.33E–07	0.0009	–0.0012	2.64E–07	3.80E–07	6.82E–07	0.0008
Male x Works in Ile–de–France	0.0465	1.29E–05	5.12E–07	1.35E–05	0.0037	0.0512	1.32E–05	1.67E–06	1.51E–05	0.0039	0.0690	2.45E–05	1.17E–05	3.74E–05	0.0061
Female x Works in Ile–de–France	0.0391	4.52E–05	8.91E–07	4.62E–05	0.0068	0.0430	4.67E–05	9.50E–06	5.71E–05	0.0076	0.0840	8.76E–05	5.97E–05	1.53E–04	0.0124
Log Sales	0.0024	1.21E–05	3.02E–05	4.54E–05	0.0067	0.0100	1.25E–05	5.17E–05	6.93E–05	0.0083	0.1337	1.40E–04	5.43E–03	6.11E–03	0.0782
(Log Sales)^2	0.0005	1.86E–08	6.72E–08	9.25E–08	0.0003	0.0004	1.89E–08	1.15E–07	1.45E–07	0.0004	–0.0051	2.96E–07	8.53E–06	9.68E–06	0.0031

Table 7. Summary of Coefficient Estimates for the Analysis of a Linear Model Predicting Log Real Annual Wage Rates with Fixed Person and Firm Effects (Continued)

	Completed Data					Masked Data					Simulated Data				
	Average Coeffi-cient	Average Variance of Coefficient	Between-Implicate Variance of Coefficient	Total Variance of Coeffi-cient	Standard Error of Average Coefficient	Average Coeffi-cient	Average Variance of Coeffi-cient	Between-Implicate Variance of Coefficient	Total Variance of Coeffi-cient	Standard Error of Average Coefficient	Average Coeffi-cient	Average Variance of Coeffi-cient	Between-Implicate Variance of Coefficient	Total Variance of Coeffi-cient	Standard Error of Average Coefficient
Time-Invariant Observables															
Male	–0.067	2.41E–05	1.79E–04	2.21E–04	0.0149	0.046	3.41E–05	5.22E–04	6.08E–04	0.0247	0.244	2.53E–05	6.69E–03	7.39E–03	0.0859
Male x Elementary School	0.002	1.90E–05	1.54E–04	1.88E–04	0.0137	0.017	2.24E–05	2.98E–04	3.50E–04	0.0187	0.016	2.56E–05	6.32E–05	9.52E–05	0.0098
Male x Middle School	0.179	2.72E–05	7.24E–04	8.24E–04	0.0287	0.155	3.49E–05	4.99E–04	5.84E–04	0.0242	0.157	3.01E–05	5.27E–04	6.10E–04	0.0247
Male x High School	0.315	4.31E–05	8.10E–04	9.34E–04	0.0306	0.316	5.26E–05	3.12E–04	3.96E–04	0.0199	0.274	5.30E–05	5.31E–04	6.37E–04	0.0252
Male x Basic Vocational School	0.149	1.68E–05	2.85E–04	3.30E–04	0.0182	0.130	1.96E–05	2.16E–04	2.57E–04	0.0160	0.051	1.83E–05	1.21E–04	1.51E–04	0.0123
Male x Advanced Vocational School	0.291	3.92E–05	2.19E–04	2.80E–04	0.0167	0.268	4.09E–05	9.53E–04	1.09E–03	0.0330	0.162	4.09E–05	3.48E–04	4.23E–04	0.0206
Male x Technical College or University	0.490	6.97E–05	6.33E–04	7.66E–04	0.0277	0.483	9.26E–05	4.79E–04	6.20E–04	0.0249	0.382	1.04E–04	6.21E–04	7.87E–04	0.0281
Male x Graduate School	0.760	7.02E–05	2.57E–04	3.53E–04	0.0188	0.716	7.89E–05	8.85E–04	1.05E–03	0.0324	0.511	1.01E–04	6.37E–04	8.01E–04	0.0283
Female x Elementary School	–0.108	3.55E–05	1.18E–04	1.65E–04	0.0129	–0.061	4.89E–05	2.66E–04	3.41E–04	0.0185	0.008	4.72E–05	8.52E–05	1.41E–04	0.0119
Female x Middle School	0.142	3.89E–05	2.29E–04	2.91E–04	0.0171	0.127	5.04E–05	2.54E–04	3.30E–04	0.0182	0.082	4.08E–05	1.00E–04	1.51E–04	0.0123
Female x High School	0.233	6.29E–05	6.11E–04	7.35E–04	0.0271	0.260	8.05E–05	2.12E–04	3.14E–04	0.0177	0.169	7.51E–05	1.55E–04	2.45E–04	0.0157
Female x Basic Vocational School	0.110	3.93E–05	1.40E–04	1.93E–04	0.0139	0.109	5.14E–05	2.61E–04	3.38E–04	0.0184	0.058	4.93E–05	6.10E–05	1.16E–04	0.0108
Female x Advanced Vocational School	0.187	6.65E–05	2.88E–04	3.83E–04	0.0196	0.199	9.28E–05	4.03E–04	5.36E–04	0.0232	0.104	7.99E–05	1.23E–04	2.15E–04	0.0147

Table 7. Summary of Coefficient Estimates for the Analysis of a Linear Model Predicting Log Real Annual Wage Rates with Fixed Person and Firm Effects (Continued)

	Completed Data					Masked Data					Simulated Data				
	Average Coeffi-cient	Average Variance of Coefficient	Between-Implicate Variance of Coefficient	Total Variance of Coeffi-cient	Standard Error of Average Coefficient	Average Coeffi-cient	Average Variance of Coeffi-cient	Between-Implicate Variance of Coefficient	Total Variance of Coeffi-cient	Standard Error of Average Coefficient	Average Coeffi-cient	Average Variance of Coeffi-cient	Between-Implicate Variance of Coefficient	Total Variance of Coeffi-cient	Standard Error of Average Coefficient
Female x Technical College or University	0.364	8.89E–05	1.72E–04	2.78E–04	0.0167	0.361	1.05E–04	5.26E–04	6.83E–04	0.0261	0.226	1.21E–04	1.94E–04	3.34E–04	0.0183
Female x Graduate School	0.460	2.23E–04	2.52E–04	5.00E–04	0.0224	0.484	2.85E–04	8.73E–04	1.25E–03	0.0353	0.351	2.72E–04	9.32E–04	1.30E–03	0.0360

Notes: The model is a linear analysis of covariance with covariates listed in the table, unrestricted fixed person, and firm effects. The dependent variable is the logarithm of annual full-time, full-year compensation.
Sources: Authors' calculations are based on INSEE DADS, EDP, and EAE data.

Figure 1. Experience Profile in Wage Regression With Fixed Worker and Firm Effects

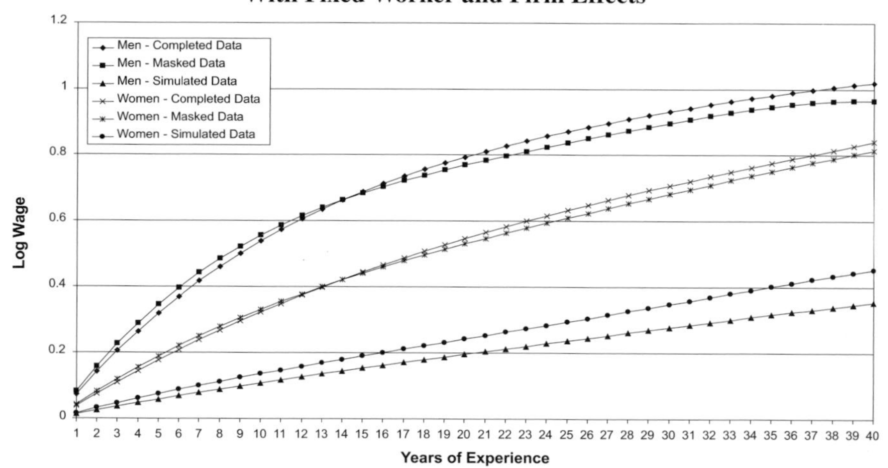

The comparison of the log sales effect reveals that the masked data are once again quite close to the completed data and that the standard errors of the masked data coefficients are larger than those of the completed data by a magnitude that allows the completed data estimate to fall within the usual posterior interval ranges of the masked data. The simulated data analysis of the log sales effect is substantially larger than the effect measured in either the completed or masked data. In contrast to the experience coefficients, however, there is plenty of evidence in the simulated data that the log sales effect has been unreliably estimated. Both the standard errors and the between-implicate component of variation indicate that this effect has been unreliably estimated in the simulated data.

Comparison of the estimated education effects for men and women reveal that the masked data yield reliable estimates of these effects. The simulated data yield acceptable results, with estimation uncertainty comparable to the masked data.

Table 8 compares correlations between the estimated effects in the completed, masked, and simulated data. The correlations are computed over all observations in the work history file that enter the data analysis. The correlations in Table 8 are used to help understand the extent to which time-varying characteristics, time-invariant characteristics, unobservable person effects, and unobservable firm effects contribute to the explanation of log wage rates. Correlations between the estimated effects and log wages in the completed data indicate that person effects are somewhat more important than firm effects in explaining log wages. This is the usual result for such analyses using French data, and it is accurately reproduced in both the masked and simulated data. The person and firm effects are negatively correlated, again the usual result for French data and reliably reproduced by the

Table 8. Correlation of Estimated Effects from the Log Wage Rate Regression Model

	Log Wage	Time-Varying Observables	Person Effect	Time-Invariant Observables	Rest of Person Effect	Firm Effect	Residual
Average Correlations in Completed Data (Between-Implicate Variance Above Diagonal)							
Log Real Annual Compensation	1	*0.000021*	*0.000023*	*0.000079*	*0.000042*	*0.000006*	*0.000000*
Time-Varying Observables	0.417	1	*0.000017*	*0.000110*	*0.000022*	*0.000004*	*0.000000*
Person Effect	0.517	–0.140	1	*0.000087*	*0.000017*	*0.000069*	*0.000000*
Time-Invariant Observables	0.336	–0.146	0.383	1	*0.000005*	*0.000011*	*0.000000*
Rest of Person Effect	0.430	–0.095	0.934	0.029	1	*0.000096*	*0.000000*
Firm Effect	0.258	0.042	–0.466	0.105	–0.545	1	*0.000000*
Residual	0.393	0.000	0.000	0.000	0.000	0.000	1
Average Correlations in Masked Data (Between-Implicate Variance Above Diagonal)							
Log Real Annual Compensation	1	*0.000067*	*0.000608*	*0.000083*	*0.000344*	*0.000036*	*0.000001*
Time-Varying Observables	0.351	1	*0.000060*	*0.000539*	*0.000029*	*0.000048*	*0.000000*
Person Effect	0.503	–0.172	1	*0.000411*	*0.000062*	*0.001582*	*0.000000*
Time-Invariant Observables	0.359	–0.086	0.339	1	*0.000015*	*0.000049*	*0.000000*
Rest of Person Effect	0.409	–0.153	0.944	0.011	1	*0.001336*	*0.000000*
Firm Effect	0.258	0.044	–0.510	0.105	–0.579	1	*0.000000*
Residual	0.402	0.000	0.000	0.000	0.000	0.000	1

Table 8. Correlation of Estimated Effects from the Log Wage Rate Regression Model (Continued)

	Log Wage	Time-Varying Observables	Person Effect	Time-Invariant Observables	Rest of Person Effect	Firm Effect	Residual
Average Correlations in Simulated Data (Between-Implicate Variance Above Diagonal)							
Log Real Annual Compensation	1	*0.008058*	*0.003133*	*0.000524*	*0.001743*	*0.000161*	*0.000032*
Time-Varying Observables	0.109	1	*0.003557*	*0.049421*	*0.002011*	*0.000176*	*0.000000*
Person Effect	0.479	–0.173	1	*0.004107*	*0.000425*	*0.004680*	*0.000000*
Time-Invariant Observables	0.399	–0.153	0.380	1	*0.000154*	*0.000130*	*0.000000*
Rest of Person Effect	0.369	–0.120	0.936	0.035	1	*0.003206*	*0.000000*
Firm Effect	0.202	0.030	–0.598	0.065	–0.670	1	*0.000000*
Residual	0.573	0.000	0.000	0.000	0.000	0.000	1

Notes: Based on statistics produced by the model summarized in Table 7.
Sources: Authors' calculations are based on INSEE DADS, EDP, and EAE data.

masked and simulated data. All correlations with log wage rates are somewhat attenuated in the simulated data, where the estimated effects explain wage variation less well than in either the completed or masked data.

Modeling Full-Time/Part-Time Status. Our second substantive model predicts whether an employee in a particular job has full-time status based on characteristics of the individual (gender, labor force experience, and education), the employer (sales, capital stock, and employment), and the job (real wage rate, occupation, and location of employment). This example was chosen for several reasons. First, the dependent variable is always observed and is among the non-confidential characteristics of the job in our masked and simulated data. Thus, the dependent variable has not been manipulated statistically in the completed, masked, or simulated data. Second, the model we use is a generalized linear model (logistic regression), which could be more sensitive to the linearity assumptions used to mask and simulate the continuous variables in the confidential data. Third, this variable is often not available in linked longitudinal employer-employee data and must be imputed. Thus, it is of substantial interest to estimate a statistical model for full-time status using a linked employer-employee dataset in which the variable is measured.

Table 9 shows that the masked data do an excellent job of preserving inferences about the effects of individual, job, and employer characteristics for predicting full-time status with very little increase in estimation error. The simulated data perform substantially less well; however, both the standard errors of the coefficients and the between-implicate component of variance signal the poorer performance of these data. The effects themselves are tricky to interpret because the equation is conditioned on the actual full-time wage rate in the job. Thus, the other effects must be interpreted as marginal effects, given the wage rate. For this reason we place more emphasis on the comparison across the completed, masked, and simulated datasets—leaving the assessment of the reasonableness of any particular set of estimated effects to the reader.

5. Conclusions

Our goal was to provide a complete description of masking and data simulation algorithms that could be used to preserve the confidentiality of longitudinal linked data while still providing data analysts with substantial information about the relationships in those data. We provided full implementation details for the application of these techniques to prototypical longitudinal linked employer-employee data. The data completion model is a full implementation of sequential regression multivariate imputation based on generalized linear models for all data with missing values. Our procedures generalize the existing methods to preserve the dynamic links among the individual, job, and employer characteristics. Our masking technique preserves confidentiality by replacing the confidential data with a draw from the predictive distribution of those data, given the values of the other

Table 9. Summary of Coefficient Estimates for the Analysis of a Model Predicting Full-Time Employment

	Completed Data					Masked Data					Simulated Data				
Predictable Variable	Average Coefficient	Average Variance	Between-Implicate Variance	Total Variance	Standard Error of Average Coefficient	Average Coefficient	Average Variance	Between-Implicate Variance	Total Variance	Standard Error of Average Coefficient	Average Coefficient	Average Variance	Between-Implicate Variance	Total Variance	Standard Error of Average Coefficient
Intercept	–6.977	0.00192	0.05725	0.06489	0.255	–7.418	0.00218	0.05747	0.06539	0.256	–10.550	0.00349	0.24928	0.27771	0.527
Male	2.216	0.00251	0.00251	0.00528	0.073	1.978	0.00292	0.00828	0.01202	0.110	0.978	0.00269	0.39920	0.44181	0.665
Male x Experience	–0.001	1.6755E–07	6.2557E–07	8.5567E–07	0.001	–0.002	1.9644E–07	1.3467E–06	1.6778E–06	0.001	–0.003	1.7804E–07	6.6510E–07	9.0965E–07	0.001
Male x Works in Ile-de-France	–0.203	0.00011	0.00001	0.00011	0.011	–0.220	0.00011	0.00005	0.00017	0.013	–0.183	0.00011	0.00017	0.00030	0.017
Male x Elementary School	0.327	0.00019	0.00065	0.00090	0.030	0.323	0.00021	0.00128	0.00162	0.040	0.011	0.00023	0.00434	0.00500	0.071
Male x Middle School	–0.236	0.00024	0.00163	0.00203	0.045	–0.202	0.00029	0.00170	0.00216	0.046	–0.101	0.00029	0.00284	0.00342	0.058
Male x High School	–0.269	0.00041	0.00201	0.00262	0.051	–0.251	0.00046	0.00199	0.00266	0.052	–0.201	0.00056	0.00249	0.00330	0.057
Male x Basic Vocational School	0.253	0.00016	0.00086	0.00111	0.033	0.280	0.00017	0.00197	0.00234	0.048	0.115	0.00017	0.00292	0.00339	0.058
Male x Advanced Vocational School	–0.074	0.00038	0.00112	0.00161	0.040	–0.011	0.00039	0.00157	0.00211	0.046	–0.114	0.00039	0.00177	0.00233	0.048
Male x Technical College or University	–0.146	0.00075	0.00107	0.00193	0.044	–0.122	0.00096	0.00255	0.00377	0.061	–0.083	0.00121	0.00288	0.00438	0.066
Male x Graduate School	–0.403	0.00085	0.00220	0.00328	0.057	–0.447	0.00084	0.00284	0.00396	0.063	–0.375	0.00104	0.00311	0.00446	0.067
Male x Engineer, Professional, or Manager	–0.565	0.00050	0.00036	0.00090	0.030	–0.706	0.00053	0.00026	0.00082	0.029	–0.974	0.00053	0.00990	0.01142	0.107
Male x Technical or Technical White Collar	0.362	0.00030	0.00004	0.00034	0.018	0.294	0.00032	0.00011	0.00044	0.021	0.237	0.00032	0.00312	0.00375	0.061
Male x Skilled Blue Collar	0.389	0.00020	0.00025	0.00047	0.022	0.370	0.00021	0.00038	0.00063	0.025	0.504	0.00022	0.00131	0.00166	0.041
Male x Unskilled Blue Collar	0.091	0.00020	0.00034	0.00057	0.024	0.099	0.00022	0.00036	0.00061	0.025	0.090	0.00022	0.00029	0.00054	0.023
Male x Log Real Annual Compensation	1.949	0.00006	0.00005	0.00012	0.011	2.129	0.00007	0.00017	0.00025	0.016	2.180	0.00006	0.00776	0.00860	0.093
Female x Experience	–0.012	2.0232E–07	3.2116E–07	5.5560E–07	0.001	–0.011	2.3378E–07	4.6906E–07	7.4974E–07	0.001	–0.009	2.2073E–07	4.4407E–07	7.0920E–07	0.001
Female x Works in Ile-de-France	–0.046	0.00012	0.00001	0.00013	0.011	–0.065	0.00013	0.00004	0.00018	0.013	0.063	0.00013	0.00008	0.00022	0.015
Female x Elementary School	0.368	0.00021	0.00065	0.00093	0.030	0.352	0.00027	0.00276	0.00331	0.058	–0.002	0.00026	0.00333	0.00392	0.063
Female x Middle School	–0.038	0.00024	0.00149	0.00189	0.043	–0.006	0.00029	0.00244	0.00298	0.055	–0.032	0.00024	0.00265	0.00315	0.056
Female x High School	–0.128	0.00042	0.00138	0.00194	0.044	–0.121	0.00047	0.00215	0.00283	0.053	–0.084	0.00046	0.00363	0.00445	0.067
Female x Basic Vocational School	0.239	0.00023	0.00058	0.00087	0.029	0.261	0.00028	0.00196	0.00243	0.049	–0.005	0.00027	0.00204	0.00252	0.050
Female x Advanced Vocational School	0.040	0.00040	0.00105	0.00156	0.039	0.085	0.00051	0.00409	0.00500	0.071	–0.122	0.00044	0.00359	0.00438	0.066
Female x Technical College or University	–0.173	0.00063	0.00223	0.00309	0.056	–0.090	0.00067	0.00496	0.00613	0.078	–0.082	0.00080	0.00600	0.00740	0.086
Female x Graduate School	–0.566	0.00151	0.00137	0.00302	0.055	–0.400	0.00167	0.00871	0.01125	0.106	–0.331	0.00164	0.00762	0.01002	0.100
Female x Engineer, Professional, or Manager	–0.758	0.00077	0.00019	0.00098	0.031	–0.882	0.00080	0.00074	0.00162	0.040	–0.905	0.00081	0.01085	0.01274	0.113
Female x Technical or Technical White Collar	0.182	0.00024	0.00022	0.00048	0.022	0.135	0.00025	0.00030	0.00059	0.024	0.086	0.00024	0.00053	0.00082	0.029

Table 9. Summary of Coefficient Estimates for the Analysis of a Model Predicting Full-Time Employment (Continued)

	Completed Data					Masked Data					Simulated Data				
	Average Coeffi-cient	Average Variance	Between-Implicate Variance	Total Variance	Standard Error of Average Coefficient	Average Coeffi-cient	Average Variance	Between-Implicate Variance	Total Variance	Standard Error of Average Coefficient	Average Coeffi-cient	Average Variance	Between-Implicate Variance	Total Variance	Standard Error of Average Coefficient
Female x Skilled Blue Collar	1.116	0.00034	0.00055	0.00094	0.031	1.112	0.00036	0.00058	0.00100	0.032	1.172	0.00036	0.00099	0.00145	0.038
Female x Unskilled Blue Collar	0.625	0.00014	0.00066	0.00087	0.030	0.643	0.00015	0.00073	0.00095	0.031	0.590	0.00015	0.00213	0.00250	0.050
Female x Log Real Annual Compensation	2.385	0.00010	0.00005	0.00015	0.012	2.504	0.00011	0.00026	0.00039	0.020	2.328	0.00009	0.02349	0.02592	0.161
Log Sales	–0.245	0.00002	0.00190	0.00210	0.046	–0.240	0.00002	0.00188	0.00209	0.046	0.082	0.00003	0.00051	0.00059	0.024
Log Capital Stock	0.338	0.00000	0.00004	0.00004	0.007	0.330	0.00000	0.00005	0.00006	0.008	0.182	0.00001	0.00043	0.00049	0.022
Log Average Employment	–0.285	0.00001	0.00287	0.00317	0.056	–0.282	0.00002	0.00294	0.00325	0.057	0.038	0.00004	0.00132	0.00149	0.039

Notes: The model was estimated with maximum likelihood logistic regression. The dependent variable is full-time employment status (not full-time is the reference category).
Sources: Authors' calculations are based on the INSEE DADS, EDP, and EAE data.

confidential and non-confidential variables, where the predictive distribution exploits the modeling techniques used to complete the data. Finally, our simulation technique preserves confidentiality by replacing the confidential data with a draw from the predictive distribution of those data, given only disclosable summary statistics.

We apply our techniques to longitudinal linked data from the French national statistical institute. We show that our masking and simulating techniques do an excellent job of preserving first and second moments for the individual and work history variables. The performance on the firm-level data is not as good, but this result is probably due to the way we specialized our methods to focus on the analysis of variables in the work history file. We believe that focusing our techniques on employer-level data, without insisting on links to the work history or individual data, would substantially improve their performance. The masked data did an excellent job of reproducing the important statistical features of analyses of the wage rate and full-time employment status variables in the work history file. The simulated data did not perform as well as the masked data but did provide many useful statistical results. The simulated data results were reliable enough to be combined with restricted access to the confidential completed data as a part of a full research program.

References

Abowd, J. M., and F. Kramarz (2000, June) 'Interindustry and Firm-Size Wage Differentials: A Comparison of the United States and France', Working Paper.

Abowd, J. M., F. Kramarz, and D. N. Margolis (1999, March) 'High Wage Workers and High Wage Firms', *Econometrica,* 67(2), pp.251-333.

Bethlehem, J., W. Keller, and J. Pannekoek (1990) 'Disclosure Control of Microdata', *Journal of the American Statistical Association,* 85, pp.38-45.

Boudreau, J-R. (1995, November) 'Assessment and Reduction of Disclosure Risk in Microdata Files Containing Discrete Data'. Presented at Statistics Canada Symposium 95.

Duncan, G. T., and S. Mukherjee (1998, June) 'Optimal Disclosure Limitation Strategy in Statistical Databases: Deterring Tracker Attacks Through Additive Noise'. Heinz School of Public Policy and Management, *Working Paper 1998-15.*

Evans, B. T., R. Moore, and L. Zayatz (1996) 'New Directions in Disclosure Limitation at the Census Bureau'. U.S. Census Bureau Research Report No. LVZ96/01.

Evans, T., L. Zayatz, and J. Slanta (1998) 'Using Noise for Disclosure Limitation of Establishment Tabular Data', *Journal of Official Statistics,* 14(4), pp.537-551.

Fienberg, S. E. (1994, December) 'A Radical Proposal for the Provision of Micro-Data Samples and the Preservation of Confidentiality'. Carnegie Mellon University Department of Statistics Technical Report No. 611.

Fienberg, S. E. (1997, September) 'Confidentiality and Disclosure Limitation Methodology: Challenges for National Statistics and Statistical Research', Presented at the

Committee on National Statistics 25th Anniversary Meeting. Carnegie Mellon Department of Statistics Technical Report, *Working Paper 668*.

Fienberg, S. E. and U. E. Makov (1998) 'Confidentiality, Uniqueness, and Disclosure Limitation for Categorical Data', *Journal of Official Statistics,* 14(4), pp.385-397.

Fienberg, S. E., U. E. Makov, and R. J. Steele (1998) 'Disclosure Limitation Using Perturbation and Related Methods for Categorical Data', *Journal of Official Statistics,* 14(4), pp.485-502.

Franconi, L. (1999, March) 'Level of Safety in Microdata: Comparisons Between Different Definitions of Disclosure Risk and Estimation Models', Presented at the Joint ECE/ Eurostat Work Session on Statistical Data Confidentiality, Thessaloniki, Greece, *Working Paper 4*.

Fries, G., B. W. Johnson, and R. L. Woodburn (1997, September) 'Analyzing Disclosure Review Procedures for the Survey of Consumer Finances', SCF Working Paper, presented at the 1997 Joint Statistical Meetings, Anaheim, CA.

Fuller, W. A. (1993) 'Masking Procedures for Microdata Disclosure Limitation', *Journal of Official Statistics,* 9(2) pp.383-406.

Greenberg, B. (1987) 'Rank Swapping for Masking Ordinal Microdata', U.S. Census Bureau, unpublished manuscript.

Hundepool, A. and L. Willenborg (1999, March) 'ARGUS: Software from the SDC Project'. Presented at the Joint ECE/Eurostat Work Session on Statistical Data Confidentiality, Thessaloniki, Greece, *Working Paper 7*.

Jabine, T. B. (1993) 'Statistical Disclosure Limitation Practices of the United States Statistical Agencies', *Journal of Official Statistics,* 9(2), pp.427-454.

Kennickell, A. B. (1991, October) 'Imputation of the 1989 Survey of Consumer Finances: Stochastic Relaxation and Multiple Imputation'. SCF Working Paper, prepared for the Annual Meeting of the American Statistical Association, Atlanta, GA, August 1991.

Kennickell, A. B. (1997, November) 'Multiple Imputation and Disclosure Protection: The Case of the 1995 Survey of Consumer Finances'. SCF Working Paper.

Kennickell, A. B. (1998, September) 'Multiple Imputation in the Survey of Consumer Finances'. SCF Working Paper, prepared for the August 1998 Joint Statistical Meetings, Dallas, TX.

Kennickell, A. B. (2000, May) 'Wealth Measurement in the Survey of Consumer Finances: Methodology and Directions for Future Research'. SCF Working Paper, prepared for the May 2000 annual meeting of the American Association for Public Opinion Research, Portland, OR.

Kim, J. J., and W. E. Winkler (1997) 'Masking Microdata Files', U.S. Census Bureau Research Report No. RR97/03.

Kooiman, P. (1998) 'Comment on Disclosure Limitation for Categorical Data', *Journal of Official Statistics,* 14(4) pp.503-508.

Lambert, D. (1993) 'Measures of Disclosure Risk and Harm', *Journal of Official Statistics,* 9(2), pp.313-331.

Little, R. J. (1993) 'Statistical Analysis of Masked Data'. *Journal of Official Statistics,* 9(2), pp.407-426.

Little, R. J. and D. B. Rubin (1987) *Statistical Analysis of Missing Data.* New York: Wiley.

Mayda, J., C. Mohl, and J. Tambay (1997) 'Variance Estimation and Confidentiality: They are Related!' Unpublished Manuscript, Statistics Canada.

Moore, R. A., Jr (1996a) 'Analysis of the Kim-Winkler Algorithm for Masking Microdata Files—How Much Masking is Necessary and Sufficient? Conjectures for the Development of a Controllable Algorithm'. U.S. Census Bureau Research Report No. RR96/05.

Moore, R. A., Jr (1996b) 'Controlled Data-Swapping Techniques for Masking Public Use Microdata Sets'. U.S. Census Bureau Research Report No. RR96/04.

Moore, R. A., Jr (1996c) 'Preliminary Recommendations for Disclosure Limitation for the 2000 Census: Improving the 1990 Confidentiality Edit Procedure'. U.S. Census Bureau Statstical Research Report Series, No. RR96/06.

Nadeau, C., E. Gagnon, and M. Latouche (1999) 'Disclosure Control Strategy for the Release of Microdata in the Canadian Survey of Labour and Income Dynamics'. Presented at the 1999 Joint Statistical Meetings, Baltimore, MD.

National Research Council (2000) *Improving Access to and Confidentiality of Research Data: Report of a Workshop.* Committee on National Statistics, Christopher Mackie and Norman Bradburn, Eds. Commission on Behavioral and Social Sciences and Education, National Academy Press, Washington, D.C.

Nordholt, E. S. (1999, March) 'Statistical Disclosure Control of the Statistics Netherlands Employment and Earnings Data', Presented at the Joint ECE/Eurostat Work Session on Statistical Data Confidentiality, Thessaloniki, Greece, *Working Paper 2*.

Pursey, S. (1999, March) 'Disclosure Control Methods in the Public Release of a Microdata File of Small Businesses'. Presented at the Joint ECE/Eurostat Work Session on Statisical Data Confidentiality, Thessaloniki, Greece, *Working Paper 5*.

Raghunathan, T. E., J. M. Lepkowski, J. Van Hoewyk, and P. Solenberger (1998) 'A Multivariate Technique for Multiply Imputing Missing Values Using a Sequence of Regression Models'. Survey Research Center, University of Michigan.

Rubin, D. B. (1976) 'Inference and Missing Data', *Biometrika,* 63(3), pp.581-592.

Rubin, D. B. (1978) 'Bayesian Inference for Causal Effects: The Role of Randomization', The Annals of Statistics, 6, pp.34-58.

Rubin, D. B. (1987) *Multiple Imputation for Nonresponse in Surveys,* New York: Wiley.

Rubin, D. B. (1993) 'Discussion of Statistical Disclosure Limitation', *Journal of Official Statistics,* 9(2), pp.461-468.

Schwarz, G. (1978) 'Estimating the Dimension of a Model', *Annals of Statistics,* 6, pp.461-464.

Skinner, C., and D. Holmes (1993) 'Modelling Population Uniqueness', *Proceedings of International Seminar on Statistical Confidentiality*, Luxembourg, pp.175-199. EUROSTAT.

Subcommittee on Disclosure Avoidance Techniques (1978) 'Statistical Policy Working Paper No. 2: Report on Statistical Disclosure and Disclosure Avoidance Techniques'. Federal Committee on Statistical Methodology, Office of Federal Policy and Standards, U.S. Department of Commerce, Washington, D.C.

Subcommittee on Disclosure Avoidance Techniques (1994) 'Statistical Policy Working Paper No. 22: Report on Statistical Disclosure Limitation Methodology', Federal Committee on Statistical Methodology, Statistical Policy Office, Office of Information and Regulatory Affairs, U.S. Office of Management and Budget, Washington, D.C.

Willenborg, L., and J. Kardaun (1999, March) 'Fingerprints in Microdata Sets', Presented at the Joint ECE/Eurostat Work Session on Statistical Data Confidentiality, Thessaloniki, Greece, *Working Paper 10*.

Winkler, W. E. (1997) 'Views on the Production and Use of Confidential Microdata'. U.S. Census Bureau Research Report No. RR97/01.

Winkler, W. E. (1998) 'Producing Public-Use Files That are Analytically Valid and Confidential', U.S. Census Bureau Research Report No. RR98/02.

Appendix A: Recent Research on Disclosure Limitation

In recent years, statistical agencies have seen an increasing demand for the data they collect, coupled with increasing concerns about confidentiality. This presents new challenges for statistical agencies, which must balance these concerns. Doing so requires techniques to allow dissemination of data that is analytically useful and preserves the confidentiality of respondents.

This appendix presents recent research on disclosure limitation methods and concepts. Our primary interest is in methods appropriate to longitudinal linked data, but this topic has not been well addressed in the literature. Hence, we review disclosure limitation methods and concepts appropriate to microdata in general. Because there are a number of reviews that provide a good summary of early research (*e.g.*, Subcommittee on Disclosure Avoidance Techniques 1978, Subcommittee on Disclosure Avoidance Techniques 1994, and Jabine 1993), we concentrate on recent research only.

The review is divided into four parts. The first presents general research on the disclosure limitation problem. The second presents research on measures of disclosure risk and harm. The third part discusses recent research into disclosure limitation methods for microdata, and the final part discusses the analysis of disclosure-proofed data.

General Research and Survey Articles

Evans, Moore, and Zayatz (1996)—This paper summarizes recent applications of a variety of disclosure limitation techniques to Census Bureau data, and outlines current research efforts to develop new techniques. The paper briefly discusses methods for microdata, (including the two-stage additive noise and data-swapping technique of Kim and Winkler (1997), the rank-based proximity swapping method of Moore (1996b), and the development of synthetic data based on log-linear models; methods under consideration for the 2000 Decennial Census; and methods for establishment tabular data. (See the more detailed discussion of Evans, Zayatz, and Slanta 1998 below.)

Fienberg (1997)—This paper presents an excellent review of the disclosure limitation problem and recent research to address it. The paper is organized in eight sections. The first section is introductory. In the second, the author defines notions of confidentiality and disclosure, presents the debate between limited access versus limited data, and describes the role of the intruder in defining notions of disclosure and methods of disclosure limitation. The third section presents two detailed examples to illustrate the issues, namely issues surrounding the Decennial Census and the Welfare Reform Act. The fourth section classifies various disclosure limitation methodologies that have been proposed and illustrates them in some detail. The fifth section considers notions of uniqueness in the sample and uniqueness in the population and their role in defining notions of disclosure risk (see the discussion of Fienberg and Makov 1998, Boudreau 1995, and Franconi 1999, below). The sixth section presents two integrated proposals for disclosure limitation: the ARGUS project (see the discussion of Hundepool and Willenborg 1999 and Nordholt 1999 below) and proposals for the release of simulated data (see Section A.3.2). The seventh section presents a brief discussion of issues pertaining to longitudinal data, and Section 8 concludes.

Winkler (1997)—This paper briefly reviews modern record-linkage techniques and describes their application in reidentification experiments. Such experiments can be used to determine the level of confidentiality protection afforded by disclosure limitation methods. The author stresses the power of such techniques to match records from disclosure-proofed data to other data sources. Emerging record-linkage techniques will allow reidentification in many existing public use files, even though these files were produced by conscientious individuals who believed they were using effective disclosure limitation tools.

National Research Council (2000)—This book describes the proceedings of a workshop convened by the Committee on National Statistics (CNSTAT) to identify ways of advancing the often conflicting goals of exploiting the research potential of microdata and preserving confidentiality. The emphasis of the workshop was on linked longitudinal data—particularly longitudinal data linked to administrative data. The workshop addressed four key issues: (1) the trade-off between increasing data access on the one hand, and improving data security on the other; (2) the ethical and legal requirements associated with data dissemination; (3) alternative approaches for limiting disclosure risks and facilitating data access—primarily the debate between restricting access and altering data; and (4) a review of current agency and organization practices. Some interesting observations in the report include the following:

- Researchers participating in the workshop indicated a preference for restricted access to unaltered data over broader access to altered data. However, these researchers also recognized the research costs associated with the former option.
- Although linking databases can generate new disclosure risks, it does not necessarily do so. In particular, many native databases are already sensitive, and hence require confidentiality protection. Linking may create a combined dataset that increases disclosure risk; however, a disclosure incident occurring from a linked data source is not necessarily caused by the linking. The breach of disclosure may have occurred in the native data as well.
- An appropriate measure of disclosure risk is a measure of the *marginal* risk. In other words, rather than comparing risks under various schemes with disclosure probability zero, one might consider the change in probability of disclosure as a result of a specific data release or linkage, or for adding or masking fields in a dataset—the marginal risk associated with an action.

- In defense of data alteration methods, Fienberg noted that all datasets are approximations of the real data for a group of individuals. Samples are rarely representative of the group about which a researcher is attempting to draw inferences; rather, they represent those for whom information is available. Even population data are imperfect, due to coding and keying errors, missing data, and the like. Hence, Fienberg finds the argument that perturbed data are not useful for intricate analysis not altogether compelling.

Measures of Disclosure Risk and Harm

Lambert (1993)—This paper considers various definitions of disclosure, disclosure risk, and disclosure harm. The author stresses that disclosure is in large part a matter of perception—specifically, what an intruder *believes* has been disclosed, even if it is false, is key. The result of a false disclosure may be just as harmful (if not worse) than the result of a true disclosure. Having distinguished between disclosure risk and disclosure harm, the author develops general measures of these.

The author defines two major types of disclosure. In an *identity disclosure* (or *identification*, or *reidentification*), a respondent is linked to a particular record in a released data file. Even if the intruder learns no sensitive information from the identification, it may nevertheless compromise the security of the data file and damage the reputation of the releasing agency. To distinguish between true identification and an intruder's beliefs about identification, the author defines *perceived identification*, which occurs when an intruder believes a record has been correctly identified, whether or not this is the case. An *attribute disclosure* occurs when an intruder believes new information has been learned about the respondent. This may occur with or without identification. The *risk of disclosure* is defined as the risk of identification of a released record, and the *harm from disclosure* depends on what is learned from the identification.

Suppose the agency holds N records in a data file $\mathbf{Z}$ and releases a random sample $\mathbf{X} = (\mathbf{x}_1, ..., \mathbf{x}_n)$ of n masked records on p variables. Lambert (1993) defines several measures of perceived disclosure risk. A 'pessimistic' risk of disclosure is given by

$$\begin{aligned} D(\mathbf{X}) &= \max_{1 \leq j \leq N} \max_{1 \leq i \leq n} \Pr[i^{th} \text{released record is } j^{th} \text{respondent's record} | \mathbf{X}] \\ &= \max_{1 \leq j \leq N} \max_{1 \leq i \leq n} \Pr[\mathbf{x}_i \text{ is } j^{th} \text{respondent's record} | \mathbf{X}]. \end{aligned} \tag{19}$$

Minimizing the measure in (19) protects against an intruder looking for the easiest record to identify. Alternate measures of disclosure risk can also be defined on the basis of (19), for example,

$$D_{average}(\mathbf{X}) = \frac{1}{N}\sum_{j=1}^{N} \max_{1 \le i \le n} \Pr[\mathbf{x}_i \text{ is } j^{th} \text{respondent's record} | \mathbf{X}] \quad (20)$$

$$D_{total}(\mathbf{X}) = ND_{average}(\mathbf{X}). \quad (21)$$

Equation (20) is a measure of data vulnerability based on the average risk of perceived disclosure, whereas (21) is a measure of the cumulative risk. An alternate measure of the total risk of perceived identification can be defined as the number of cases for which the risk of perceived disclosure exceeds a threshold τ:

$$D_{\tau}(\mathbf{X}) = \#\left\{j : \max_{1 \le i \le n} \Pr[\mathbf{x}_i \text{ is } j^{th} \text{respondent's record} | \mathbf{X}] \ge \tau\right\}.$$

The author proceeds to develop several detailed examples and provides a general measure of disclosure harm, which is not presented here.

Fienberg and Makov (1998)—This paper reviews several concepts, namely uniqueness in the sample, uniqueness in the population, and some notions of disclosure. The main contribution is a proposed approach for assessing disclosure potential as a result of sample uniqueness, based on log-linear models. A detailed description of this method follows.

Suppose a population is cross-classified by some set of categorical variables. If the cross-classification yields a cell with an entry of '1', then the individual associated with this entry is defined as a *population unique*. Population uniqueness poses a disclosure risk, because an intruder with matching data has the potential to match his or her records against those of the population unique. This creates the possibility of both reidentification and attribute disclosure.

A *sample unique* is defined similarly—an individual associated with a cell count of '1' in the cross-classification of the sample data. Population uniques are also sample uniques if they are selected into the sample, but being a sample unique does not necessarily imply being a population unique. The focus of the Fienberg and Makov (1998) approach for assessing disclosure potential is to use uniqueness in the sample to determine the probability of uniqueness in the population. Note that sample uniqueness is not necessarily required for such an endeavor—'small' cell counts may also pose a disclosure risk. For example, a count of '2' may allow an individual with almost unique characteristics to identify the only other individual in the sample with those characteristics. If the intruder did not also possess these characteristics, then a cell count of '2' could allow the individuals to be linked to the intruder's data with probability 1/2. The extension to larger yet still 'small' cell counts is obvious.

Let N denote the population size, n the size of the released sample, and K the maximum number of 'types' of individuals in the data, as defined by the cross-classifying variables (*i.e.*, the total number of cells). Let F_i and f_i, $i = 1, ..., K$, denote the counts in the cells of the multiway table summarizing the entire population and sample, respectively. Then a crucial measure of the vulnerability of the data is given by

$$\sum_{i=1}^{K} \Pr(F_i = 1 | f_i = 1) \quad (22)$$

Most prior attempts to estimate (22) assumed distributions for F_i and f_i (*e.g.*, Bethlehem, Keller, and Pannekoek 1990 and Skinner and Holmes 1993). The Fienberg and Makov (1998) approach differs by assuming the released sample is drawn from a population with cell probabilities $\{\pi_i^{(N)}\}$ that follow a log linear model (including terms such as main effects interactions), of the form

$$\log(\pi_i^{(N)}) = g_N(\boldsymbol{\theta}_i) \quad (23)$$

where θ_i are parameters. The authors propose to fit (23) to the observed counts $\{f_i\}$. Denote the estimated cell probabilities $\{\pi_i^{(n)}\}$. Ideally, one would like to develop analytical formulae for $\Pr(F_i = 1 | f_i = 1)$, but this will frequently be infeasible since many of the log-linear models that result from the estimation process will not have a closed-form representation in terms of the minimal marginal sufficient statistics. Instead, the authors propose the following simulation approach. First, use the records on the n individuals in the sample $(x_1, ..., x_n)$ to generate $(N - n) \times H$ records from $\{\pi_i^{(n)}\}$. This results in H populations of size N, each containing $(N - n)$ 'new' records obtained by some form of imputation (*e.g.* see Little and Rubin (1987)) or multiple imputations from some posterior distribution (*e.g.* see Rubin (1987)). Next, let $\bar{F}_i(j) = \bar{F}_i(x_1, ..., x_N, j)$ be the count in cell i of the jth imputed population. Similarly, let $\bar{f}_i = \bar{f}_i(x_1, ..., x_N)$ be the count in cell i of the released data (the sample). Clearly, $\bar{f}_i \neq 1 \Rightarrow F_i(j) \neq 1$. We can estimate (22) by

$$\sum_{i=1}^{K} \hat{\Pr}(F_i = 1 | f_i = 1) = \sum_{i=1}^{K} \sum_{j=1}^{H} \frac{\mathbf{1}[(\bar{F}_i(j) = 1) \cap (\bar{f}_i = 1)]}{H} \quad (24)$$

where the function $\mathbf{1}[A] = 1$ if A is true, and zero otherwise. Equation (24) can be used to assess the disclosure risk of the released data for a given release size n. Because (24) is likely to decrease as $(N - n)$ increases, the statistical agency is motivated to reduce n to the point that (24) indicates disclosure is infeasible. Note that if we remove the summation over i in (24), we can obtain a cell-specific measure of disclosure risk.

Fienberg and Makov (1998) do not address the sample error of the estimate in (24). They also do not address the inherent trade-off that an agency faces when choosing n based on (24) between reduced disclosure risk and increased uncertainty in the released data.

Boudreau (1995)—This paper presents another measure of disclosure risk based on the probability of population uniqueness given sample uniqueness. For the case of microdata containing discrete key variables, the author determines the exact relationship between unique elements in the sample and those in the population. The author also gives an unbiased estimator of the number of population uniques, based on sample data. Because this estimator exhibits great sampling variability for small sampling fractions, the author models this relationship. After observing this conditional probability for a number of real populations, the author provides a parametric formulation of it. This formulation is empirical only—it has no theoretical justification. However, the empirical formulation is much more flexible than earlier measures of disclosure risk based on uniqueness that required distributional assumptions (*e.g.*, the Poisson-Gamma model of Bethlehem, Keller, and Pannekoek 1990 or the Poisson-Lognormal model of Skinner and Holmes 1993).

Willenborg and Kardaun (1999)—This paper presents an alternative measure of disclosure risk appropriate to microdata sets for research (as opposed to public use files). In such files there is generally no requirement that all records be absolutely safe, because their use is usually covered by a contractual agreement that includes a non-record-matching obligation. The approach is to define a measure of the 'degree of uniqueness' of an observation, called a *fingerprint*. A fingerprint is a combination of values of identifying (key) variables that are unique in the dataset at hand and contain no proper subset with this property (so it is a minimum set with the uniqueness property). The authors contend that records with 'many' 'short' fingerprints (*i.e.*, fingerprints composed of a small number of variables) are 'risky' and should not be released. Appropriate definitions of 'many', 'short', and 'risky' are at the discretion of the data collection/dissemination agency. In this way, defining disclosure risk in terms of fingerprints is very flexible. The authors propose that agencies use the fingerprinting criterion to identify risky records and then apply disclosure-limitation measures to these records. The paper contains a discussion of some design criteria for an implementation of the fingerprinting criterion and stipulates some useful heuristics for algorithm design.

Franconi (1999)—This paper reviews recent developments in measures and definitions of disclosure risk. The author stresses differences between methods appropriate to social data and business data. These differences are due to differences in the underlying data. Namely, social data are generally from large populations, have an inherent dependent structure (*i.e.,* groups such as families or households exist in the data), and are characterized by key variables of a categorical nature. These characteristics allow one to tackle the disclosure limitation problem via concepts of uniqueness. Business data, on the other hand are generally from small populations with a skewed distribution and have key variables that are primarily continuous. Uniqueness concepts are generally not useful here, because nearly all cases would be considered unique. In both cases, the author stresses the need to take account of hierarchies in the data, such as the grouping of cases into families and households. These hierarchies provide additional information to an intruder attempting to identify records; hence, they should be incorporated into measures of disclosure risk.

Disclosure Limitation Methods for Microdata

Additive Noise Methods

Fuller (1993)—This paper considers a variety of masking methods in which error is added to data elements prior to release. These fall generally within the class of measurement error methods. The author stresses that to obtain consistent estimates of higher-order moments of the masked data and functions of these moments such as regression coefficients, measurement error methods and specialized software are required. Other techniques, such as data switching and imputation, can produce biased estimates of some sample covariances and other higher-order moments. The approach is related to that of Kim and Winkler (1997), but is also applicable to data that are not necessarily multivariate normal.

Kim and Winkler (1997)—This paper presents a two-stage disclosure limitation strategy applied to matched CPS-IRS data. The disclosure concern in this data arises from the match: the CPS data are already masked, but the IRS tax data are not. The IRS data need to be sufficiently well masked to prevent reidentification, either alone or in conjunction with unmasked key variables from the CPS. The procedure is as follows.

The data in question are known to be approximately multivariate normal. Hence, in the first stage, noise from a multivariate normal distribution with mean zero and the same correlation structure as the unmasked data is added to the IRS income variables. As discussed in Little (1993) and Fuller (1993), such an approach is currently the only method that preserves correlations. Following the addition of noise to the data, the authors determine the reidentification risk associated with the data by matching the raw linked data to the masked file. In cases where the reidentification risk is deemed too great, the authors randomly swap quantitative data within collapsed (age $\times$ race $\times$ sex) cells. This approach preserves means and correlations in the subdomains on which the swap was done and in unions of these subdomains. However, the swapping algorithm may severely distort means and correlations on arbitrary subdomains. Finally, the authors assess both the confidentiality protection offered by their method and the analytic usefulness of the resulting files, and conclude that both are good.

Moore (1996a)—This paper provides a critical examination of the degree of confidentiality protection and analytic usefulness provided by the Kim and Winkler (1997) method. The author concludes that the method is both useful and highly feasible. The author also considers some particular aspects of the algorithm, such as optimal parameter values that generate 'sufficient' masking with minimal distortion to second moments. Finally, the author considers how much masking is 'sufficient', given reasonable assumptions on intruder knowledge, tools, and objectives.

Winkler (1998)—This paper compares the effectiveness of a number of competing disclosure limitation methodologies to preserve both confidentiality and analytic usefulness. The methods considered include the additive noise and swapping techniques of Kim and Winkler (1997), the additive noise approach of Fuller (1993), and μ-ARGUS suppression as described in Hundepool and Willenborg (1999) and Nordholt (1999). The author arrives at several conclusions. First, the Fuller (1993) additive noise method may not provide as much protection as that author had originally suggested. In particular, sophisticated matching techniques may allow for a significantly higher reidentification rate than previously thought. Second, a naive application of μ-ARGUS to the linked CPS-IRS data described in Kim and Winkler (1997) did little to preserve either confidentiality or analytic usefulness. More sophisticated methods, including a variant of the Kim and Winkler (1997) method that included a μ-ARGUS pass on the masked data, were much more successful. The authors conclude that additive noise methods can produce masked files that allow some analyses to approximately reproduce the results obtained with unmasked data. When additional masking procedures are applied, such as limited swapping or probability adjustment (Fuller 1993), disclosure risk is significantly reduced, though analytic properties are somewhat compromised.

Duncan and Mukherjee (1998)—This paper derives an optimal disclosure limitation strategy for statistical databases—in other words, microdatabases that respond to queries with aggregate statistics. As in all disclosure limitation problems, the aim is to maximize legitimate data access while keeping disclosure risk below an acceptable level. The particular confidentiality breach considered is called a *tracker attack*: a well-known intruder method in databases with query set size (QSS) control. QSS control is a query restriction technique in which a query is disallowed if the number of records satisfying the query is too small (or too large, by inference from the complementary query). A tracker attack is a finite sequence of legitimate queries that yields the same information as a query precluded under QSS control. The authors show that the optimal method for thwarting tracker attacks is a combination of query restriction and data masking based on additive noise. The authors also derive conditions under which autocorrelated noise is preferable to independent noise or 'permanent' data perturbation.

Evans, Zayatz, and Slanta (1998)—This paper presents an additive noise method for disclosure limitation that is appropriate to establishment tabular data. The authors propose adding noise to the underlying microdata prior to tabulation. Under their approach, 'more sensitive' cells receive more noise than less sensitive cells. There is no attempt to preserve marginal totals. This proposal has numerous advantages over the cell-suppression approach that is usually applied to such data. In particular, it is far simpler and less time-consuming than cell-suppression techniques. It also eliminates the need to coordinate cell suppressions between tables and eliminates the need for secondary suppressions, which can seriously reduce the amount of information in tabular releases. The authors also contend that an additive noise approach may offer more protection than cell suppression, although suppression may give the appearance of offering more protection.

Pursey (1999)—This paper discusses the disclosure control methods developed and implemented by Statistics Canada to release a public use microdata file (PUMF) of financial data from small businesses. This is a fairly unique enterprise—in most cases, statistical agencies deem it too difficult to release public use microdata on businesses that preserve confidentiality. The paper discusses the five steps taken to create the PUMF: (1) make assumptions about an intruder's motivation, information, and tools; (2) make disclosure control goals based on these assumptions; (3) translate these goals into mathematical rules; (4) implement these rules to create the PUMF; and (5) measure the data quality of the PUMF. These are discussed briefly below.

It is assumed that an intruder seeks to identify any record in the PUMF and has access to the population data file from which the PUMF records are drawn. It is assumed that identification is achieved via nearest-neighbor matching to the population file. Given these assumptions, the following disclosure control goals were set:

- Ensure a low probability that a business from the population appears in the PUMF (less than $r\%$) and that an intruder cannot determine that a particular business is in the PUMF
- Ensure that each continuous variable is perturbed and that an intruder cannot undo the perturbation
- Ensure a low probability that a PUMF record can be correctly linked to itself in the population file (less than $p\%$) and that an intruder cannot determine whether a link has been correctly or incorrectly made
- Remove unique records

Continuous variables were perturbed according to methods similar to those of Kim and Winkler (1997). First, independent random noise was added to each datum, subject to the constraints that the minimum and maximum proportion of random noise is constant for each datum and that within a record the perturbations are either always positive or always negative. Next, the three highest data values of each variable in each cell were replaced with their average. Finally, all data values were rounded to the nearest $1,000. Because a less than $p\%$ linkage rate was deemed necessary, in industry cells with a correct linkage rate greater than $p\%$, the data was further perturbed by data swapping with the second-nearest neighbor until a $p\%$ linkage rate was achieved.

After implementing the above disclosure control methods, the resulting data quality was analyzed. The general measure used was one of relative distance: $Rd = (x_a - x_b)/(x_a + x_b)$, where x_a is data or a sample statistic after disclosure control and x_b is the same data or sample statistic before disclosure control. All variables in the PUMF and a variety of sample statistics were analyzed according to this distance measure. The results indicated that the resulting data quality was good to fair for unincorporated businesses, fair to poor for incorporated businesses.

Multiple Imputation and Related Methods

Rubin (1993)—Rubin (1993) was the first author to suggest the use of multiple-imputation techniques for disclosure limitation for microdata analyses. His radical suggestion—to release only synthetic data generated from actual data by multiple imputation—is motivated by the forces outlined at the outset of this review—namely, an increase in the demand for public use microdata, and increasing concern about the confidentiality of such data.

Rubin's (1993) approach has a number of advantages over competing proposals for disclosure limitation, such as microdata masking. For example, valid statistical analyses of masked microdata generally require 'not only knowledge of which masking techniques were used, but also special-purpose statistical software tuned to those masking techniques' (Rubin 1993, p. 461). In contrast, analysis of multiply-imputed synthetic data can be validly undertaken using standard statistical software simply by using repeated applications of complete-data methods. Furthermore, an estimate of the degree to which the disclosure proofing techniques influence estimated model parameters can be obtained from between-imputation variability. Finally, since the released data are synthetic, (*i.e.*, contain no data on actual units), they pose no disclosure risk.

The details of Rubin's (1993) proposal are as follows. Consider an actual microdata sample of size n drawn using design D from a much larger population of N units. Let X represent background variables (observed, in principle, for all N units), Z represent outcome variables with no confidentiality concerns, and Y represent outcome variables with some confidentiality concerns. Note that Z and Y are only observed for the n sampled units and are missing for the $N - n$ unsampled units. A multiply-imputed population consists of the actual X data for all N units, the actual $[\,Z\;\;Y\,]$ data for the n units in the sample, and M matrices of $[\,Z\;\;Y\,]$ data for the $N - n$ unsampled units, where M is the number of multiple imputations. The multiply-imputed values of $[\,Z\;\;Y\,]$ are obtained from some model with predictors X. Given such a multiply-imputed population and a new survey design D^* for the microdata to be released (possibly the same as D), the statistical agency can draw a sample of $n^* \ll N$ units from the multiply-imputed population, which is structurally like an actual microdata sample of size n^* drawn from the actual population using design

D^*. This can be done M times to create M replicates of the $[\,Z \;\; Y\,]$ values. To ensure that no actual data are released, the statistical agency could draw the samples from the multiply-imputed population excluding the n actual units.

Rubin (1993) recognizes the information loss inherent in the multiple-imputation technique. However, some aspects of this information loss are subtle, and he presents these as the following two facts. First, although the actual $[\,Z \;\; Y\,]$ and the population values of X contain more information than the multiply-imputed population, if the imputation model is correct, then as M increases, the information in the latter is essentially the same as in the former. Second, the information in the original microdata sample of size n may be greater than, less than, or equal to the information in the multiply-imputed sample of size n^*; the relationship will depend on the estimand under investigation, the relative sizes of n and n^*, the magnitude of M, the designs D and D^*, and the ability of X to predict $[Z \;\; Y]$.

Fienberg (1994)—Fienberg (1994) proposes a method of confidentiality protection in the spirit of Rubin (1993). Whereas Rubin (1993) suggests generating synthetic microdata sets by multiple imputation, Fienberg (1994) suggests generating synthetic microdata by bootstrap methods. This method retains many of the desirable properties of Rubin's (1993) proposal—namely, disclosure risk is reduced because only synthetic data are released, and the resultant microdata can be analyzed using standard statistical methods.

To discuss the details of his proposal, let us restate the statistical agency's problem. As before, suppose the agency collects data on a random sample of size n from a population of size N (ignore aspects of the sample design). Let F be the true p-dimensional c.d.f. of the data in the population, and let $\bar{F}$ be the empirical c.d.f. based on the sample of size n. The disclosure problem arises because researchers request, in essence, access to the full, empirical p-dimensional c.d.f., $\bar{F}$. Because of guarantees of confidentiality, the agency believes it cannot release $\bar{F}$, as an intruder may be able to identify one or more individuals in the data.

Fienberg's (1994) proposal is as follows. Suppose the statistical agency has a 'smoothed' estimate of the c.d.f., $\hat{F}$, derived from the original sample c.d.f. $\bar{F}$. Rather than releasing either $\bar{F}$ or $\hat{F}$, the agency could sample from $\hat{F}$ and generate a synthetic bootstrap-like sample of size n. Denote the empirical c.d.f. of the synthetic microdata file as $\bar{G}$. Fienberg (1994) notes some technical details surrounding $\bar{G}$ that have yet to be addressed. Namely, under what conditions would replicates of $\bar{G}$, say $\bar{G}_i$ for $i = 1, \ldots, B$, be such that as $B \to \infty, \frac{1}{B}\sum_{i=1}^{B} \bar{G}_i \to \hat{F}$? Is a single replicate sufficient, or would multiple replicates be required for valid analyses, or possibly the average of multiple replicates? Bootstrap theory may provide some insight into these issues.

Fienberg, Makov, and Steele (1998)—The authors reiterate Fienberg's (1994) proposal for generating synthetic data via bootstrap methods and present a related application to the case of categorical data. Categorical data can be represented by a contingency table, for which there is a direct relationship between a specific hierarchical log-linear model and a set of marginal tables that represent the minimal sufficient statistics of the model. The authors present an example of a three-way table, for which they obtain maximum likelihood estimates of the expected cell values under a log-linear model. The suggestion is to release the MLEs, rather than the actual data, as a public use product. They then generate 1 million tables with the same two-way margins and perform a goodness-of-fit test based on the MLEs. They find that the sparseness of the table in their example presents some problems for accurate log-linear modeling.

In a comment to this article, Kooiman (1998) expresses doubt as to the feasibility of the Fienberg (1994) and Fienberg, Makov, and Steele (1998) proposal for generating synthetic data. He makes a connection between the proposed method and a data-swapping exercise subject to fixed margins. Kooiman (1998) shows that for large datasets with many categorical variables and many categories, such an exercise is likely impossible. He also finds the relationship between the synthetic data proposal and the categorical data example tenuous, at best.

Kennickell (1991, 1997, 1998, 2000)—In a series of articles, Kennickell (1991, 1997, 1998, and 2000) describes the Federal Reserve Imputation Technique Zeta (FRITZ), used for both missing value imputation and disclosure limitation in the Survey of Consumer Finances (SCF). The SCF is a triennial survey administered by the Federal Reserve Board to collect detailed information on all household assets and liabilities. Because holdings of many types of assets are highly concentrated in a relatively small fraction of the population, the SCF heavily oversamples wealthy households. Because such households are likely to be well-known, at least in their localities, the data collection process presents a considerable disclosure risk. As a first step toward implementing the proposal of Rubin (1993), the SCF simulates data for a subset of sample cases, using the FRITZ multiple imputation algorithm. This approach is highly relevant for our current research, and hence we discuss it in some detail here.

Using the FRITZ Algorithm for Missing Data Imputation—As mentioned above, the FRITZ algorithm is used both for missing value imputation and for disclosure limitation in the SCF. The algorithm is most easily understood in the context of missing data imputation. We return to the issue of its application to disclosure limitation below.

The FRITZ model is sequential in the sense that it follows a predetermined path through the survey variables, imputing missing values one (occasionally two) at a time. The model is also iterative in that it proceeds by filling in all missing values in the survey dataset, using that information as a basis for imputing the following round, and continuing the process until key estimates are stable. Five imputations are made for every missing value; hence, the method is in the spirit of Rubin's

(1993) proposal. The following describes the FRITZ technique for imputing missing continuous variables.

For convenience, suppose the iterative process has completed $l-1$ rounds, and we are currently somewhere in round l, with a data structure as given below: (reproduced from Kennickell 1998)

$$
\underset{\text{Iteration } l-1}{\begin{bmatrix}
y_1 & X_{11}^{l-1} & x_{12} & x_{13} \\
\Psi_2^{l-1} & X_{21}^{l-1} & x_{22} & X_{23}^{l-1} \\
y_3 & x_{31} & x_{32} & x_{33} \\
\dots & & & \\
y_{n-2} & x_{n-2,1} & x_{n-2,2} & x_{n-2,3} \\
y_{n-1} & x_{n-1,1} & X_{n-1,2}^{l-1} & x_{n-1,3} \\
\Psi_n^{l-1} & X_{n1}^{l-1} & X_{n2}^{l-1} & x_{n3}
\end{bmatrix}}
\underset{\text{Iteration } l}{\begin{bmatrix}
y_1 & \cdot & x_{12} & x_{13} \\
\cdot & r_{21} & x_{22} & \cdot \\
y_3 & x_{31} & x_{32} & x_{33} \\
\dots & & & \\
r_{n-2} & x_{n-2,1} & x_{n-2,2} & x_{n-2,3} \\
y_{n-1} & x_{n-1,1} & X_{n-1,2}^{l} & x_{n-1,3} \\
\cdot & r_{n1} & X_{n2}^{l} & x_{n3}
\end{bmatrix}}
$$

Here, y indicates complete (non-missing) reports for the variable currently the subject of imputation; ψ^p represents round p imputations of missing values of y; x represents complete reports of the set of variables available to condition the imputation; X^p represents completed imputations of x from iteration p. Variables that were originally reported as a range but are not currently imputed are represented by r, and $\cdot$ represents values that are completely missing and not yet imputed. Every x variable becomes a y variable at its place in the sequence of imputations within each iteration. Note that no missing values remain in the stylized $l-1$ dataset.

Ideally, one would like to condition every imputation on as many variables as possible, as well as on interactions and higher powers of those terms. Of course there are always practical limits to such a strategy due to degrees of freedom constraints, and some judgment must be applied in selecting a 'maximal' set of conditioning variables, X. Of that maximal set, not every element may be non-missing at a given stage of imputation. For each variable to be imputed, the FRITZ algorithm determines the set of non-missing variables among the maximal set of conditioning variables for each observation, denoted $X_{(i)}$ for observation i. Given the set of available conditioning variables $X_{(i)}$, the model essentially regresses the target imputation variable on the subset of conditioning variables using values *from the previous iteration of the model.* This process is made more efficient by estimating a maximal normalized cross-product matrix for each variable to be imputed, denoted $\sum(X,Y)_{l-1}$, and then subsetting the rows and columns corresponding to the non-missing conditioning variables for a given observation, denoted $\sum(X_{(i)},Y)_{l-1}$. The imputation for observation i in iteration l is thus given by

$$\Psi_{il} = \beta_{(i)l} X_{(i)il} + e_{il} \tag{25}$$

where $X_{(i)il}$ is the rows of $X_{(i)l}$ corresponding to i; $X_{(i)l}$ is the subset of X that is available for i in iteration l; $\beta_{(i)l} = \sum(X_{(i)}X_{(i)})^{-1}_{l-1} \sum(X_{(i)}Y)_{l-1}$, and e_{il} is a random error term. Once a value is imputed, its imputed value is used (along with reported values) in conditioning later imputations.

The choice of error term e_{il} has been the subject of several experiments (see Kennickell 1998). In early releases of the SCF, e_{il} was taken to be a draw from a truncated normal distribution. The draw was restricted to the central 95 percent of the distribution, with occasional supplementary constraints imposed by the structure of the data or respondent-provided ranges for the variable under imputation. More recently, e_{il} has been drawn from an empirical distribution.

The FRITZ algorithm for imputing multinomial and binary variables works similarly, with an appropriate 'regression' substituted for (25).

Using The FRITZ Algorithm for Disclosure Limitation—The FRITZ algorithm is applied to the confidentiality protection problem in a straightforward manner. In the 1995 SCF, all dollar values for selected cases were simulated. The procedure is as follows. First, a set of cases that present excessive disclosure risk are selected (see Kennickell 1997). These are selected on the basis of having unusual levels of wealth or income, given other characteristics or other unusual combinations of responses. Second, a random set of cases is selected to reduce the ability of an intruder to determine even the set of cases determined to present an excessive disclosure risk. Then, a new dataset is created for all the selected cases, and shadow variables (which detail the 'type' of response given for a particular case-variable pair, *e.g.*, a complete report, a range report, or non-response) are set so that the FRITZ model interprets the responses as range responses. The type of range mimics one where the respondent volunteered a dollar range—a dollar amount of $\pm p$ percent (where p is an undisclosed number between 10 and 20 percent) is stored in a dataset normally used to contain unique range reports. Finally, the actual dollar values are set to missing, and the FRITZ algorithm is applied to the selected cases, using the simulated range reports to constrain the imputed values. Subsequent evaluation of the 1995 SCF (Fries, Johnson, and Woodburn 1997) indicates that while the imputations substantially masked individual cases, the effect on important distributional characteristics was minimal.

Other Methods

Moore (1996b)—This paper presents a brief overview of data-swapping techniques for disclosure limitation and presents a more sophisticated technique than found elsewhere in the literature. The author presents an algorithm for a controlled data swap based on the rank-based proximity swap of Greenberg (1987). The contribution of this paper is that it provides a technique that preserves univariate and bivariate relationships in the data. Based on a simulation using the 1993 Annual Housing Survey Public Use File, the author concludes that the algorithm preserves the desired moments to an acceptable degree (and hence retains some degree of analytic usefulness), while providing a level of confidentiality protection comparable to simple additive noise methods.

Moore (1996c)—This paper suggests modifications to the Confidentiality Edit, the data-swapping procedure used for disclosure limitation in the 1990 Decennial Census. The suggested improvements are based on the ARGUS system for determining high-risk cases (see Hundepool and Willenborg 1999 and Nordholt 1999 below), and the German SAFE system for perturbing data. The author also presents two measures of the degree of distortion induced by the swap and an algorithm to minimize this distortion.

Mayda, Mohl, and Tambay (1997)—This paper examines the relationship between variance estimation and confidentiality protection in surveys with complex designs. In particular, the authors consider the case of the Canadian National Population Health Survey (NPHS), a longitudinal survey with a multistage clustered design. To prepare a public use file, it was deemed necessary to remove specific design information such as stratum and cluster identifiers due to the extremely detailed level of geography they represented. Furthermore, providing cluster information could allow users to reconstitute households, increasing the probability of identifying individuals. However, specific design information is necessary to correctly compute variances using jackknife or other methods. This highlights yet another aspect of the conflict between providing high-quality data and protecting confidentiality. The authors describe the approach taken to resolve this conflict. Specifically, strata and clusters are collapsed to form 'super-strata' and 'superclusters' in the public use file, which protect confidentiality while providing enough information for researchers to obtain unbiased variance estimates under certain conditions. The drawback of this approach is that it does not generate the exact variance corresponding to the original design, and that collapsing reduces degrees of freedom and hence the precision of variance estimates.

Nadeau, Gagnon, and Latouche (1999)—This paper presents a discussion of confidentiality issues surrounding Statistics Canada's Survey of Labour and Income Dynamics (SLID) and presents the release strategy for microdata on individual and family income. SLID is a longitudinal survey designed to support studies of economic well-being of individuals and families. With the demise of the Canadian Survey of Consumer Finances (SCF) in 1998, SLID became the official source of information for both longitudinal *and* cross-sectional income data on individuals and families. This presented some rather unique issues for disclosure limitation.

Prior to integrating SLID and SCF, Statistics Canada did not release sufficient information in the SLID PUMFs to allow household reconstitution. It was considered too difficult to protect confidentiality at the household level in a longitudinal microdata file. However, since integrating SLID and SCF, it has become a priority to release cross-sectional PUMFs that meet the needs of former SCF users. In particular, the cross-sectional PUMFs now contain household and family identifiers, which allow household and family reconstitution. This compromises the release of longitudinal PUMFs. Instead, Statistics Canada has opted to explore other options for the release of longitudinal data—namely, release of synthetic files and creation of research data centers. In the meantime, a number of disclosure limitation methods have been explored for the cross-sectional PUMFs to limit the ability of intruders to link records dynamically (constructing their own longitudinal file, considered 'too risky' for reidentification) and/or reidentify records by linking to the Income Tax Data File (ITDF).

The disclosure control methods applied in the cross-sectional PUMFs include both data reduction and data modification methods. The data reduction methods include dropping direct identifiers, aggregating geographic variables, and categorical grouping for some occupational variables. Data modification methods are applied to numeric variables. In particular, year of birth is perturbed with additive noise, income variables are both bottom- and top-coded, and the remaining values are perturbed with a combined random-rounding and additive noise method.

Finally, the authors assess how successful these measures are at protecting confidentiality and maintaining analytical usefulness. To address the former, they consider linking consecutive cross-sectional PUMFs and linking to the ITDF. In both cases, they consider both direct matches and nearest-neighbor matches. They find that the ability of an intruder to match records in either consecutive PUMFs or to the ITDF is severely limited by the disclosure control measures. As for the usefulness of the data, they find little difference in the marginal distribution of most variables at highly aggregated levels (*i.e.*, the national level) but more significant differences at lower levels of aggregation (*i.e.*, the province × sex level).

Hundepool and Willenborg (1999), Nordholt (1999)—These papers describe the τ-ARGUS and μ-ARGUS software packages developed by Statistics Netherlands for disclosure limitation. Nordholt (1999) describes their specific application to the Annual Survey on Employment and Earnings (ASEE). The τ-ARGUS software tackles the problem of disclosure limitation in tabular data. It automatically applies a series of primary and secondary suppressions to tabular data on the basis of a dominance rule: A cell is considered unsafe if the n major contributors to that cell are responsible for at least p percent of the total cell value. The μ-ARGUS software is used to create a public use microdata file from the ASEE. The public use microdata have to satisfy two criteria, which are implemented with μ-ARGUS: First, every category of an identifying variable must occur 'frequently enough' (200,000 times is the default for ASEE); second, every bivariate combination of values must occur 'frequently enough' (1,000 times is the default for ASEE). These objectives are achieved via global recoding and local suppression.

Analysis of Disclosure-Proofed Data

Little (1993)—This paper develops a model-based likelihood theory for the analysis of masked data. His approach is to formally model the mechanism whereby case-variable pairs are selected for masking, to describe the masking method, and to derive an appropriate model for analysis of the resulting data. His method is sufficiently general to allow for a variety of masking selection mechanisms and such diverse masking methods as deletion, coarsening, imputation, and aggregation. The formal theory follows.

Let $\mathbf{X} = \{x_{ij}\}$ denote an $(n \times p)$ unmasked data matrix of n observations on p variables. Let $\mathbf{M} = \{m_{ij}\}$ denote the masking indicator matrix, where $m_{ij} = 1$ if x_{ij} is masked and $m_{ij} = 0$ otherwise. Let $\mathbf{Z} = \{z_{ij}\}$ denote the masked data, in other words, z_{ij} is the masked value of x_{ij} if $m_{ij} = 1$, and $z_{ij} = x_{ij}$ if $m_{ij} = 0$. Model the joint distribution of $\mathbf{X}$, $\mathbf{Z}$, and $\mathbf{M}$ with the density function:

$$f(\mathbf{X},\mathbf{Z},\mathbf{M}|\boldsymbol{\theta}) = f_X(\mathbf{X}|\boldsymbol{\theta})\, f_Z(\mathbf{Z}|\mathbf{X})\, f_M(\mathbf{M}|\mathbf{X},\mathbf{Z})\,. \tag{26}$$

Here $f_X(\mathbf{X}|\theta)$ is the density of the unmasked data given unknown parameters θ, which would be the basis for analysis in the absence of masking; $f_Z(\mathbf{Z}|\mathbf{X})$ formalizes the masking treatment; and $f_M(\mathbf{M}|\mathbf{X}, \mathbf{Z})$ formalizes the masking selection mechanism. If the analyst knows which values are masked and the masking method, then the analyst knows $\mathbf{M}$, as well as the distributions of $\mathbf{M}$ and $\mathbf{Z}$. If not, then $\mathbf{M}$ is unknown. A more general specification would also index the distributions of $\mathbf{M}$ and/or $\mathbf{Z}$ by unknown parameters, and a full likelihood analysis would then involve both θ and these unknown masking parameters.

Let $\mathbf{X} = (\mathbf{X}_{obs}, \mathbf{X}_{mis})$ and $\mathbf{Z} = (\mathbf{Z}_{obs}, \mathbf{Z}_{mis})$ where *obs* denotes observed components and *mis* denotes missing components of each matrix. Analysis of the masked data is based on the likelihood for θ given the data $\mathbf{M}$, $\mathbf{X}_{obs}$, and $\mathbf{Z}_{obs}$. This is obtained formally by integrating the joint density in (26) over the missing values $\mathbf{X}_{mis}$ and $\mathbf{Z}_{mis}$:

$$L(\boldsymbol{\theta}|\mathbf{M},\mathbf{X}_{obs},\mathbf{Z}_{obs}) = \int f_X(\mathbf{X}|\boldsymbol{\theta})\, f_Z(\mathbf{Z}|\mathbf{X})\, f_M(\mathbf{M}|\mathbf{X},\mathbf{Z})\, d\mathbf{X}_{mis}\, d\mathbf{Z}_{mis}\,. \tag{27}$$

Because the distribution of $\mathbf{M}$ in (27) may depend on $\mathbf{X}$ and $\mathbf{Z}_{obs}$, but should not depend on $\mathbf{Z}_{mis}$, we can write $f_M(\mathbf{M}|\mathbf{X},\mathbf{Z}) = f_M(\mathbf{M}|\mathbf{X},\mathbf{Z}_{obs})$. Thus we can rewrite (27) as

$$L(\boldsymbol{\theta}|\mathbf{M},\mathbf{X}_{obs},\mathbf{Z}_{obs}) = \int f_X(\mathbf{X}|\boldsymbol{\theta}) f_Z^*(\mathbf{Z}_{obs}|\mathbf{X})\, f_M(\mathbf{M}|\mathbf{X},\mathbf{Z}_{obs})\, d\mathbf{X}_{mis} \tag{28}$$

where $f_Z^*(\mathbf{Z}_{obs}|\mathbf{X}) = \int f_Z(\mathbf{Z}|\mathbf{X})\, d\mathbf{Z}_{mis}$.

The author notes that the likelihood in (28) can be simplified if the masking selection and treatment mechanisms satisfy certain ignorability conditions, in the sense of Rubin (1976) and (1978). Specifically, if the masking selection mechanism is ignorable, then $f_M(\mathbf{M}|\mathbf{X},\mathbf{Z}) = f_M(\mathbf{M}|\mathbf{X}_{obs},\mathbf{Z}_{obs})$ for all $\mathbf{X}_{mis}$, $\mathbf{Z}_{mis}$. In this case, the density of $\mathbf{M}$ can be omitted from (28). Similarly, the masking treatment mechanism is ignorable if $f^*_Z(\mathbf{Z}_{obs}|\mathbf{X}) = f^*_Z(\mathbf{Z}_{obs}|\mathbf{X}_{obs})$ for all $\mathbf{X}_{mis}$. In this case, the density of $\mathbf{Z}_{obs}$ can be omitted from (28). Finally, if both mechanisms are ignorable, then the likelihood reduces to

$$L(\boldsymbol{\theta}|\mathbf{M},\mathbf{X}_{obs},\mathbf{Z}_{obs}) = \int f_X(\mathbf{X}|\boldsymbol{\theta})d\mathbf{X}_{mis},$$

which is proportional to the marginal density of $\mathbf{X}_{obs}$.

Confidentiality, Disclosure, and Data Access: Theory and Practical Application for Statistical Agencies
Pat Doyle, Julia I. Lane, Jules J.M. Theeuwes and Laura M. Zayatz (Eds)

Chapter 11
Licensing*

Marilyn M. Seastrom
National Center for Education Statistics

1. Introduction

This chapter discusses the definition and rationale for data licensing or data use agreements to provide external researchers access to individually identifiable confidential data. The discussion is accompanied by a list of the federal agencies that are using these mechanisms. Section 2 provides a detailed description of the licensing system at one agency, the National Center for Education Statistics (NCES). Section 3 takes a broader look at how a number of federal agencies use data agreements. It identifies the laws and federal regulations that provide a basis for the data agreements and then discusses the similarities and differences in the implementation and enforcement of the provisions in the agreements. Section 4 looks at how well the systems are working, and Section 5 makes some recommendations for improvement.

A number of federal studies require the collection of individually identifiable data to fulfill the mandates of the sponsoring agencies. While federal agencies typically report out the basic descriptive data, there are usually additional research questions that could be addressed with the respondent-level data. To share these individually identifiable confidential data with qualified nongovernmental researchers, some agencies use data licenses or data use agreements that extend the legal responsibilities for handling and using confidential data to external data users.

What Is a Data License?

A data license or a data use agreement allows external researchers access to individually identifiable confidential data covered under federal statutes and regulations, by extending the legal responsibilities for handling and using confidential

* In addition to my predecessors who conceived and started the NCES licensing system, I would like to thank all of the anonymous authors of the information found on federal websites. Much of the information included in this chapter from agencies other than NCES would not otherwise have been accessible.

data to the external researchers through a data loan program. In particular, by executing the license agreement and supporting materials, the researcher agrees to assume the responsibilities and penalties for violations that apply to agency staff. Further, the researcher sometimes agrees to participate in unannounced security inspections. Licenses are granted under the authority of the agency head to permit qualified requestors to examine respondent records for statistical or research purposes consistent with the purpose for the data collection.

Why Are Data Licenses Needed?

Federal agencies are required by law to develop and enforce standards designed to protect the confidentiality of individually identifiable respondents. Yet many federal agencies are also required by law to collect and disseminate data on specific topics. The first response to this tension is to develop public use data files. The public use files are derived from restricted use data files that include individually identifiable confidential data. In the public use data files, the restricted use data are altered by deleting individual identifiers and other sensitive data items, by aggregating data, or by altering the original data in some way. While these alterations do not affect the analyses of most data users, there are times when a specific analysis requires some of the information that has been altered or deleted. A data license or a data use agreement allows qualified researchers access to the unaltered restricted use data files (McMillen 2000).

Which Agencies Use Data Licenses?

The National Center for Education Statistics pioneered the data license approach. In 1989, in consultation with the chief statistician at the Office of Management and Budget (OMB), NCES began work developing a protocol for the licensing system. NCES staff collaborated with lawyers in the Department of Education's Office of the General Counsel in the preparation of legal documents needed to implement the licensing system. On February 6, 1991, NCES issued its first restricted use data license.

A number of other agencies are now using similar approaches that can be described generically as data use agreements. The exact name of the agreement and some of the particulars of the agreement vary by agency. For example, the Division of Science Resources Studies of the National Science Foundation follows the same procedures used by NCES. The Social Security Administration, the National Cancer Institute, and the Health Care Financing Administration all have 'Data Use Agreements'. Several agencies within the Department of Justice use 'Data Transfer Agreements'—Bureau of Justice Statistics, Bureau of Justice Assistance, Office of Juvenile Justice and Delinquency Prevention, National Institute of Justice, and Office of Justice Programs. The Bureau of Labor Statistics enters into 'Letters of Agreement', and the National Institute for Child Health and Human Development has 'Agreements for the Use of Sensitive Data'.

Several of the Institutes at the National Institutes of Health provide access to both data records and biological materials. The National Heart, Lung, and Blood Institute uses 'Data and Materials Distribution Agreements', the National Institute on Drug Abuse uses 'Access Requests', and the National Institute of Mental Health and the National Institute on Alcohol Abuse and Alcoholism use 'Distribution Agreements'.

Table 1. Names of Data Use Agreements and Relevant Datasets, by Agency

	Type of Agreement	Datasets
National Center for Education Statistics	License	11-NELS:88, SASS, TFS, NHES, NPSAS, RCG, NSOPF, B&B, BPS, HS&B, NAEP
National Science Foundation	License	Survey of Earned Doctorates, Survey of Doctorate Recipients
Department of Justice	Data Transfer Agreement	Datasets from 5 Agencies—Bureau of Justice Statistics, Bureau of Justice Assistance, Office of Juvenile Justice and Delinquency Prevention, National Institute of Justice, Office of Justice Programs
Health Care Financing Administration	Data Use Agreement	6-Enrollment Database, HISKEW, NCH, SAF, MEDPAR, MCBS
Social Security Administration	Data Use Agreement	The Health and Retirement Study
HCFA-National Cancer Institute	Data Use Agreement	Surveillance, Epidemiology, and End Results-Medicare Database (SEER-Medicare)
Bureau of Labor Statistics	Letters of Agreement	National Longitudinal Survey of Youth, Census of Fatal Occupational Injuries)
National Institute of Child Health & Human Development	Agreement for the Use of Sensitive Data	The National Institute of Child Health and Human Development Study of Early Child Care
National Heart, Lung, and Blood Institute	Data and Materials Distribution Agreement	Framingham Heart Study

Table 1. Names of Data Use Agreements and Relevant Datasets, by Agency (Continued)

	Type of Agreement	Datasets
National Institute of Mental Health	Distribution Agreement	Genetic Analysis of Schizophrenia, Bipolar Disorder, Alzheimer's Disease, and Other Mental Disorders
National Institute on Drug Abuse	Access Requests	Genetics of Drug Addiction Vulnerability
National Institute on Alcohol Abuse and Alcoholism	Distribution Agreement	Collaborative Studies on Genetics of Alcoholism

The common theme across all data licenses or use agreements is that they are used to extend the provisions of the relevant laws and regulations to qualified researchers. In each case, the researcher seeking access to confidential data must acknowledge his or her responsibilities under the relevant law, and must present a research plan that justifies the need for restricted use data. In many cases the researcher must also include a security plan that describes the measures to be taken to protect the confidential data while they are in the possession of the researcher.

2. How the Licensing System Works at One Agency: NCES

Who Needs a Restricted Use Data License?

NCES staff members' oath of office serves in lieu of a license document, and NCES staff members are required to sign the Data Security Office log book to obtain restricted use data. NCES is the exception; virtually every other organization requires a license document to authorize individual access to restricted use data. The specific license document varies by type of organization. Researchers in other parts of the Department of Education, Congress, and other federal agencies must enter into a memorandum of understanding with NCES to gain access to restricted use data from NCES. Researchers in Department of Education Research Laboratories, state government or local education agencies, or other nonfederal agencies, groups, or organizations must enter into a restricted use data license with NCES. Researchers who work as NCES data collection contractors or subcontractors are licensed under terms included in the contractor license. Finally, a data collection contractor who wants to use NCES restricted use data for independent, but related, research must submit a formal written request. If the purpose of the independent research is different from the use stated in the contract, the contractor must follow the standard application process to obtain a license.

Components of the NCES License Application

Each NCES data license applicant must submit a formal letter of request, a license document, a security plan, and notarized affidavits of nondisclosure. These materials are reviewed for accuracy and completeness by the NCES security staff before each application is forwarded to the commissioner for approval (*Judicial Administration* 1997).

Formal Letter of Request. The letter of request must name the requested data files, identify the exact location of the proposed licensed site, specify the time period of the requested loan of the restricted use data, and demonstrate the need for the specified data file. More specifically, the requestor must provide assurance that the data will be used only for statistical purposes consistent with the purpose for which the data were collected. And the proposed project must be explained in sufficient detail to be deemed an appropriate statistical use of the data and to justify the need for the restricted use data.

The letter of request must also include the names and titles of the principal project officer, the senior official, the system security officer, and additional authorized users for the license. The principal project officer oversees the daily operations; the senior official must have the authority to legally bind the organization to the provisions of the license; and the system security officer implements and maintains the computer security plan.

License Document. An executed license document is a legally binding agreement that names the agency or organization being licensed, and includes signatures from the senior official, the principal project officer, and the commissioner of NCES.

Security Plan. Restricted use data must be protected from unauthorized disclosure, use, or modification at all times. Strict security procedures are required to protect individually identifiable confidential data. The security plan contains the detailed procedures for protecting the restricted use data. A copy of the security plan signed by the senior official, the principal project officer, and the system security officer must be submitted as a part of the application package.

Affidavits of Nondisclosure. Each person who will have access to the data must complete an affidavit of nondisclosure that names the data files requested. The affidavit includes an oath or affirmation not to disclose individually identifiable information to any person not licensed to access the data, the penalties for disclosure, a personal signature, and the signature and imprint of a notary public.

Required Security Procedures

Individually identifiable confidential data are sensitive and require high levels of protection to prevent unauthorized disclosure, use, or modification. Restricted use data on loan under a license agreement must be maintained at the licensed site in a safe environment 24 hours a day for the period of the license.

The principal project officer has full responsibility for the security of loaned restricted use data; this includes supervision of the preparation and implementation of the restricted use data security plan. The system security officer is responsible for the implementation, maintenance, and periodic update of the security plan.

The security plan must describe the arrangements for the secure storage of the data and printouts when not in use. The plan must also specify the location where the data will be used and describe the physical security arrangements for that location. Finally, the plan must describe the computer security requirements that will be implemented when the data are in use.

The security plan must be maintained as part of the license file, along with the license and its attachments, any amendments to the license document, a current list of all authorized users, and copies of their notarized affidavits of nondisclosure.

Storage. The restricted use data and any resulting printouts or tables must be secured in a locked cabinet when not in use. Only authorized users may have access to the storage facility.

Physical Security. The restricted use data may be used only at the licensed site. They may not be used at any other location or accessed from a remote site. The room where the data are used must be reasonably secure during business hours and locked after close of business.

Computer Security Requirements. The safest computer environment is based on the use of a stand-alone personal computer that has no active connections to another computer or network. The users must be password-protected, with passwords changed at least every three months. The computer must include a notification warning either on the machine or displayed during login. The warning states, 'Unauthorized access to licensed individually identifiable information is a violation of federal law and will result in prosecution'. Either the computer should have an automatic shutdown feature set to three to five minutes, or the computer or room should be locked when the authorized user is away. One backup copy is permitted; other than that, restricted use data may not be backed up on a routine basis. After each session with restricted use data, the hard disk data must be overwritten.

Who May Use the Licensed Data

Only individuals who have a signed affidavit of nondisclosure that is on file with NCES as a part of an approved license may have access to the licensed data specified in their affidavit. Before they access licensed data, all authorized users must read and understand the NCES security procedures and the security plan for their site. The principal project officer must notify NCES of any changes in personnel. Any new members of the research team must submit notarized affidavits of nondisclosure to NCES for inclusion on the amended license. Similarly, if any authorized user leaves the research project, NCES must be notified so that user can be deleted from the license.

Requirements for Publishing Results From Restricted Data

While analyzing the data, the licensee should edit all printouts, tabulations, and reports for any possible disclosures of the data. The general rule is to not publish a cell in which there are fewer than three respondents or where such a small cell could be obtained by subtraction. Care must also be taken to not disclose information through subsequent uses of the same data. The licensee also agrees to submit to NCES a final copy of each publication containing information based on an NCES restricted use data file. If the licensee suspects a publication might disclose individually identifiable confidential information, the licensee must forward an advance copy of that publication to NCES for review and delay public release until formally notified by NCES that no potential disclosures were found.

Security Inspection Procedure

The license authorizes representatives of NCES to make unannounced and unscheduled inspections of the licensee's facilities to evaluate compliance with the terms of the license and security procedures. NCES contracts with an information security firm that employs trained security officers for the on-site inspections.

The security officer's first contact with the license holder occurs through a phone call before the inspection. During the phone call, the security officer identifies himself or herself as an agent of the U.S. government and arranges an appointment, usually for the next day. Once on-site, security officers introduce themselves and explain that they are there for an unannounced inspection as agreed to in the license granting access to an NCES database that contains individually identifiable confidential information. The security officers note that the main goal of the inspection is to ensure that licensees are in compliance with the statutes, as outlined in the security procedures and the license. The security officers also explain that if a licensee is not in compliance, the objective is to provide advice and assistance to achieve compliance.

During an inspection, the security officers review the license holder's file for a copy of the license and a list of persons authorized to access the data. This list is compared to a current list of all project personnel to ensure that all personnel are authorized users. The security officers also confirm whether all project staff have reviewed a copy of the license and the security procedures.

The security officers review the security procedures with the license holder, comparing the security procedures to the submitted security plan. The security officers use an on-site inspection guideline to ensure that compliance with the agreement is enforced uniformly across license holders and to that all appropriate questions and topics are covered by the interview. The guideline includes questions pertaining to license procedures; physical handling, storage, and transportation of the data; and computer security requirements.

At the conclusion of a field inspection, the security officers submit each completed inspection report to the NCES security staff for review. NCES staff review the form and determine whether there are any violations. The license holder learns about the outcome of the inspection through a follow-up letter from the NCES security program.

If no problems were found, the license holder receives a letter acknowledging full compliance. If violations were found, the license holder receives a letter enumerating the areas in need of correction. The license holder is instructed to submit a letter describing the steps adopted to correct the violations. NCES security staff monitor these cases until a written response is received. Once the correction letter is received, the security officer adds that license site to the inspection schedule for inclusion in the next trip to that region.

How a License Is Terminated

When the proposed research is complete, the licensee is required to notify NCES that the project is complete, and return the original restricted use data and additional data file documentation to NCES by certified mail. At this time the licensee is required to overwrite the restricted use data from the computers where the data were analyzed.

3. How Are Federal Agencies Using Licensing Systems?

Laws and Regulations Covering Data Licenses

Two laws pertain to individually identifiable or confidential data at any federal agency: the Privacy Act of 1974 and the Computer Security Act of 1987. In addition, federal agencies use the Code on Crimes and Criminal Procedures, the federal regulation on the 'Federal Policy for the Protection of Human Subjects', and agency-specific laws to protect individually identifiable data.

The Privacy Act of 1974. This act protects the privacy of personal data maintained by the federal government. It imposes specific requirements on federal agencies to safeguard the confidentiality and integrity of personal data, and limits the uses of these data. Federal Information Processing Standard Publication 41, *Computer Security Guidelines for Implementing the Privacy Act of 1974,* offers guidance to ensure that government-provided individually identifiable information is adequately protected in accordance with federal statutes and regulations. Unlawful disclosure is a misdemeanor and is subject to a fine up to $5,000.

The Computer Security Act of 1987. The second law relates to sensitive information, defined as any unclassified information that could adversely affect the national interest, the conduct of federal programs, or individual privacy covered by the Privacy Act of 1974. This law requires each federal agency to identify all federal computer systems that contain sensitive information and implement security plans to protect these systems against loss, misuse, disclosure, or modification. Unlawful disclosure is a misdemeanor and is subject to a fine up to $5,000.

Crimes and Criminal Procedure. The Code on Crimes and Criminal Procedure (U.S. Code: Title 18, Section 641) does not relate directly to confidentiality, but it includes penalties for misusing government property. The Bureau of Labor Statistics currently uses this code for violations of the bureau's confidentiality agreement with external researchers; like the first two laws, it could be used by any federal agency. According to the law, anyone who knowingly converts to his or her use or the use of another person or who, without authority, conveys or disposes any record of any department or agency of the U.S. government is in violation of the law and is subject to penalties that include imprisonment up to 10 years and fines up to $10,000.

Federal Policy for the Protection of Human Subjects. In addition to these three laws, ten federal departments and five additional agencies or administrations have federal regulations identified as the 'Federal Policy for the Protection of Human Subjects'. These regulations mandate and govern the formation and operation of institutional review boards (IRBs) and define research that is subject to regulation, human subjects' rights, and the concepts of minimal risk and informed consent. IRBs approve and monitor research on human subjects research that is conducted under federal jurisdiction (Lowrance 1977).

Research institutions are required to have IRBs that comply with federal regulations to be eligible to receive federal funding for human-subjects research. Because these approvals are already in place, several agencies have tied their data use agreements for individually identifiable data to proof of an IRB approval. In addition, some agencies have specific confidentiality legislation or regulations.

National Center for Education Statistics. The National Education Statistics Act of 1994 authorizes NCES to collect and disseminate information about education in the United States. The act requires NCES to develop and enforce standards to protect the confidentiality of persons in the collecting, reporting, and publication of data. The act also protects the confidentiality of individual schools in the National Assessment of Educational Progress data. Under this law, no person may use any individually identifiable information for nonstatistical purposes, make any publication in which data furnished by any particular person can be identified, or permit anyone other than the individuals authorized by the NCES commissioner to examine the individual reports. A confidentiality violation of this law is a class E felony, punishable by up to five years in prison and/or a fine up to $250,000.

Department of Justice. Title 28 of the Code of Federal Regulations specifies the requirements for a transfer agreement that allows five different data collection components of the Department of Justice to share individually identifiable research and statistical data with external researchers for the sole purpose of research and statistical analysis. The legal authority for the regulations in Section 28 is from the

Omnibus Crime Control and Safe Streets Act of 1968, as amended, the Juvenile Justice and Delinquency Prevention Act of 1974, as amended, and the Victims Crime Act of 1984. These regulations specify that the data may not be disseminated in any format that permits identification of private persons. The data must be kept secure, with access limited to authorized employees. The knowing and willful use, publication, or dissemination of individually identifiable information constitutes a violation punishable by a fine up to $10,000.

Health Care Finance Administration-Medicare Data. Section 1306 of the Public Health Service Act (42 U.S.C.) guards against disclosure of individually identifiable information from a tax return or any portion of a return. Under this act, the disclosure of such data that have been filed with the Commissioner of Internal Revenue and transmitted to the head or staff of another agency is a felony, subject to a fine up to $10,000 or imprisonment up to five years, or both.

Public Health Service. Section 903 of the Public Health Service Act (42 U.S.C. 299a-1) specifies that data collected under this provision from individuals or organizations for research purposes may be used only for research purposes. The act also states that no individually identifiable information may be released or published without the respondent's consent. Using both the Public Health Service Act and the Privacy Act, several agencies within the Public Health Service have developed guidelines for loaning data and materials to external researchers. In each case, the researcher agrees to use the data and materials only for statistical purposes and to not attempt to identify any individuals. The information must be used in a secure environment, with access limited to authorized users. Violations of the terms of the agreement may result in the revocation of the agreement and the imposition of unspecified monetary damages.

How Agencies Compare in Implementation

An extensive search of government websites identified 13 different license or data use agreements that are designed to allow qualified researchers access to confidential or restricted access data. Several of these agreements are applicable to one single data collection. Others may be used for multiple data collections within one agency. In the case of the Department of Justice, the same regulations cover data from five separate agencies. There are several stipulations that are common across all of these agreements (including licenses). The agency-specific features are shown in Table 2.

Research Proposals. Each application starts with a requirement for a research proposal that describes the data requested, describes the analysis proposed, and explains why public use data (if available) are not adequate. It must also be clear that the proposed analysis is consistent with the statistical/research purpose for which the data were supplied.

Universal Terms of the Agreement. If an agreement is executed, each authorized data user must agree to release data only in statistical summaries so as to not disclose information about any individual, and to share the individually identifiable data only with members of the immediate authorized research team. Researchers are also prohibited from using the data to learn the identity of any person or other legally protected entity. Researchers may not transfer an agreement to another individual or move the data to another institution without the written consent of the federal agency that is party to the agreement. At the completion of the approved research project, the researchers must either return the data to the agency or destroy the data under terms specified in the specific agreement.

Additional Terms of Agreements. A number of other features of licenses or data agreements are shared across many, but not all, agreements. First, as part of the research proposal phase, three of the five agreements identified at the National Institutes of Health, the agreements at the Department of Justice and the Health Care Financing Administration, and those involving the Social Security Administration require IRB approval of the proposed research project. In the remaining agreements, the staff approving the agreements assume the responsibility for reviewing the proposals. Also, all but four of the agreements examined require institutional concurrence from the researcher's institution as part of the application process. This is usually accomplished by having an official who has the legal authority to bind the institution sign the license.

Data Security. In the area of data security, the principal investigator who signs the license or agreement makes a written commitment to protect the confidentiality of the data. In seven of the thirteen agreements examined, there is an additional requirement for signed security pledges or nondisclosure affidavits from each authorized user on an approved research project; and six of these seven require the reporting of any identified disclosures. Nine of the thirteen agreements include a security plan, either as prescribed in the agreement or as submitted by the applicant as a component of the agreement or license. The remaining four agencies distribute biomaterials along with data with no personal identifiers; however, because of the sensitive nature of the genetic materials, they include nondisclosure provisions in their agreements. Finally, as an additional security precaution, seven of the thirteen agreements require the researchers to agree to on-site inspections of their data security arrangements.

Table 2. Agency-Specific Features of Data Use Agreements and Licenses

	IRB Approval Required	Institutional Concurrence	Security Pledges All Users	Report Disclosures	Security Plan	Security Inspections	Cell Size Restrictions	Prior approval-Reports	Notification of Reports
National Center for Education Statistics		X	X	X	X	X	X	X	X
National Science Foundation		X	X	X	X	X	X	X	X
Department of Justice	X	X	X						
Health Care Financing Administration					X	X		X	
Social Security Administration	X	X	X	X			X		X
Health Care Financing Administration-National Cancer Institute					X	X		X	
Bureau of Labor Statistics-National Longitudinal Survey of Youth			X	X	X	X	X		X
Bureau of Labor Statistics-Census of Fatal Occupational Injuries			X	X	X	X	X	X	
National Institute of Child Health and Human Development	X	X	X		X	X	X		X
National Heart, Lung, and Blood Institute	X								X
National Institute of Mental Health	X	X							X
National Institute on Drug Abuse	X	X							X
National Institute on Alcohol Abuse and Alcoholism		X							X

Reporting Standards. Standards for reporting are not as well delineated. While all agreements require researchers to release data only in statistical summaries in order to avoid disclosure of information about any individual entity, only four agreements include any detailed cell size or other reporting requirements. Three agreements require that any written releases of data be approved before release, and two additional agreements specify that the researcher must submit any written releases that could raise reasonable questions about disclosure for review before release. These two agreements and seven others require notification of publication, with no review function. Only one agreement includes no provisions for review or notification of publications resulting from the data in the agreement.

Penalties. Penalties for violations of the confidentiality provisions of the license are also variable. Each agreement or license includes provisions covering the termination or revocation of the agreement. Because all the agreements involve data from a federal agency, violations of confidentiality in each case are, at a minimum, subject to the penalties specified in the Privacy Act of 1974—a misdemeanor subject to a fine up to $5,000. The agreements for the Retirement History Study and the agreements with the Health Care Financing Administration (HCFA) include a provision that the researcher must be a current recipient of a federal grant for the approved research, with the understanding that a disclosure may result in a recommendation for the revocation of research funding. Laws governing the violation of confidential data from HFCA include fines up to $5,000 and felony prison terms up to 10 years. The federal regulations for the agencies covered in the Department of Justice include penalties up to $10,000 for a violation of confidential data. The law used by the Bureau of Labor Statistics includes fines up to $1,000 and prison terms up to 10 years. The law governing the use of restricted data from NCES makes a confidentiality violation a Class E felony, with fines up to $250,000 and prison terms up to five years.

How Agencies Compare in Enforcement

Strong enforcement starts with the strength of the license or agreement. There are five elements of an agreement that can potentially influence compliance, if not enforcement.

Conditions Influencing Compliance. First, the requirement that the parent institution either signs on to the agreement or gives IRB approval of the proposed research ensures that senior officials at the researcher's institution are aware of the responsibilities of projects using confidential data. Second, the requirement that each person who will have access to the confidential data submit a signed, preferably notarized, security pledge or nondisclosure affidavit ensures that everyone authorized to use the data is aware of the restrictions and responsibilities associated with using confidential data. Third, requiring the researcher to submit a security plan for review and approval as part of the agreement increases the researcher's awareness of the necessary security precautions and provides assurances to the federal agency that the data will be secured appropriately. Fourth, having prescribed guidelines for publishing tables with confidential data to avoid disclosures is likely to go further toward ensuring against disclosures in reporting than simply telling the researchers that they should not disclose any data in any report. Fifth, requiring the researcher to submit reports for review before release provides another mechanism to guard against disclosures.

Security Inspections. A key element of enforcement involves on-site security inspections of the researcher's facilities. Six of the thirteen data licenses and agreements included a provision for on-site inspections. Three of them are for individual surveys: the National Institute of Child Health and Human Development study of 'Early Child Care', and two Bureau of Labor Statistics studies, the National Longitudinal Survey of Youth and the Census of Fatal Occupational Injuries. The National Institute of Child Health and Human Development study has only recently initiated a data agreement program, so although the Institute reserves the right to inspect researchers' security arrangements, it had not yet made plans to do so. The Bureau of Labor Statistics conducts seven to ten inspections a year, by identifying a geographic location and sending in-house professional staff to inspect researcher sites at the selected location.

The remaining three agreements that include an inspection provision are agreements that are used agencywide. HCFA lists seven major data files on its website that are available under 'Identifiable Data Use Agreements'. While its documentation includes a provision for on-site security inspections, the agency is not currently conducting site inspections. It does, however, expect to start inspections in the near future. NCES and the National Science Foundation (NSF) both license external researchers to use restricted data. NSF currently lists four restricted access databases and NCES lists eleven. These two agencies collaborate in the management of their on-site security inspections. Rather than use in-house professional staff to conduct site inspections, these agencies contract with an information security firm to have security site inspections performed by trained security officers. Each licensee can be assured of receiving at least one inspection over the lifetime of the license period. Two additional agencies are exploring the feasibility of joining NCES and NSF in the use of the information security firm for on-site security inspections.

Annual Reporting. Six of the agreements for individual databases require annual reporting on the part of the researcher. While this requirement is not strictly direct enforcement, periodic contact between the agency and the researcher can serve to keep the researcher mindful of the responsibilities associated with the confidential data.

Close-Out. Finally, each of the licenses or agreements includes a requirement that the data must be either returned or destroyed at the end of the agreement. All but one of the thirteen set specific time limits, and that one reports plans to add time limits. In each case, the agency contacts the researcher to give notification that the time period is ending, and to negotiate the return or destruction of the data.

4. How Well Does Licensing Work?

Review of the Security Inspection Process

Absent an external audit, security inspections are the best way to determine whether researchers are complying with the terms of the license or agreement. A review of some of the security inspection reports from NCES provides some insight into the difficulties that researchers may have in successfully executing the data use agreements or licenses.

In the spring of 1998 the NCES security staff reviewed the site inspection process. By that time, NCES had issued 391 restricted use data licenses, and one-third of the licensed sites had been inspected. To evaluate the reasons for the failures, a more in-depth analysis was performed focusing on 54 inspections conducted between April 1997 and April 1998. The site visits analyzed were conducted roughly in proportion to the distribution of license holders across different sectors of licensees (e.g., universities, research organizations, other government agencies).

Minor Violations

A review of the violations revealed that a number of them were less serious in nature in that they did not pose a direct risk of unauthorized users accessing restricted use data. Some of the minor violations represent poor record keeping and failure to update NCES on changes in personnel. For example, they included incomplete or missing license documentation, licenses where the senior official on the license was no longer at the institution, or cases where one or more of the authorized users had left the project. These violations involved missing license documentation or someone leaving the institution, and thus did not increase the risk of an unauthorized user accessing the data.

In addition, minor procedural violations occurred when there were inadequate displays of warning notices. Because authorized machines are all password-protected and maintained in limited access environments, the absence of the warning sign is not likely to result in unauthorized access to restricted use data.

Major Violations

More serious violations involved cases where there was a risk that unauthorized users might access the data. One clear example of a serious reporting problem occurred when an unauthorized user was actually accessing restricted use data, because the project director failed to have a new staff member execute and submit an affidavit of nondisclosure. As a result, the new user may not have been informed of the security precautions required when one is using restricted data.

Another serious reporting problem occurred when the project director left the institution without notifying NCES. If the researcher left the data behind, there was no longer an authorized project director responsible for the security and proper use of the confidential data. And if the researcher moved the data to a new institution without first notifying NCES and submitting an application for a new license, the safeguards associated with the institutional commitment and the approved security plan were lost.

Serious procedural problems occurred when a researcher was found using the data off-site or accessing the data from an off-site location. The security plan is specific to the licensed site, and the use of the data in a location with no approved security plan poses a risk to the security of the data; similarly, accessing the data from a remote location may place the data at risk during transmission. Additional cases were discovered in which the data were lost. These cases usually involved a situation where the project director had died or retired and the data were not locatable, but they still create a concern over the potential misuse of the missing data.

Corrections

As a result of the review of the inspection process, NCES developed and implemented computer-assisted personal interview (CAPI) site inspection reports that include programmed follow-up letters. This reduces unnecessary processing time and the resulting delay in providing feedback to licensees. The fully implemented system allows the security inspectors in the field to transmit the inspection letters to NCES electronically for review and signature. It will also provide NCES with a more immediate and more accessible database to monitor the licensing system.

An External Audit

One federal agency with a data use agreement system in place was recently audited by the General Accounting Office (GAO) for its compliance with the Privacy Act of 1974 and its ability to protect the confidentiality of individually identifiable data. The GAO found that weaknesses in the implementation of the agency's policies could potentially compromise the confidentiality of the individually identifiable data (U.S. Government 1999b).

One of the factors cited was a failure on the part of the agency to routinely monitor contractors and researchers who use individually identifiable data. In fact, the agency officials reported that they do not have a system for monitoring whether organizations with data use agreements have established safeguards for protecting the individually identifiable confidential data received from the agency. Instead, the agency relies on data users to monitor their own compliance with the commitments made in the agreement to establish appropriate administrative, technical, and physical safeguards not below the level of security established by OMB.

Another factor cited was a failure to track and monitor the return or destruction of data upon completion of the project as outlined in the data use agreement. The absence of this control could result in subsequent uses other than those approved in the agreement or in the data drifting into unauthorized hands.

The important fact here is not which agency was inspected, but rather the fact that the same criticisms could be leveled against any agency that enters into data use agreements or licenses without putting a monitoring system in place, or any agency that does not diligently follow up at the end of the agreement period to ensure the return or destruction of the data.

5. How Can Licensing Systems Be Improved?

Security Inspections

Any agency that is entering into data use agreements or licenses without an active monitoring system should seriously consider implementing one. A periodic inspection cannot catch all potential problems. However, the knowledge that each researcher will be inspected at least once during the life of the agreement sends a strong signal to researchers that the agency takes its responsibilities seriously and expects the same from authorized data users.

The inspections can also help inform the agency of potential problems or vulnerabilities in the agreement or license system that can be corrected by modifications in procedures. For example, as a result of the review of NCES inspection reports, minor and serious violations were identified and separated, allowing a more intensive focus on the serious problems. In addition, the security inspectors were empowered to resolve correctable problems during the course of the inspection—problems such as identifying authorized users who have left the project, supplying the warning notice for terminals where it is missing, and advising on computer security problems.

Tracking Database

It is important for each licensing or agreement system to develop and maintain a database application that allows the agency to keep accessible records of the restricted use data files and all authorized users for each agreement. The database should be developed to produce periodic confirmation letters including this information. The letters would serve both as a reminder of the importance of protecting the confidential data and as a mechanism for updating any changes in the status of the authorized data users. A database with this information could also be used to identify the subset of researchers with approved access to specific data files. The information used in the interim confirmation letters could be used to send notification letters at the end of the approved time period for the agreement and to monitor agreements through closeout.

References

Computer Security Act of 1987 (1987) Public Law 100-235, 100th Congress, Washington, D.C.: U.S. Government Printing Office.

Computer Security Guidelines for Implementing the Privacy Act of 1974 (1974) Federal Information Processing Standard Publication (FIPSPUB 41), Washington, D.C.: U.S. Government Printing Office.

Crimes and Criminal Procedures (2000) 18, U.S. Code, 641, Washington D.C.: U.S. Government Printing Office.

Hawkins-Stafford Elementary and Secondary School Improvements Amendments of 1988 (1988) Public Law 100-297, 100th Congress, Washington, D.C.: U.S. Government Printing Office.

Judicial Administration (1997) 28CFR22, Washington, D.C.: U.S. Government Printing Office.

Lowrance, W.W. (1977) *Privacy and Health Research,* Washington, D.C.: U.S. Department of Health and Human Services.

McMillen, M.M. (2000) 'Data Access: National Center for Education Statistics', *Of Significance,* 2000, 2(1), pp.51-6.

National Center for Education Statistics (1995) *Restricted-Use Data Procedures Manual,* Washington, D.C.: Author.

National Education Statistics Act of 1994 (1994) Public Law 103-382, 103rd Congress, Washington, D.C.: U.S. Government Printing Office.

Privacy Act of 1974 (1974) 5, U.S. Code Section 522a, as amended, Washington, D.C.: U.S. Government Printing Office.

Public Health Service Act (1999) 42, U.S. Code, 299a-1 and 1306, Washington, D.C.: U.S. Government Printing Office.

U.S. General Accounting Office (1999) *Medicare: Improvements Needed to Enhance Protection of Confidential Health Information,* 1999, Letter Report, HEHS-99-140, Washington, D.C.: Author.

Confidentiality, Disclosure, and Data Access: Theory and Practical Application for Statistical Agencies
Pat Doyle, Julia I. Lane, Jules J.M. Theeuwes and Laura M. Zayatz (Eds)

Chapter 12

Issues in the Establishment and Management of Secure Research Sites*

Timothy Dunne
University of Oklahoma

1. Introduction

Modern statistical agencies collect a vast amount of data on individuals and households, firms and establishments, and other organizations such as governments and schools. These data represent incredibly rich sources of information to social scientists and statisticians and form the basis of the statistical infrastructure for most countries. While statistical agencies make a concerted effort to allow as much data as possible into the public domain, a vast archive of data exists that is not accessible by the public. The restrictions on the dissemination of most of these data stem from the fact that confidentiality concerns limit the amount, detail, and type of data that can be released to the public. For example, most statistical agencies rarely release individual record data from business surveys. This is because disclosure avoidance techniques are difficult to design for business data, where the size of the business unit in a particular sector often is sufficient information to reveal the identity of the respondent[1]. This situation is unlike many demographic surveys, where a public use sample of microdata is made available for analysis. However, even public use files may seriously constrain the data available for research. Data from demographic collection programs often suppress detailed geography, top-code income items, or swap out data for individuals in small areas. Hence, depending on the specific circumstance, the information loss to the data user can vary from relatively minor to quite substantial.

Given that it is well understood that the privacy of survey respondents and administrative records must be maintained, statistical agencies often face the question of how to increase access to the rich data archives they house while main-

* The author thanks Julia Lane and a set of reviewers for comments on an earlier draft. The author also thanks Gustav Goldman for information on the Statistics Canada secure site program.

1 In particular, in some instances, large businesses are known to be in the sample based on the sampling criteria. See McGuckin and Nguyen (1990) for a discussion of disclosure issues and public use files for business data.

taining the confidentiality of the data. This point is emphasized in a recent publication of the Social Sciences and Humanities Research Council of Canada and Statistics Canada (1999):

> Unfortunately, a great deal of the valuable data is underutilized because of Canada's lack of national capacity to fully analyze them. Compounding this problem…is the difficulty of providing access for researchers to highly detailed files, because of Statistics Canada's legal responsibility to protect the confidentiality of individually identifiable responses. Against this backdrop lies a pressing need for current social statistics to sustain the increasing demand for evidence-based decision-making.

This demand for access to the microdata archives of statistical agencies is not new. Flaherty (1979) analyzes the discussions that occurred on this same topic in the late 1960s and the 1970s. As is true today, the demands by researchers, policy makers, and analysts for increased access were driven by the inadequacies of the analysis that utilized aggregate data or limited public use files. Flaherty (p.21) quotes Charles L. Schultze of the Brookings Institution in 1970 as saying that 'matching data on individuals and organizations from different surveys and administrative files are required to construct measures of social performance by various population groups'.

The situation is little different today, and the demand for access to microdata at statistical agencies remains strong, for several reasons. First, it is increasingly apparent that many public policy issues require the analysis of panel microdata on individuals and/or firms. These requirements stem from the recognition that statistical and econometric analyses must often use panel data on individual agents to model and evaluate the impact of government programs and policies on economic outcomes[2]. Second, the usefulness of certain datasets can often be greatly enhanced by the ability to match the data with another dataset. There is a growing body of research using matched data that come from household, establishment, and administrative data[3]. Typically such matching occurs at the microdata level and may require access to very detailed matching keys available only at the statistical agency. Third, technological improvements in data storage, data access, and computing power have made it more feasible and less costly to access the data archives

[2] In addition, recent research looking at aggregate fluctuations in the economy argues that it is critical to understand the microadjustment process in order to understand how the macroeconomy responds to a variety of shocks. An excellent discussion of the importance of microdata in understanding macroeconomic fluctuations is provided in Haltiwanger (1997).

[3] The papers of a recent symposium on efforts to utilize linked employer-employee data are reported in Haltiwanger, Lane, Spletzer, Theeuwes, and Troske (1999). The papers in that volume detail efforts to use matched employer-employee data in various countries around the world. Important in this effort is the Linked Employer-Household Database (LEHD) project at the U.S. Census Bureau, which is creating large-scale databases of workers-establishment matched data.

of statistical agencies. Finally, while statistical agencies collect a vast amount of data for public policy research, it is not unusual for statistical agencies to keep an arm's-length relationship with direct public policy analysis.

For all these reasons as well as others, there has been strong demand for access to the microdata archives at statistical agencies. Agencies and researchers around the world have dealt with this demand in a variety of ways. Projects such as the Luxembourg Income Study build data archives that protect the data files in a secure location. Researchers submit programs to be run by the project but do not have direct access to the microdata. Alternatively, some statistical agencies have allowed restricted access to certain microdata files by external researchers at the statistical agency. Groot and Citteur (1997) document the approaches taken at a variety of national statistical agencies to provide access to establishment and enterprise data. In a small number of statistical agencies, these efforts have led to the establishment of secure research sites—physical locations that provide researchers restricted access to data files that the statistical agency does not release to the general public. This chapter examines the establishment and management of such secure research sites. It focuses on the conditions for establishing secure sites; the operational aspects of secure sites, including security and management issues; and the benefits both to the statistical agency and to the research community of establishing secure sites. In addition, the chapter highlights issues posed when statistical agencies establish secure research sites at other institutions such as universities.

2. Establishing Secure Sites

For a statistical agency to establish a secure site, housed either at the agency or at an alternative site, the agency must have the legal authority to allow access to the microdata. Access to the confidential microdata is described in either the laws under which the data are collected or the laws under which the statistical agency operates. The specific laws and regulations vary widely across agencies and countries, and hence, the potential availability of restricted data to external researchers varies by source and across countries[4]. Moreover, the laws and regulations governing access to microdata usually do not speak directly to the establishment of secure sites but often focus on the confidentiality provisions, penalties associated with noncompliance with the confidentiality provisions, legitimate use of the data, and sometimes a description of the classes of authorized users. However, the laws and regulations governing the collection of the data and

[4] In the United States, the Privacy Act of 1974 provides overarching regulation covering how federal agencies handle the information they collect on individuals. In addition, specific federal agencies may have laws or titles that regulate their specific data collection and data dissemination procedures. Furthermore, individual states have their own legislation that governs the use of state-level data—in particular, state-level administrative record data. In Canada, the Statistics Act governs the access and use of data collected by Statistics Canada.

operations of data collection agencies generally provide a good description of the type of data access agencies can provide at secure sites.

Duncan *et al.* (1993), Flaherty (1979), and Groot and Citteur (1997) provide detailed descriptions of the legal frameworks that allow access to nonpublic data resources at various statistical agencies around the world. The key questions an agency must address are these:

(1) Do the laws governing the data collection programs allow access to non-publicly available data?

(2) If external researchers may use certain data, under what conditions is authorization appropriate?

(3) Who are valid users of these non-publicly available data?

(4) Can secure sites be established to allow access?

The first question that must be addressed before a secure site is established is whether the data provider has the authorization to allow access to the microdata for analysis. The data collection laws must allow analysis of the data for research purposes. In most cases, the agencies collecting the data are allowed access to the data for 'statistical uses'. Such analyses must adhere to confidentiality provisions proscribed by the relevant law and/or regulation. In situations where the potential use may involve data from a variety of sources, the agency may have to consider the regulations governing the use of all the data sources including administrative data sources[5]. This is often the case when the project involves matching employer and employee data or when the proposed research uses data from different sponsoring agencies. Under certain regulations, the analyses of the data may be undertaken only if they benefit directly the statistical programs of the agency (improve quality, design better statistical methods, *etc.*). Alternatively, in some cases, legitimate uses may be defined more broadly (*e.g.*, improving public policy decisions). It is important to reemphasize that use provisions may vary by the specific data source, as well as across agencies and countries. Hence, agencies interested in establishing secure sites must first identify the authorized uses of their data.

Once the authorized uses have been defined, the next question is, Under what conditions is authorization appropriate? Here, the agency needs to define the scope of the types of research that may be carried out with the restricted data. For example, in the United States, Title 13 of the U.S. Code allows the use of the data col-

[5] The U.S. Census Bureau collects survey data under its own main title, Title 13, and under Title 15 when it carries out surveys for other agencies (reimbursable work). However, the confidentiality provisions for a Title 15 survey are established by the survey sponsors' confidentiality provisions, not by the Census Bureau's. In addition, the Census Bureau uses administrative data from both the Internal Revenue Service and the Social Security Administration in some of its programs, as provided in Title 13. Authority to use these administrative data for statistical purposes must be formally agreed upon with the relevant data-supplying agency.

lected by the Census Bureau for 'statistical purposes'. The Census Bureau requires that the use of the data by an external researcher provide a benefit to the statistical programs of the Census Bureau. However, while a specific data use may meet the conditions of authorized use, they may not sufficiently meet the conditions of appropriate use. Here, agencies must use judgment and, perhaps, a detailed review process (see below) to define the potential appropriate uses of their microdata files[6]. The agency must weigh both the costs (*e.g.*, feasibility, required resources, and the potential for disclosure) versus the benefits (*e.g.*, improvements in a specific data collection program) when assessing appropriate use. Moreover, any potential use and the resulting statistical analysis must satisfy the agency's confidentiality restrictions.

Obviously, the premise for establishing a secure site is to allow researchers who are not full-time employees of the statistical agency controlled access to generally unavailable data resources. Therefore, mechanisms must be established that allow researchers to gain specific security clearances and authorization. In the case of the U.S. Census Bureau, individual researchers must obtain 'special sworn status'. To obtain this status, an individual must apply for a Census Bureau security clearance and take the Census Bureau oath protecting the confidentiality of the data. At Statistics Canada, external researchers granted access are referred to as 'deemed employees'. Certain other statistical agencies that allow access to their internal data files use taking an oath and receiving special status[7]. By taking an oath, the researchers pledge to comply with the confidentiality provisions imposed by the agency and they become subject to the penalties (often including both fines and prison terms) if they are caught violating the confidentiality provisions.

A key question that an agency should address when considering making such research appointments is: What is the status of the individual researcher vis-à-vis the agency? Is the researcher a contractor or an employee? If the researcher reimburses the agency for access, what is the status of that researcher in this situation? In many circumstances, only employees (in the traditional sense) have been given access to internal datasets for research purposes. This has been generally true in the case of Statistics Canada for accessing internal Statistics Canada establishment and firm data files. Alternatively, the U.S. Census Bureau has allowed broader access to include employees of other government agencies and academics. Again in the case of the U.S. Census Bureau, these individuals must adhere to all the laws and regulations associated with the use of the data and gain special sworn status.

If an agency decides that it can provide authorized access to internal data files for research and statistical purposes, and that external researchers can be given ac-

---

[6] Obviously, certain uses of the data will be precluded. For example, a statistical agency would not allow the analysis of a firm's data by an agency that regulated the commerce or industry of the firm even if the proposed analysis were not regulatory in nature. In addition, the Luxembourg Income Study explicitly prohibits commercial use of the data.

[7] Groot and Citteur (1997) detail the process of oath-taking and associated penalties for breaking confidentiality rules for 16 national statistical agencies.

cess under the appropriate conditions, the agency then can entertain establishing a secure site to allow access to the data. In the most basic sense, one can think of a secure site as a physical location where researchers can go to access specific microdata resources of the statistical agency. This is the type of access that is considered in this chapter[8]. However, with changing technology and increased sensitivity to privacy issues, alternative models of less direct access are being considered. We will not discuss these alternatives at length, but recognize that they exist and may have advantages over the establishment of physical locations (*e.g.,* remote access over virtual private networks or access to simulated data but only indirect access to confidential data).

The most straightforward type of secure site to establish is the secure site at the statistical agency. Here the agency sets up a facility that lies inside the agency's security environment. For these sites, it is easier for the agency to establish the required security arrangements, to manage the facility, and to monitor the activity at the site. A related type of site is one that is established at a regional office of the statistical agency. Typically, such offices already follow the security regulations of the statistical agency involved and thus offer an attractive location for a potential secure site. Both Denmark and the United States use regional offices of statistical agencies to provide researchers access to restricted data[9]. The third alternative is the establishment of a secure site at a third party institution such as a university or research institute. The U.S. Census Bureau has pioneered these efforts and has established four sites at four U.S. universities. Much of the remainder of the chapter will discuss the mechanics of establishing and operating secure sites.

3. Structure of Secure Sites

This section examines the structure of the secure sites, focusing on security, data, and management issues. It draws heavily on the experience of the U.S. Census Bu-

---

[8] While the chapter focuses on the experience of the U.S. Census Bureau's Center for Economic Studies (CES) in operating secure sites, the Census Bureau effectively operated an additional secure site program through the Census Fellows program. In the middle 1980s and early 1990s, the American Statistical Association, the National Science Foundation, and the Census Bureau sponsored a program that allowed researchers restricted access to non-publicly available data files at the Bureau's offices in Suitland, Maryland. The program had many of the basic features of CES's research data center at the Census Bureau.

[9] In Denmark, researchers at the Aarhus School of Business access microdata files through a regional office of Danmarks Statistik. In this case, the data files are not maintained at the regional office; researchers access computer files at the Danmarks Statistik in Copenhagen through secure computers in the regional office in Aarhus. In the United States, the Boston regional data center was the first secure site located away from the Census Bureau's main Suitland, Maryland, office. In this case, the data were and still are housed directly at the Census Bureau's regional office in Boston, Massachusetts.

reau's Center for Economic Studies (CES) in operating secure sites at the statistical agency, at the Census Bureau's regional office in Boston, and at universities[10].

The authors of the National Research Council's book entitled Private Lives and Public Policies discuss the use of both 'restricted data' and 'restricted access' in providing greater access to the microdata collected by statistical agencies[11]. 'Restricted data' refers to various approaches used to ensure that the microdata, as well as some tabular data, cannot be used to identify individuals. Typically the techniques used include such procedures as suppressing geographic detail, top-coding data items, and swapping data items across individuals. However, as information technology has improved, external data resources have developed, and data matching algorithms have been enhanced, questions have been raised about the release of traditional public use microdata files. 'Restricted access' refers to allowing the use of the statistical agency's microdata under controlled conditions. Through careful review of potential projects and secure management of the research facility, the statistical agency increases access but strictly controls the disclosure of information. The restricted access vehicle considered in this chapter is the secure site.

Security Issues

Before allowing researchers to access specific microdata, the agency must establish the physical location and the security of the site. First, controlled physical access must be maintained. Whether the site is at the agency or at an alternative site, only authorized personnel may be allowed into the area. This restriction can create difficulties in university settings, where such secure access areas may not be readily available. The physical security of the area must meet the requirements of the statistical agency's security plans. They may include alarm systems, monitored access through the use of security badges or security codes, and even the use of security cameras that monitor the restricted access areas. In addition, mundane but important issues must be addressed, such as how a site disposes of its confidential printed matter or how computer maintenance will be undertaken when needed. If the site resides in the statistical agency, many of these issues are easily resolved. At a university, the secure site operators will have to develop the appropriate solutions.

Second, the secure site must develop a secure computer and network infrastructure. Again, the agency must define necessary security provisions. In secure sites within agency walls, the research computers already reside behind firewalls. How-

---

[10] Reznek *et al.* (1997) discuss the U.S. Census Bureau's Center for Economic Studies research data center network. Currently, the network has secure sites at the main office (Suitland, Maryland), the Boston regional census office, Carnegie Mellon University, the University of California at Berkeley and at Los Angeles, and Duke University. Statistics Canada had opened a secure site at McMaster University, and it planned to open offices at the University of Toronto and the University of Montreal in March 2001.

[11] See Chapter 6, pp.141-79, for a detailed discussion of the 'restricted data' and 'restricted access' approaches to protecting the confidentiality of the data.

ever, computing equipment used by external researchers typically will not have the networking features of other machines in the agency. For example, it is not unusual to have very restricted access or even no access to the agency network where the secure site is housed. In effect, the external researcher's computers become stand-alone machines. The agency staff may be able to place data on the machine over the network, but the researcher cannot use the network to transfer data, to access the Internet, or to use e-mail services. In addition, some agencies require that the external disk drives and serial and parallel ports be disabled on the computers that external researchers use. In the case of secure sites not at the agency, the secure site must have an approved firewall. In certain installations, the secure sites elect to or are required to physically isolate their network from the larger university or research institution network so that no data traffic can flow either into or out of the site. In the secure sites operated by CES for the Census Bureau, all facilities must undergo a security evaluation by the Census Bureau's security office before confidential data can be housed at the site and before research can commence.

Third, the site must have skilled personnel who can help external researchers carry out their research in such an environment. Many researchers will be unfamiliar with the workings and security arrangements of a statistical agency. These researchers must be trained in the 'culture of confidentiality' that typically exists at statistical agencies. This is a particularly true when secure sites are set up outside the agency. The agency considering setting up a secure site must establish a detailed security training program for both staff at the site and researchers working at the site. In addition, employees need to be trained to ensure that the security provisions are well understood and are appropriately executed. This may be challenging because it may involve staff in complex computer security issues as well as in managing and monitoring the physical site.

Fourth, all statistical output (including printed and electronic versions) must be analyzed for disclosure before it can be removed from the secure site. This last issue is discussed at greater length below.

It is clear that in setting up a secure site, the agency must develop an environment where it can maintain strict control of the information housed at the site and released from the site. Besides these fundamental security measures that include both physical security and computer security features, the agency can improve its security by implementing a restricted data access policy. Agencies can reduce security risks by carefully restricting data to external researchers. For most research projects, researchers rarely require access to the most sensitive fields in a data record—name, address, Social Security numbers, and so on. In data extracts provided to researchers, these fields should be carefully controlled. At the Census Bureau, projects that need detailed identifiers often link various data sources. The sensitive data are required not for the analysis but for the match. In these situations, the Census Bureau typically supervises or directly carries out the match and then provides a dataset for the researcher with the sensitive match keys removed[12]. In this way the statistical agency can use both restricted access and restricted data policies to minimize disclosure risk and at the same time provide access to its unique data resources.

In summary, security issues are paramount. To establish a secure site either at a statistical agency or at another location, the security of the data must be ensured. This can be accomplished by the establishment of computer and physical security systems, the development of skilled staff, and the careful release of data extracts to external researchers. Each of these elements is important in establishing the overall security of a site. The data products made available and the management of the sites are also critical to the success of endeavor. It is to these issues that we turn next.

Data Issues

Presuming that the security issues can be addressed, the agency needs to consider what data to make available to researchers. Because of the very sensitive nature of some data or because of authorization issues that limit use, a statistical agency may forgo the opportunity of making certain data available to external researchers. At the same time, it is likely that the statistical agency will be asked to consider various microdata files for access as researchers make specific requests. The agency must be prepared to handle a range of requests for its internal data files, which raises several issues dealing with data documentation, data problems, and data management.

The microdata collections of statistical agencies around the world represent a valuable resource to data analysts and researchers. As mentioned above, data on firms and establishments are rarely made public. Detailed geographic data on households are often suppressed. However, the microdata files at statistical agencies often contain these types of data plus a wealth of other potentially useful datasets and data items. In making these data available to researchers at secure sites, statistical agencies must be prepared to provide researchers with detailed documentation for the specific files. Unfortunately, user-friendly documentation for internal files may not always exist for a specific data file. Constructing such documentation can be both time-consuming and expensive, and statistical agencies must be aware of these potential costs. These costs can be particularly high when there is no analogous public use file available (this is typically true for establishment- and firm-level data). For some demographic data, small modifications to the existing public use documentation may suffice. This would be the case if the only

---

[12] The LEHD project, along with the administrative records staff of the U.S. Bureau of the Census, has developed a set of detailed procedures for the appropriate use of sensitive identifier fields. This involves the creation of secure data management systems that restrict access to such sensitive data to specific computers that have high security systems installed. The matching process occurs on these very restricted access computers, and the matched data are then stripped of sensitive information or restricted use data items for use by researchers on other computing platforms. In this way, all uses of administrative record data follow elements of both a restricted access and a restricted data approach to ensure that privacy conditions are met.

difference between a public use sample and the internal agency files were items such as geographic detail or top-coding.

A second issue for statistical agencies is that their internal files are usually not 'edited' and/or 'cleaned up' as public use files are. It is not unusual for external researchers to identify anomalies and measurement errors in the microdata files. This is to be expected because the files may not have been originally constructed with the idea that they would be used as microdata files. This can be both a blessing and a curse for the agency involved. It represents a blessing because much of the gain to the agency from allowing use of the data is in the form of feedback on its collection programs and datasets. The agency can use this feedback to improve its programs. It represents a curse because some of the problems stem from the agency's belief that the benefit of additional data editing was not worth the cost when the data were originally edited. When the microdata are accessed by researchers, it is likely that some data problems will emerge. This may be particularly true when the researcher is constructing panel data with series of linked cross sections[13]. The editing routines of a statistical agency may focus on generating accurate cross-section statistics but not on creating accurate panel datasets. In these cases, the researchers may uncover inconsistencies and problems in the panel that are not problems in the cross section. The statistical agency needs to be able to deal with such situations and questions from researchers.

A third data issue centers around data management. One of the great benefits of research analysts poring over the raw microdata is that they will find errors and often suggest corrections. Ideally, some corrections and suggested improvements should find their way back into the data. The agency will need to decide first how to incorporate such corrections into the data files, and second how to update the existing files in the field. Similarly, the agency will need to decide how it will treat changes in the structure of the files, collection mechanisms, or data items. For example, industry coding and geographic coding typically change over time as coding systems are updated. The experience at the Center for Economic Studies is that researchers expect or (at least) desire consistent coding in data files. It is rare that the programs that produce the original data will go back and recode the historical data as coding systems change. Thus, it is likely left up to the organization that is operating the secure sites to decide whether it will actively reconstruct historical data files to provide consistent coding. Such recoding efforts, while important in making the data more user-friendly, can be quite costly. Given these data management issues, it is probably wise to start a secure site program at the outset with a

[13] For example, in the creation of the Longitudinal Research Database at CES a huge amount of time was spent finding errors and correcting the plant linkage variables. Because of mergers/acquisitions or changes in organizational structure, plant identifiers would change. These errors would have no impact on the tables that were constructed from these data (the intended use), but they had a large impact on the microdata panels constructed from these data (secondary use).

predefined list of data files that are in reasonable shape with respect to documentation and data consistency.

Project Management Issues

Once the security issues and data issues have been addressed, an agency will find itself with a wide range of project requests from a variety of sources. The agency needs to develop project management schemes to select, monitor, and manage the myriad projects that will come its way. It is important to remember that a secure site will typically provide data to researchers from a number of different disciplines. Secure sites must be prepared to deal with this variety.

The first project management issue is project selection. The review process serves two purposes at the Census Bureau. One purpose is the obvious one of screening proposals to establish which projects are feasible. The second purpose is to offer feedback to enable potential users to redesign projects that have flaws or to provide detailed explanations of why a project is not feasible. The Census Bureau has developed a set of procedures that review all project requests and bases its approval or rejection of a project on a specific set of criteria, which include the following:

(1) Analysis of a project for disclosure problems.

(2) Evaluation of whether a project falls under appropriate use conditions.

(3) Evaluation of the benefit to the Census Bureau.

(4) Analysis of the feasibility of the project.

(5) Analysis of the costs to the Census Bureau of supporting the project.

(6) Review of the scientific merit of the project.

While certain of these criteria are typical in any review of scientific projects, the Census Bureau requires that the project present no inordinate disclosure problems, provide benefits for the Census Bureau programs, and represent appropriate use. Once these conditions are met, the Census Bureau evaluates a proposal based on scientific merit, feasibility, and cost. A researcher receives feedback on what is feasible and, in some instances, suggestions to modify the project to make it fall within appropriate use guidelines, meet disclosure requirements, or satisfy the benefits conditions of the Census Bureau. In all these facets, the researcher typically needs to work with the staff of the secure sites to develop proposals that address the Census Bureau evaluation criteria. This can be challenging, because outside researchers may not fully understand how their proposed research benefits the Census Bureau. By working with secure site staff, researchers can tailor their project proposals to better articulate these messages and avoid disclosure problems through careful design[14].

Once a project has been selected, formal agreements between the researcher and the agency need to be established. These may involve writing contracts and setting fees for the proposed work. Contractual relationships can be more complex when the secure site is located at a university or other research institution. In these cases, the researchers may need to form agreements with the secure site operator, as well as with the statistical agency. In addition, security clearances and the taking of the required oaths will be required before the researcher commences the project.

Operators of secure sites must be prepared to train the researchers in security and data issues when the researchers arrive at the site. While many of the security provisions are common sense, external researchers may be relatively unfamiliar with the appropriate methods for disposal of printed matter, the use of secure computing facilities, the restrictions on the use of mobile computer platforms, the monitoring activities (badges, sign in sheets, *etc.*), and the disclosure regulations. Secure sites should set up orientation sessions with each researcher.

In terms of the computing environment, access can be provided through a centralized workstation where user privileges are tightly controlled, or data can be placed on individual workstations where network access is restricted. The choice depends on the computer configuration of the secure site, the computer expertise available, and the costs of providing centralized versus standalone computers. If data are placed on individual workstations, some agencies disable the floppy disk drives and communications ports. Secure sites that use centralized workstations for computing and statistical analysis should be prepared to provide access to standard office suite software as well. This is necessary because researchers will often construct tables and write research papers at the secure site.

Throughout the research project, staff of the secure site should be prepared to provide help with data and computer problems and, more important, help the researcher develop a strategy to deal with potential disclosure problems. While the proposal review process should provide researchers with a reasonable view of disclosure constraints they will face, in practice, constraints become real when the project is beginning to generate output. For example, it is not unusual for a researcher to ask to take some preliminary output back to the researcher's home institution. The specific output the researcher requests for release may pass the disclosure requirements of the statistical agency and secure site. However, releasing a certain set of preliminary output may preclude the researcher from taking out a different slice of the data later because of complementary disclosure problems. The researcher must be made aware that decisions on what preliminary output to take out at the beginning of the project may affect any subsequent requests for disclosure analysis.

---

[14] In addition to the formal review process, an active prescreening of potential projects and discussions with potential researchers are important. Early discussions with researchers often identify projects that either are not feasible or that will not be authorized well before the researcher submits a proposal.

As the project finishes, the disclosure analysis on the final research results must be carried out, and a summary report of the data used and the research output (papers, presentations, *etc.*) should be submitted. The secure site should be prepared to entertain subsequent data requests by the researcher in certain circumstances. This is particularly true when the researcher is submitting the papers developed in the project to peer-review journals. Such journals often request revisions to the empirical analysis. Secure sites need to be prepared and flexible in handling such requests, and such eventualities should be discussed before the researcher completes the initial project.

System Management Issues

In establishing secure sites at non-agency locations, the agency is entering into a partnership with another party. In the United States and Canada, these arrangements have been with universities. Currently, the Census Bureau operates research data centers with four separate universities. One issue that must be addressed is how such secure sites are awarded. In the United States, the Center for Economic Studies at the Census Bureau works with the National Science Foundation (NSF) to award sites through a formal review process that includes input from the Census Bureau, external reviewers, and the NSF. The review process is organized by the NSF, and interested parties submit proposals directly to the NSF. The NSF obtains external reviews of the proposals, and the Census Bureau also reviews the proposals. It is important that the award process be viewed as unbiased. The secure sites are self-funding and pay for the equipment, the staff, the Census Bureau personnel at the site, and the office space. The NSF provides some start-up funds to the sites, but these funds do not cover the annual costs of operation. The agency, in this case the Census Bureau, provides the data, the security plans, and oversight, and employs a staff person at the location. The Census Bureau maintains project approval authorization at all sites.

This partnership relationship can be a challenging one. Ideally, one would like the secure sites to operate as 'branch offices' of the statistical agency, with the objectives and goals of the branch office the same as those of the main office. But at times some natural tension can exist. The secure site at the university has an interest in promoting research opportunities at the location. While the agency has a similar goal, it is concerned first and foremost that the data be used in an appropriate and authorized fashion. Hence the agency may reject research proposals or require substantial revisions in projects that the university partners are promoting. Nonetheless it is important that the agency make all attempts to establish a 'branch office' organizational culture. The alternative organizational structure is the 'franchise' model. In this model, the individual units are more independent in their objectives (though they must abide by the contractual arrangements) and there can be increased friction between the franchisee (university) and the franchiser (statistical agency).

In addition to creating the appropriate organizational culture, operators of secure sites have to develop an overall disclosure strategy for their sites. While statistical agency rules and disclosure review boards will provide overall guidance, the disclosure problems faced by the secure site are very different from those faced by the statistician deciding what should be released from a particular dataset. Consider the case of disclosure analysis on tabular data. The statistician usually defines a set of rules that tables must follow so that the risk of disclosure meets the agency's standards. These rules will include the analysis of primary and complementary disclosure. The statistician will predefine the tables under consideration and the disclosure analysis will commence. Cells will be suppressed or combined to make sure that no disclosures are present in the tables.

The secure site faces a more difficult task. It must envision a range of requests for data releases from a number of researchers over an extended period. A particular release of data to a researcher may preclude the further release of data from that project or even from another project. With complex projects, the problems of complementary disclosure can become quite difficult to manage. One solution (which the Center for Economic Studies employs in managing its research data center network) is to try to minimize the amount of tabular output that the secure site will consider for disclosure analysis. Researchers are encouraged at the initial stages of their projects to restrict their tabular output requests and to focus on model-based output. This is important for two reasons. First, tabular output is more likely to generate complementary disclosure problems, both with other projects and with statistical agency publications. Second, it is quite costly for the statistical agency to perform detailed disclosure analysis on tables with large numbers of cells. In practice, it can take much less time to judge whether the estimated parameters and standard errors from a statistical or econometric model have disclosure problems[15]. Another solution that the CES has adopted is to set very strict (high) disclosure standards for tabular output[16]. This discourages researchers from submitting extensive tabular data that they know will not pass some predefined cell count thresholds.

Regardless of the specifics of the disclosure procedure, an agency must establish a workable strategy. The strategy must include relatively straightforward rules that can be applied by the staff who interact with the researchers. It must recognize that detailed disclosure analyses of certain types of data are costly and should be avoided. It should be communicated early and clearly to the researchers at the secure site so that they can design their research to minimize disclosure issues. And most important, the disclosure process should err on the side of caution.

---

[15] Model-based output still requires disclosure analysis, and certain types of models and parameter estimates raise disclosure issues. In particular, discrete variables that identify the mean difference of small numbers of observations and models with individual effects incorporated can present problems. Reznek and Nucci (2000) discuss these problems.

[16] The issues of disclosure analysis in secure sites operated by CES are discussed in greater length in Cooper et al. (1998).

4. Benefits of Secure Sites

This section will focus on two main benefits of secure sites: benefits to the statistical agencies involved and benefits to the research community. The benefits to the statistical agency are multifaceted; they include direct benefits to the statistical programs, the development of skilled human resources, and the development of new data and statistical products. The benefits to the research community are substantial and are evidenced by the observed demand for such data access services. The available data at statistical agencies allow researchers to undertake analyses of detailed geographic areas, allow the creation of unique data sources by enabling researchers to match across microdata sets, and allow researchers access to whole classes of microdata where there are no public use substitutes (*e.g.,* business establishment data). In addition, the creation of secure sites allows the research community to benefit from close interaction with the skilled staff of statistical agencies. In this way, details on the creation of various datasets can be communicated to the external researchers.

The statistical programs of the agency that develops secure sites can benefit in a number of ways. First, by developing a community of scholars who use the data, the agency can receive feedback on methodological, measurement, and data quality issues. In the case of certain microdata collection programs, detailed feedback is often lacking. This is particularly true in situations where public use microdata are not released to the public. For example, as discussed above, statistical agencies often have extensive data collection programs dealing with establishments and firms. These programs generally release aggregated tabulations for industry and geography cells. Microdata from these programs are generally not released to the public. Thus, the agency rarely receives feedback on methodological aspects or data quality issues relating to these kinds of data. Second, by allowing restricted access to the microdata, the agency builds up a community of skilled data analysts. These individuals represent, in part, a pool of potential future employees to the agency. The experience of the U.S. Census Bureau suggests this is important. Many research assistants who initially worked on research projects at secure sites have become Census Bureau employees. Statistics Canada's secure site program specifically discusses training of data analysts and researchers as an overall goal of the program. Also, secure sites can be viewed as supporting the overall mission of the statistical agency. The mission of a statistical agency is typically to provide data to society to shed light on important demographic, sociological, and economic problems and to provide quality data for public policy analysis. Secure sites further that mission by providing access to rich data resources that are generally unavailable but that have great value to researchers.

With respect to the benefits to the research community, secure sites provide access to incredibly rich data housed at the statistical agency. In the United States, secure sites at the Bureau of the Census and its university partners have been used over the past 15 years to provide access to a broad range of researchers[17]. These researchers have developed new data series that provide new descriptions of the economy and have analyzed fundamental issues in the areas of productivity, entrepreneurship, the interaction of workers and firms, and industry dynamics[18]. In addition, secure sites enable researchers to increase the information yield of traditional data sources by the ability to match across sources at secure sites. The recent volume by Haltiwanger, Lane, Spletzer, Theeuwes, and Troske (1999) discusses at length the benefits of using matched employer-employee data. Such matched data can address a range of issues that individual datasets simply cannot address[19]. Under the strict supervision of the statistical agency authorities, data sources can be matched based on detailed keys. The most sensitive identifier data (which are often contained in the keys) can then be removed from the data and the researcher can gain access to matched files without access to the sensitive key fields. Secure sites are ideal environments for such matching to take place. Secure sites and increased access to microdata in various agencies have already had a significant impact on social science research. The ability to create and analyze even more detailed datasets at secure sites suggests that this impact will continue for some time to come.

5. Conclusions

The development of secure sites offers statistical agencies a mechanism to increase access to their unique data resources and at the same time maintain strict control over how the data are used. Secure sites enable statistical agencies to satisfy (in part) the increasing demand for microdata on individuals and business units when these data cannot be made publicly available because of confidentiality concerns. This is a reasonable and prudent strategy, employing elements of both restricted data and restricted access approaches to microdata provision. The restricted access is provided through the secure sites, and it limits use to authorized researchers. Within the secure site, a restricted data approach is used. While researchers are given access to data that are not publicly available, only the minimum data required are made available. Sensitive information such as names and addresses are removed from the data files before a researcher can use the data. In addition, secure

---

[17] McGuckin (1995) provides a review of a large number of studies that have been undertaken at the U.S. Census Bureau's network of research data centers.

[18] New data series on employment flows were developed by Davis et al. (1996), and new diversification indices were constructed by Gollop and Monahan (1991). Baily et al. (1992) and Olley and Pakes (1996) provide detailed analyses of the role of plant heterogeneity in the evolution of industry productivity. Bernard and Jensen (1997) illustrate the usefulness of such data to international economists.

[19] Papers by Abowd *et al.* (1999), Haltiwanger, Lane, and Spletzer (1999), and Hellerstein *et al.* (1999) are examples of recent papers that use matched employee-employer data to analyze issues in labor economics.

sites allow researchers to gain access to new datasets that use a variety of sources in their creation. Important in this area are the newly created employer-employee matched datasets.

The management of secure sites, especially sites located away from the statistical agency, is not without challenges. First, agencies need to impart the 'culture of confidentiality' throughout the secure site network. This requires a trained and dedicated staff. Second, secure sites typically require investments in documentation development, data management, project oversight mechanisms, disclosure infrastructure, and staff training, in addition to the physical infrastructure investments involved in establishing a secure location. While this chapter has focused on the model used by Statistics Canada and the U.S. Census Bureau (*i.e.*, establishing of physical locations with statistical agency employees and data on-site), it is likely that technological change will lead to different modes of access. As network speed and network security features advance, remote access to a centralized data resource will represent an alternative access model. In this case, data and computing facilities could remain centralized, and access to the data would be through secure network transmissions. Danmarks Statistik already operates such a secure site through its regional office in Aarhus. Finally, as networks of secure sites evolve, it is critical that the statistical agency, secure site operators (universities, research institutions), and the researchers become real partners in the process. This will help to ensure that secure sites function well and provide the benefits discussed throughout the chapter.

References

Abowd, John M., Francis Kramarz, and David Margolis (1999) 'High Wage Workers and High Wage Firms', *Econometrica,* 67(2), pp.251-333.

Baily, Martin N., Charles Hulten, and David Campbell (1992) 'Productivity Dynamics in Manufacturing Plants', in *Brookings Paper on Economic Activity: Microeconomics Annual,* pp.187-249, Washington, D.C.: The Brookings Institution.

Bernard, Andrew B., and J. Bradford Jensen (1997) 'Exporters, Skill Upgrading and the Wage Gap', *Journal of International Economics,* 42(1), pp.3-31.

Cooper, Joyce M.R., David R. Merrell, Alfred R. Nucci, and Arnold P. Reznek (1998) 'Protecting Confidential Data at Restricted Access Sites: Lessons Learned from Census Bureau Research Data Centers', 1998 Proceedings of the Section on Government Statistics and Section on Social Statistics, pp.91–96, Alexandria, Va.: American Statistical Association.

Davis, Steven, John C. Haltiwanger, and Scott Schuh (1996) *Job Creation and Destruction,* Cambridge, Mass.: MIT Press.

Duncan, George T., Thomas B. Jabine, and Virginia A. de Wolf (eds.) (1993) *Private Lives and Public Policies: Confidentiality and Accessibility of Government Statistics,* Washington, D.C.: National Academy Press.

Flaherty, David H. (1979) *Privacy and Government Data Banks: An International Perspective,* London: Mansell Publishing.

Gollop, Frank M., and James L. Monahan (1991) 'A Generalized Index of Diversification: Trends in U.S. Manufacturing', *Review of Economics and Statistics,* 73(2), pp.318-30.

Groot, Andrea, and Cor A.W. Citteur (1997) 'Accessibility of Business Microdata', *Netherlands Official Statistics,* Volume 12, Winter, pp.18-32.

Haltiwanger, John C. (1997) 'Measuring and Analyzing Aggregate Fluctuations: The Importance of Building from Microeconomic Evidence', *Federal Reserve Bank of St. Louis Review,* 79(3), pp.55-77.

Haltiwanger, John C., Julia I. Lane, and James R. Spletzer (1999) 'Productivity Differences Across Employers: The Role of Employer Size, Age and Human Capital', *American Economic Review: Papers and Proceedings,* 89(2), pp.94-8.

Haltiwanger, John C., Julia I. Lane, James R. Spletzer, Jules Theeuwes, and Kenneth Troske (1999) *The Creation and Analysis of Employer-Employee Matched Data,* Amsterdam: Elsevier Science B.V.

Hellerstein, Judith K., David Neumark, and Kenneth R. Troske (1999) 'Wages, Productivity and Worker Characteristics: Evidence from Plant-Level Production Functions and Wage Equations', *Journal of Labor Economics,* 17(3), pp.409-46.

McGuckin, Robert H. (1995) 'Establishment Microdata for Economic Research and Policy Analysis: Looking Beyond the Aggregates', *Journal of Business and Economic Statistics,* 13(1), pp.121-6.

McGuckin, Robert H., and Sang V. Nguyen (1990) 'Public Use Microdata: Disclosure and Usefulness', *Journal of Economic and Social Measurement,* 16(1), pp.19-39.

Olley, G. Steven, and Ariel Pakes (1996) 'The Dynamics of Productivity in the Telecommunications Equipment Industry', *Econometrica,* 64(6), pp.1,263-97.

Reznek, Arnold P., Joyce M. Cooper, and John B. Jensen (1997) 'Increasing Access to Longitudinal Survey Microdata: The Census Bureau's Research Data Center Program', *American Statistical Association 1997 Proceedings of the Section on Government Statistics and Section on Social Statistics,* pp.243-8.

Reznek, Arnold P., and Alfred R. Nucci (2000) 'Protecting Confidential Data at Restricted Access Sites: Census Bureau Research Data Centers of Significance,' 2(1), pp.42–50.

Social Sciences and Humanities Research Council of Canada and Statistics Canada (1999) *Canadian Initiative on Social Statistics,* Ottawa: SSHRC.

Confidentiality, Disclosure, and Data Access: Theory and Practical Application for Statistical Agencies
Pat Doyle, Julia I. Lane, Jules J.M. Theeuwes and Laura M. Zayatz (Eds)

Chapter 13

The Potential and Perils of Remote Access

Michael Blakemore
University of Durham

1. Introduction

Remote access to data seemingly has never been easier than in the era of the World Wide Web, where a wide range of e-commerce solutions offers secure access to confidential data. In late 2000 I demonstrated the use of a secure banking system to a group of civil servants in Malaysia. I was confident, 8,000 kilometers from my home, that I could access our family charge card account and that my personal details would not be intercepted. On an overhead display, we all saw what my wife and I had recently spent in Durham and, even more recently, in Kuching.

On an individual basis, I had opted into a process whereby some of our personal information could be shared, and I demonstrated the trust I have in secure information transactions. In doing so, I was making a conscious risk assessment with respect to our own data. It is important to note, however, that this decision did not impinge on the confidentiality of others, which is a very different proposition. Imagine the task of the charge card company if it had to secure the permission of every cardholder to allow the use of personal financial data for research purposes, and you get some idea of the task facing statistical agencies charged with releasing confidential data for research.

The essence of the problem is the definition of 'secure'. In the commercial environment, as with the secure Internet database facilities available commercially, the definition of secure is a 'best endeavors' type of approach—balancing the risk of disclosure against commercial benefit. For an official statistical agency, however, striking a stable balance is much more difficult. The public, upon whom statistical agencies depend for the comprehensiveness and integrity of their data collection (and therefore their reputation and ability to do their job), are hostile to the possibility of any disclosure of individual data. The U.S. Federal Statistical Confidentiality Order of 1997 (OMB 1997, p.35,047) explicitly recognizes this in its statement that 'willingness to cooperate in statistical programs substantially affects both the accuracy and completeness of statistical information and the efficiency of statistical programs'.

But absolute protection against the risk of disclosure implies no release of individual data at all—a position that is not tenable for statistical agencies because of the rising demand by analysts and policy makers for such data.

Thus, although remote access to sensitive statistical data is, on the surface, a technological issue, in reality technology is only one of the factors statistical agencies must take into account when making decisions about remote access to confidential data. Fear of disclosure, a large part of the social context, is equally important, and can include legal barriers to disclosure imposed by governments in response to public fear.

The next two sections of this chapter describe the social context and the information technology (IT) context within which agency policy must be made. The chapter ends with four case studies illustrating differences in how the tensions between the social and IT contexts are being addressed, an assessment of where we are in 2001, and a conclusion.

2. Agencies and the Social Context

Rapidly developing information technologies offer increasing potential for unrestricted information flow; and 'with every new advance in technology and every new generation of analyst', the desire for access continues to grow (Keller-McNulty 1997). For statisticians, Malaguerra (2000, p. 89) warns that 'as the methods employed by NSIs [National Statistical Institutes] undergo radical change, do we have any clear idea of the consequences which will arise from this revolution.' As the United Nations Economic Commission for Europe (UNECE) noted in a recent report, the 'need to provide customized access to statistical resources appears to be an unavoidable challenge to statistical institutes' (UNECE 2001a, p.5). For official statisticians, the Internet in particular is heightening expectations for both data collection and dissemination.

At the same time, it has never been easier to let confidential information get into the wrong hands and lose the trust of those who provide the data. Surrounded by a diversity of mass communication channels, citizens can be alerted, alarmed, informed, and misinformed about threats to their individual liberty, with the development of technology inevitably heightening the risk of breach. As Fienberg (1997, p.11) has warned, 'never before will so many have had access to so much data. In such circumstances, the opportunities for malfeasance will inevitably grow'. Hacking is an ever-growing threat to data security, to the extent that even governments participate in the practice (Lemos 2001). Surveys paint a bleak picture of increasing cybercrime threats. They also document high levels of citizen concern about the ramifications of those threats. The Pew Internet Project (Fox and Lewis 2001, p.2), for example, found high levels of concern among U.S. citizens in early 2001 about on-line credit card theft (87 percent), cyberterrorism (82 percent), fraud (80 percent), hacking government computers (78 percent), hacking business networks (76 percent), and viruses (70 percent).

And public apprehension has been rising over time (Marx, 1988). (The only countervailing pressure is fear of government nondisclosure under the Freedom of Information Act and related 'democracy' themes (Freedman 1987)). Complaints to the U.K. Data Protection Registrar increased by 36 percent in the reporting year 1999-2000, for example, with complaints relating to personal data held on computers in all sectors and reportedly overburdening the registrar staff (Hencke 2000).

Nor is public fear restricted to disclosure of individual pieces of information. It also encompasses fears of record linkage, not only by users external to government but by government itself, as articulated in the context of the U.K. 2001 Census of Population. Particular targets for criticism in that context were questions about topics on which government databases already exist, such as unemployment and employment. The census asked for the postal address of the respondent's employer, for example, which could be very useful to tax authorities cross-checking income data. One observer, admitting that 'this may sound like conspiracy', made note of the fact that the census did not ask that of anyone 'over 75 (in whom the Inland Revenue would not be so interested) about their employment' (Waterhouse 2001).

The *Statistical Program of the United States Government—Fiscal Year 2000* acknowledged the tension between data access and public privacy concerns when it agreed that a consistent confidentiality policy would be welcome across federal agencies, but warned that any such policy should not be allowed to compromise 'public confidence in the security of information respondents provide to the Federal Government' (OMB 1999, p.40). The U.S. Commerce Department has made a similar acknowledgment of these tensions in creating the post of privacy adviser, its spokesperson emphasizing that the Secretary of Commerce 'feels strongly that we should lead by example, and therefore the first thing we should do is get our own house in order' (cited in McGuire 2001).

The abuse (perceived or real) of personal citizens' data by government resonates throughout the debate on access to microdata (Abramson 2000, Hurt 2001) and is of the highest political concern. Duncan (1998, p.16) quotes Al Gore: 'Privacy is a basic American value—in the Information Age, and in every age. And it must be protected.'

This pervasive public fear makes it crucial that discussion of remote access take a pragmatic approach in assessing the technologies of disclosure control and confidentiality protection. Communicating risks to the public clearly, simply, and objectively is as essential in the case of confidentiality risks as it is in other types of risk, and plausibly as difficult. In the late 1990s, for example, the U.K. public became hysterical over 'mad-cow' disease and genetically modified (GMO) foods—hysteria generated largely by the media, not the specialists. The U.K. government's Chief Medical Office produced a balanced report on GMO (Donaldson and May 1999, pp.2–3); yet the media, with the simple label 'Frankenstein foods', were able to influence public opinion to such an extent that major retail food companies removed all genetically modified components from their products.

Thus, remote access technologies must be evaluated as part of a bigger 'package'—that of a continuing partnership between citizens and government in the provision of information, policies, and services to citizens. In this context, statistical agencies must continue to balance their data needs with not only the technological risks but also the social risks involved (Zealand 2000b).

How has the community of national statistical agencies in fact reacted to these concerns? The basic answer is, pretty conservatively. In the early days of computer use, statistical agencies were slow to use computer mechanisms to preserve confidentiality, typically continuing to ensure nondisclosure by vetting each publication before release (Felligi 1972). As IT systems developed, the basic trend toward protecting individual privacy continued (Duncan 1998, Kruskal 1981, Posner 1984). This trend has its critics, of course, among them Openshaw, who has criticized mechanisms of disclosure control as a 'set of essentially ad hoc methods' (Openshaw *et al.* 1997, p.1) and has persistently referred to the conservatism of agencies as unnecessary, given the 'implicit assumption in mapping census data that the spatial variations in colour across the map are real and not largely an artefact of the mapping process" (Openshaw and Rao 1995).

The tendency of statistical agencies to be conservative has been exacerbated by the general fact that legislation tends to lag behind technological advances. This 'legislative legacy' (Camp 1999, p.256) can often be more of a constraint than technological failings, in part because of knee-jerk types of reaction that tend to come from legislators who have to react to events rather than preempt development (Spencer 1997, p.168). Statistics Canada sees a benefit in being able to disseminate the individual records from the 1901 census, for example, but the Statistics Act as it stands forbids such dissemination. Options being considered are to amend the legislation 'to retroactively change the confidentiality provisions of the... Act', or to secure informed consent from those who are related to individuals in past censuses (Statcan 2000c). In the United Kingdom, legislation protects individual census records for 100 years, so the Public Records Office is preparing the 1901 data for public release on the Web in January 2002[1].

3. Agencies and the IT Context

The technology of access control itself encapsulates the tensions between dissemination of statistical microdata and the risk of unauthorized disclosure. The approach to data has been one that is largely skeptical of the power of technology and digital networks to minimize risk—with use of remote data centers, to which researchers must travel to use the data, the preferred approach adopted by agencies

[1] http://census.pro.gov.uk/

such as the U.S. Census Bureau and Statistics Canada. Here the technological risks are contained within the confines of the statistical agency. Any failings of networks, hardware, and software are checked by the administrative practices of the data control centers: formal contracts between users and the agency, physically secure IT installations on materials that can be taken in and out of the centers, and rigorous vetting of outputs by automated and/or human intervention. As one observer of Statistics Canada's launching of new data research centers late in 2000 put it, 'the strictly controlled environment of the centers makes it possible to perform essential social research while assuring the security of the data' (Statcan 2000a).

This form of remote access has serious limitations, however. Travel and subsistence costs can be high for researchers who must travel to a center, the number of researchers who can be hosted at any one time is limited, and the task of writing a discrete proposal for each period of research denies researchers the opportunity to experiment. More fundamentally, as an observer of U.S. practice in the mid-1990s noted, 'the current approaches to data and metadata in the Federal Statistical System make very little use of emerging new technologies' (Keller-McNulty 1997).

Moving from internally controlled data centers to remote access, however, involves ceding control of many security aspects to third parties, with failure of any of these security aspects impinging mostly on the reputation of the statistical agency, not the third party.

There are three key components to 'liberating' researchers from having to travel to the data. First is the network, primarily the Internet, along with the physical IT structure of the installation, using firewalls and related technologies. Second is the software used, which ranges from security systems, through analytical statistical software, to techniques such as encryption for the transmission of data to researchers at remote locations. Third is the organizational context within which the preparation and transmission of data take place.

Network and Physical Infrastructure

Focus on protecting the security of the network and physical infrastructure (Oppliger 2000) has intensified with the expansion of hacking and cyberterrorism. At the national level in the United States, for example, the National Infrastructure Protection Center has been set up as the 'focal point for threat assessment, warning, investigation, and response for threats or attacks against our critical infrastructures' (NIPC 2000), such as the three-year assault on the U.S. infrastructure (Abreu 2001). At an agency and organizational level, one model is to give a 'chief security officer' (Hancock 2001) the key task of generating a holistic approach to security, which Rathmell (2001, p.45) conceives of as having five dimensions: 1) vulnerability analysis, 2) definition of the critical infrastructure, 3) assessment and monitoring of the threat, 4) balancing security against liberty, and 5) coordination, control, and influence.

Given the levels of security concerns of recent years, the most sensible assumption is that any facility that is remotely accessible is a target for hackers. This places a considerable burden on the efficacy of the software and the organizational environment. As Bace (2000) has noted, the 'roots of security problems' are in system design and development, in system management, and in allocating trust appropriately.

Software

At the statistical level, software needs to ensure 1) that any data outputs that contain disclosure risks are transmitted to the recipient in a secure mode (*i.e.,* a mode that, if intercepted, cannot be constructed into identifiable records) and 2) that the recipient is the only person authorized to receive and use the data.

The specifically statistical software methodologies are covered elsewhere in this volume. Disclosure control techniques have developed radically from the random sample queries proposed by Denning (1980, p.308). In the medical sector, for example, the development of 'digital hospitals' (Delio 2001) necessitates the protection of privacy in an environment where there is a strong justification for linking and sharing patient records both for individual care and for wider epidemiological monitoring and research. Quantina and colleagues present a hash algorithm approach that produces anonymized medical records that can then be linked (Quantina *et al.* 1998, p.122); another algorithmic, Boolean, approach is adopted by Øhrn and Ohno-Machado (1999).

Ensuring that the right records are accessible by the correctly authorized users involves methodologies such as those developed in the games industry for role-based access control, where 'permissions are associated with roles, and users are assigned to appropriate roles' (Ahn 2000, p.1176). Role-based approaches such as discretionary access control (DAC) are considered inflexible by Lin, however, who extends role-based access to role-based policy-enforced access control (RBPEAC). RBPEAC maintains security policies within a role's privileges so that 'RBPEAC is capable of dealing with both Internet users and distributed policy-based access control, using PKI and role-assignment policy' (Lin and Brown 2000, p.1591). Furnell and Dowland (2000) consider intrusion monitoring, and Turega (2000, p.247) considers accessing confidential documents using certified public keys.

Authentication is vital to the growing suite of e-commerce security systems, with solutions being provided by companies such as Identrus[2]. Public-key cryptography (PKI) goes beyond conventional encryption where the sender and receiver use the same encryption algorithm. PKI uses a public key and a private key, where 'information encoded using the public key can be converted to plain text only with the private key' (Ackerman 1998, p.372). Biometric approaches are a way to overcome the unreliability of passwords and personal identification numbers (PINs),

---

[2] http://www.identrus.com/

which can be obtained by others who can impersonate an authorized user, by requiring physical characteristics of a user to be authenticated (Bennett 2000, p.23). For example, BioconX[3] can base authentication on fingerprints, hand silhouettes, facial geometry, the color pattern in the iris of the eye, and the blood vessel pattern in retina of the eye.

The application of biometrics, and its implied criticism of previous technologies, highlights the dilemma for statistical agencies. Any technology is fallible. It is hoped that the newest technology is potentially less fallible then the previous technologies. How do you use technology that can only be 'less infallible' than previous technology in applications where any failure of security can be massively damaging to the agency? This highlights the real challenge to technology: It must aim to replicate 'the nuances of the real-world contractual relationships that reflect the practical realities of access to information in our modern, inter-organisational era' (Eaton 1999, p.191).

In an extensive review of record linkage procedures, the U.S. General Accounting Office (GAO) identifies stages in the technical approaches to disclosure control (Kingsbury 2001). At the software level, the task involves building a 'privacy toolbox', which comprises techniques for masked data sharing, and reducing both the reidentification risks and the sensitivity of the data being linked (Kingsbury 2001, p.76). The next stage is 'secure transfer of data', including encryption in preference to techniques such as transferring data in subsets. The third stage involves practices such as 'safer settings', including physically secure data research centers, licensing procedures, and usage audit trails. The GAO report notes the utility of maintaining rigorous audit trails, calling them 'an effective security tool because they create a continuous log of information about system activity. This includes the user's identity, location, date, time, information accessed, and the function performed' (Kingsbury 2001, pp.98–99). The GAO even cites approvingly the strong view that 'some believe that audit trails are potentially "one of the strongest deterrents to abuse."' (Kingsbury 2001, p.99).

The Organization

In many aspects of IT, the weakest linkages are at the organizational, not the technological, level. Organizations, for example, generally view information security as largely a technological issue to be delegated to IT staff. The result is 'gaps in the corporate information security chain of defenses' (Mitchell *et al.* 1999, p. 226), with the 'chief security officer' left to confront such threats as e-mail viruses, denial of service attacks, and direct hacking (Hancock 2001).

There are plenty of horror stories about IT security failures, new and old. For example: 'I can tell you categorically that I can gain access to virtually any data bank, given enough time and money. Categorically! The system is extremely

[3] http://www.bioconx.com/

leaky (McNulty 1979, p.100). Scary stuff, and not the sort of statement that inculcates a feeling of trust and satisfaction into skeptical citizens. McNulty was managing director of General Robots, and therefore technically qualified to make such a wide-ranging warning. In the context of remote access, however, is it technology or organizational behavior (Mabuchi *et al.* 1996) that presents the biggest risk? Dinnie (1999) reports on the second annual Ernst & Young Global Information Security Survey. Data were collected on real and perceived IT threats from more than 4,200 companies in 29 countries. The most common actual problems were reported as IT system failure, occurring in 76 percent of the companies responding. Accidental errors (66 percent) were followed by viruses (49 percent). This contrasts dramatically with the most common perceived threats: hackers (53 percent); unauthorized users (49 percent); former employees (46 percent); computer terrorists (37 percent); contract workers/consultants (35 percent); competitors (32 percent); industrial spies (32 percent); and authorized users/employees (31 percent) (Dinnie 1999, p.113).

That hacking is high on the agenda of concern is not surprising, given the penetration of websites in recent years. Hawkins observes the process of leap-frog, where the industry is often barely one step ahead of, and often one step behind, the IT terrorists (Hawkins *et al.* 2000, p.142). In the statistical arena, experience of the Centers for Disease Control and Prevention (CDC) and the Agency for Toxic Substances and Disease Registry (ATSDR) indicates that the major threat to data security occurred when data were being transmitted over the Internet, but 'while the data reside on workstations and servers that are accessible from the Internet' (CDC/ATSDR 1999).

Individual and organizational behavior therefore are fundamental aspects of any successful remote access system for confidential data; and user training or the development of ethical/moral users (Siponen 2000, p.39) has received attention. Lynch proposes that a sensitive balance be reached between 'functional requirements, about roles and responsibilities…in structuring the authentication and authorization mechanisms that will support the networked information environment' (Lynch 1997, p.38). Individual behavior and perceptions of risk and trust were cited as part of the Eurostat disclosure research needs in the European Commission Fifth Framework program. For Eurostat there were issues related to cognition and the ways in which citizens comprehend and interpret the confidentiality assurances made by statistical agencies, whether citizens viewed some data as being 'more' confidential than others, and 'the extent to which data aging is felt to make certain types of data less confidential' (Thorogood 1999, p.102).

Agencywide risk management strategies help to articulate the balance of technological, organizational, and individual threat. U.K. National Statistics embrace formal information security standards (BS7999) (Statistics 2000a). The Netherlands Bureau of Statistics articulates formal staff and organization policies such as vetting visitors, minimizing computer interconnections, limiting e-mail facilities, restricting data access on a need-to-use basis, and such procedures as ensuring that those analyzing data 'only have access to the statistical variables of the linked cases, and not to the identifying variables' (Al and Altena 2000).

Formal risk assessments of disclosure are undertaken for surveys such as the Census of Population. The Scotland General Register Office did this, for example, in preparation for the 2001 census (GRO 2001). This effort focused primarily on the physical security aspects of long-term storage of the individual returns, as well as the ability of individual members of a household to fill in individual forms if they wished no other member to see their details. Referring to computer security, the document presents an inevitable conundrum, arguing that 'to preserve the high level of security guaranteed by the Registrars General it would not be prudent to publish the details' (GRO 2001).

The literature on IT and business promotes the broad issues of general awareness (Bertino 1998), organizational awareness of legislation (Henderson and Snyder 1999), and using trusted 'infomediaries' to gather (and redisseminate) information (Anon. 2000). The proof of success, as ever, of course is in the successful application of the procedures.

Integrated Solutions?

Earlier in this chapter, I referred to commercial solutions developed for the e-commerce sector and noted that secure data transmission was fundamental in my willingness to access my personal financial information over the Internet. It does not stop me checking each monthly statement with great care, however, because I still fear fraud and identity theft. There are commercially secure, remote access e-commerce solutions on the market, such as the i-Planet offering. It claims 'security similar to that of a Virtual Private Network (VPN)' and 'provides secure enterprise access on virtually any browser-enabled end user device without costly, hard-to-configure software' (SUN 1999). I-Planet and similar solutions have such clients as blue-chip companies, so the pedigree of such software is high. Applications have been developed for clients ranging from America Online (AOL) (SUN 2001) to the U.S. Department of Defense Web Portal (IPlanet 2001). But are these packages new solutions for remote access to confidential statistics, or are they more an indication of the state of the art in the sense that they are better than the 'secure' database systems sold 10 years ago? The issue is not so much absolute (is this system the best and most secure?) as relative (is this system good enough to persuade citizens that their data are being fully protected in the climate of increasing security threat on the Internet?).

In reality, the official statistical community would seem to have taken a rigorous attitude to information security. The threats of premature disclosure of time-sensitive statistics, along with the threats of compromising the privacy of individuals and organizations, are well understood. Leaks seem to be rare. Yet it is this very conservatism toward disclosure that some use to argue that statistical agencies are overprotective of information. So there are two crucial factors at issue in remote access, indeed any access, to confidential data. First, the techniques and algorithms are subject to challenge. Until there is unequivocal acceptance that they are robust and secure, the task of moving to the second stage, allowing remote access, is complex and controversial. The second stage itself is subject to challenge, and that challenge increases as the Internet security problems themselves increase. To what extent the algorithms, software, and IT infrastructure improve faster than the threats is a contentious issue.

4. Case Studies

The case studies presented here illustrate four approaches to the problem of reconciling the need to make information available with the need to protect its security.

'Soft' Disclosure

This rather pejorative term is used here to illustrate how other sectors are using access control to disseminate and protect valuable intellectual property. The 'risks' include loss of income stream and theft of intellectual property by people and organizations beyond the legal limit of national boundaries. While these risks can pose threats to individual privacy, such as the use of commercial credit referencing databases, the overriding concerns are intellectual property protection.

Access control software was used when researching this chapter. If it had been written 10 years ago, the likely alternative would have been many hours spent in the university library searching for journals and books, taking notes (either on paper or on a word processor), and using literal cut-and-paste techniques to put together material. Following that would be the inevitable anguish of chasing down the last few reference details to complete the bibliography. The task of accessing journals is now far easier through the use of e-journals via the Durham intranet and through electronic bibliographic software. Access control software enables researchers to use library resources much more effectively, and fears of mass theft of publishers' intellectual property have been largely offset by ethical use policies and electronic audit trails—the conventional carrot-and-stick approach. Any unauthorized access to journal papers damages the financial viability of the intellectual property owners but is not likely to prejudice the privacy of individuals.

The library sector therefore is strongly involved in remote access control (Willson and Oulton 2000). And for the suppliers of material to libraries, the ethical use policy is every bit as important as the technology. According to Garrison and McClellan (1997, p.53), 'our experience has shown that database vendors are more interested in the library making a good faith effort to implement secure access to their licensed databases using readily available technologies'.

An ongoing example of remote access to sensitive intellectual property is the DIGIMAP[4] project in the United Kingdom, which is providing access to a wide range of Ordnance Survey (OS) map information. U.K. academic researchers have long wanted access to OS map data. But OS protects its intellectual property very aggressively (Survey 1998) against the danger that it would be copied, transmitted, and abused by academics who too often promoted the 'public domain' view of information that is prevalent in the United States with regard to federal map data. The DIGIMAP project (EDINA 2001) uses commercially developed access control software within a geographic information system environment, and this software operates within a strong licensing, reporting, and audit environment that has been essential in persuading OS to participate in such large-scale nonintermediated dissemination.

The ethical use approach to the dissemination of official statistics has long been a practice for academic researchers, with intermediaries such as data archives playing an important role in both encouraging and policing data use. Flaherty (1979) made a comparative study of ethical use practice in the United Kingdom, Sweden, the Federal Republic of Germany, Canada, and the United States. The Essex Survey Archive (currently the ESRC Data Archive[5]) in the United Kingdom maintained a strong licensing and usage policy, while the United States and Canada disseminated public use samples with a code of ethics for users.

A service combining maximum dissemination detail with ethical use policy was detailed for the United Kingdom by Blakemore (1998). The Nomis[6] service, owned by U.K. National Statistics, is a publicly accessible (via a signed license agreement) database of labor market and demographic statistics. It contains data series that technically include confidential establishment data. At the smallest geographical areas (electoral wards) for unemployment data (disaggregated by age, gender, and occupation) it is technically feasible to identify individuals in wards with small populations and low levels of unemployment, and the Nomis database contains such data. This contravenes the 1947 Statistics of Trade Act, which guarantees that employment establishments will not be individually identifiable. The disclosure control approach used in Nomis is a combination of licensing, auditing, and pragmatism.

Licensing ensures that when using potentially disclosive establishment data, users actively acknowledge their obligations to use the data for their own research or applications but not to publish or redisseminate the data. Auditing in Nomis ensures the ability to track use and link any identified disclosure back to a user session. Pragmatism exists because there is, in fact, a legal contradiction. On the one hand, the 50-year-old Statistics of Trade Act forbids disclosure of establishments. On the other hand, many commercial databases have emerged during the past half

---

[4] http://digimap.edina.ac.uk/
[5] http://www.data-archive.ac.uk/
[6] http://www.nomisweb.co.uk/

century that identify establishments in considerable detail. This does not mean that official statisticians can override legislation, because that would threaten the high compliance levels that are so vital to the quality of official statistics. It does mean, however, that data can be disseminated in a way that increases use without leading to a risk of major disclosure. Nevertheless, it is highly unlikely that such an approach would be acceptable for data relating to individual citizens.

A related development under way in U.K. National Statistics is the Neighbourhood Statistics Service. This will be a no-charge service that 'is an innovative long-term project to make information widely available for thousands of small areas across the U.K.'[7] While it is a major improvement in public access to statistics, it will still suffer from the geographical 'cookie-cutter' approach—that is, you can have access to any geographical levels as long as they are the supplied administrative geographies. Only a remote access system linking users to a microdata processor, allowing customized geographies to be extracted, is likely to overcome this ecological fallacy. On that basis, development of remote access systems to microdata could be the biggest advance in the geographical production of data since digital databases were invented.

The commercial sector also has strategies for protecting individual information. The driver here is more often one of commercial confidentiality—to avoid competitors understanding your strategy. Thus these strategies have more to do with protecting commercial secrets than with disclosure control. Some commercial strategies, in contrast, have emerged on the Web whereby individuals knowingly and consciously sign away their data confidentiality rights, often in return for some form of financial gain. Sites here are typified by Permission Marketing[8]. In many ways, such sites are no different from the retail industry's existing practice of sending out complex questionnaires via direct mail and encouraging the recipients to respond in return for a prize, or the chance to win a prize.

The fundamental difference between the 'sophisticated' official strategy and commercial strategies tends to be that they are coming from different directions. Official statisticians are most concerned with any degradation of citizens' full compliance (hence the concern to maximize response levels in the 2000 U.S. census and the 2001 U.K. censuses), whereas commercial databases are trying to improve on historically low levels of response. Whether the two response levels will eventually meet, and if so at what level, is a debatable, but interesting, question.

Pre-prepared Restrictive Disclosure

The second category for disclosure covers restrictive access, either by limiting the spatial extent of access by making the data users travel to the data, or by preparing a prechecked dataset that, as far as can be assessed, will not infringe disclosure rules.

---

[7] http://www.statistics.gov.uk/neighbourhood/home.asp
[8] http://www.permission.com/

Restricting access spatially has been a characteristic of the research data centers set up in the United States by federal agencies such as the Bureau of the Census. Another example is the research data center (RDC) at the National Center for Health Statistics (NCHS) headquarters in Hyattsville, Maryland. The purpose of these centers is to allow access to data that otherwise 'would not be permissible to analyze because of confidentiality/disclosure rules and regulations' (NCHS 2000). The procedures run along lines similar to the ways access to confidential data is undertaken by agencies such as Eurostat, where a research project is accepted and the researcher becomes a member of staff subject to all the ethical and security regulations of the agency. The Eurostat approach to microdata access is relatively restricted, with the exception of the successful European Communities Household Panel (ECHP), although a 1999 draft European Commission regulation on 'access to confidential data for scientific purposes' does indicate the Commission's intent to address access issues (Trivellato 2000).

The NCHS undertakes rigorous screening of researchers. A proposal is fully evaluated, ideally in collaboration with an existing member of RDC staff. Confidentiality agreements are signed, work must be undertaken within RDC facilities, and any merging of NCHS data with researchers' data from other sources is undertaken by RDC staff. Any remote access to data is the result of more internal processes involving RDC staff producing of data files, limiting the analysis to particular statistical software and to restricted facilities within the software. For example, 'no cell fewer than five observations; if found, other cells also suppressed. Job log scanned for conditions that spawn case listings. Manual disclosure limitation review where necessary' (NCHS 2000). The NCHS is critically aware of the two-way relationship between data providers and users, and its phrasing is quite definite—not best endeavors but 'all steps possible to protect the confidentiality of individually identifiable information' (NCHS 2001).

A pre-prepared dataset that can be accessed remotely under license is typified by the U.K. Sample(s) of Anonymised Records (SAR or SARs). First produced from the 1991 Census of Population data, the genesis of the SARs was the result of a long process of lobbying the census agency (then Office for Population Censuses and Surveys and more recently the Office for National Statistics) and of persuading official statisticians that the sample was feasible, in terms of both disclosure limitation and user licensing and behavior. A paper on disclosure risk prepared for the Cathie Marsh Center at the University of Manchester[9] by Professor Tim Holt concludes that 'the risk of disclosure is negligible for the large majority of the population and thought to be extremely small for the remainder. International experience indicates that the levels of risk are consistent with a decision to go ahead with the release of the SARs, and I so advise' (CSSR 2000).

Remote access to the U.K. SARs has operated successfully, and work during the late 1990s focused on persuading the census agency not only that similar products

---

[9] http://les1.man.ac.uk/ccsr/

could be generated from the forthcoming 2001 Census of Population but that it may even be feasible to enhance the detail. Such behavior is typical of the cautious approach to the inclusion of more detail by official statisticians, caution that is well justified in the light of general privacy and disclosure threats. (See Dale and Elliott 1998 for a statistical justification of 2001 SARs.) A major concern was that users could obtain confidential data by differencing multiple overlapping geographies. According to an expert assessment, however, 'in virtually all of our comparisons, the risks that differenced areas with person or household numbers below the confidentiality threshold were either absent or extremely small (less than half of 1%)' (Duke-Williams and Rees 1998, p.602).

The U.S. Computer Science and Telecommunications Board (CTSB) organized a workshop in 2000 to consider the challenge of disclosure control in this geographical context. Into the debate on classifying database records it introduced a concern that the spatial element of the data has 'received relatively little attention', citing the need for research that balances confidentiality and disclosure risk with 'the benefits of using spatial information to link with a broader suite of information resources' (CSTB 2000). An international example of remote access to potentially disclosive data is the Luxembourg Income Study[10] (LIS 1997). The LIS database comprises data from household income surveys from 25 countries, covering Europe, America, Asia, and Oceania. Data are at household, person, and child levels, along themes of demographics, income, and expenditure. A key activity of LIS is to harmonize and standardize national-level disclosive data and provide remote access, through the datasets that 'can be accessed via the internet mailing system by submitting SAS, SPSS, or STATA programs' (LIS 2000).

The examples in this section are typical of the managed remote dissemination of disclosive data. Other agencies are also examining the techniques and implications of data integration and dissemination. U.K. National Statistics, for example, widened general access to Labour Force Survey (LFS) data when the geographical resolution was increased from 11 regions and 16 counties, to more than 400 Local and Unitary Authorities in Great Britain (ESRC Data Archive 2000). However, the key issue here for users was less potential disclosure than it was statistical sampling robustness once the LFS sample was dissected into such a large number of geographical areas. Statistics Denmark has taken first steps in managing confidential statistical data in its registers through a single integrated database, although it acknowledges that this is not seen as feasible by some, given concerns about confidentiality threats. Following its introduction of integrated registers, Statistics Denmark will work on a metadatabase of statistical documentation. It should be noted that this is very much an internal project, not a project to provide remote access to external users (Spieker 1999). A further step in the dissemination process is informing and involving providers of data. For example, the Australian Bureau of

[10] http://www.lis.ceps.lu/access.htm

Statistics (ABS) has consulted providers of international merchandise trade statistics on changes to confidentiality suppression (ABS 1999).

Pre-prepared Public Disclosure

The main difference between this and the preceding section at first sight is rather small. However, there are now examples of public access dissemination services that deal with disclosure in innovative ways. The electronic database for Statistics Netherlands, Statline[11] (Keller 1999), is accessible on the Web. Its design enables users to request sophisticated data extractions, while at the same time ensuring that the wholesale copying of the database is not feasible.

Statistics Canada provides remote access to the National Population Health Survey (NPHS), and the software implemented makes use of 'good survey documentation, the creation of synthetic (dummy) files for program testing, the ability to run a variety of software, and a relatively fast turn-around time' (Tambay 2001, p.5). Remote access for the census exists in Canada through pre-prepared products that have been verified through Census Products and Services System (PASS). Here, outputs are manually reviewed for validity and for residual disclosure before they are sent to users (Tambay 2001, p.6). The PASS has been decentralized to speed up access to data for the five federal departments that provided funding for the 1996 census, but the overall approach to wider dissemination is (sensibly) conservative.

On another level is the American FactFinder[12]. This is a Web-based public access facility for the U.S. decennial census of population, the American Community Survey, and the 1997 Economic Census, the integrated data being made accessible at detailed geographical levels (Census 2000). The product is a government and private sector partnership between the Census Bureau and the Environmental Systems Research Institute (ESRI), one of the leading geographic information system developers. American FactFinder generates customized outputs from a database 'covering more than 100 million U.S. households and more than twenty million U.S. businesses' (ESRI 2000).

The technology behind American FactFinder (AFF) is detailed by Hawala (2001). A firewall exists 'to permit web-service requests which originate on only AFF's external server, and which terminate on AFF's internal server' (Hawala 2001, p.3). Any other machine attempting to communicate with the internal server is blocked, and no confidential information is routed along the path allocated for user requests. Rowland and Zayatz (2001) describe this system as the software equivalent of a 'separate DMZ or demilitarized zone' whereby 'the separate DMZ provides a secure place for the external web servers to reside inside the first firewall. The internal servers will reside behind the DMZ inside the second firewall. The internal servers will execute the query and results filters'.

---

[11] http://www.cbs.nl/en/statLine/index.htm
[12] http://factfinder.census.gov

The custom-made American FactFinder tabulations from microdata files are produced via a set of query filters and results filters (Hawala 2001, p.4). A query filter, which is supposed to detect queries that will not pass such disclosure rules before they are submitted, is complemented by a results filter, which performs a final check on the resulting table before returning it to the user (Domingo-Ferrer 2001, p.3). However, there is concern that any software should have the sophistication to check current queries from one user against all previous queries, because 'an intruder can accumulate knowledge through successive queries and eventually succeed in reidentifying an individual' (Domingo-Ferrer 2001, p.3). Should this be taken further to argue that the query should be checked against all those of other users who are collaborators or colleagues? The task could become endless; perhaps here again the better approach is software combined with ethics, much along the lines argued by Gates (2001, p.7), who investigates the links between data collection, confidentiality, and privacy. He promotes a 'rationale for an integrated approach to data release that requires a commitment to research on SDL techniques and research on public perceptions' (Gates 2001, p.7).

The major innovation with services such as American FactFinder is that they no longer provide data along the Ford Model T approach—any car as long as it is black, or any data as long as they are this set of pre-prepared tables. But the issue of flexible geographies still needs to be addressed. More than any of the examples provided thus far, innovations such as American FactFinder illustrate how a partnership between data custodians and leading IT innovators can disseminate more data while still keeping within the legislative framework of disclosure control.

Sensitive Disclosure

This group of activities is not really under the remote access heading, but it does illustrate how statistical and other agencies are operating with the leading-edge research and policy making that, in the future, will have a significant impact on how remote access is developed.

The health sector, for example, has been examining the issues of data linkage, security, and disclosure in the context of integrated, or electronic, patient records. Integrated records provide medical specialists with full information about a patient, improving diagnosis and treatment possibilities. Griew *et al.* (1999) extend attention to shared case records where family doctors, hospitals, and perhaps remote specialists can access integrated patient information. The strength of integrated access is also a threat, because 'a shared electronic record is potentially available for inspection by all, in comprehensible form' (Griew *et al.* 1999, p.276).

For the treatment of diabetes in the U.K. National Health Service, Chadwick *et al.* (2000) describe a project for 'securely connecting a hospital diabetes information system across the internet, thereby providing patient specific information to general practices or even patients'. The system has strong user authentication, public key cryptography and digital signatures, commercial certification software[13], and clearly defined access rights that make sure that medical practitioners access only their own patients' data. So up to the point of access rights, this project has strong implications for remote access to confidential statistical data, but it seems to stop short of allowing access to all data for analytical purposes. Still, the Internet security issues are addressed and indicate that remote access is achievable using commercial software.

The issues of disclosure and confidentiality are shared across many sectors (for example, Internet banking, or share-trading systems). In the statistical sector, agencies maintain their own specialized units to research and apply disclosure control, including the Statistics Canada Disclosure Control Resource Center (Statcan 2000b); the 'safe custody' approach of Statistics New Zealand, which 'must be assured that the data is going into safe custody before access is granted' (Zealand 2000a); and the Social Survey Methodology Unit of U.K. National Statistics (Statistics 2000b).

A Note on Costs and Opportunity Costs

Building a secure remote access service costs money. How much? To a large extent the answer is 'how long is a piece of string'? Hardware and software prices are both volatile and rather opaque. And the IT infrastructure itself is a small part of the overall application infrastructure, along with such items as buildings and running costs and, most important of all, the training and retention of skilled human capital. It is not even easy to argue that IT applications are cost-beneficial. Stratopoulos and Dehning (2000) study the 'productivity paradox', and cite a range of studies that fail to show that IT demonstrably produces efficiency savings.

The cost justification for remote access disclosure control systems does involve real costs, but those costs are part of a larger cost-benefit appraisal, which includes perceived economic benefits from better analysis feeding into better policy. The expertise of the research community can be harnessed in the analysis, and the volume and quality of the research should have an economic impact—although the expectations of data providers that users prefer electronic to conventional printed products are not guaranteed (Laskowski 2000). How the development and running costs are covered, or recovered, is driven largely by the pricing dogma of each nation.

For developing countries, and for countries with emerging statistical services, the entry cost into the hardware and software market can be significant, and organizational behavior problems can militate against microdata dissemination. The UN-ECE reviewed practice in the European Union transition countries in the winter of 2000/2001, finding that 'several transition countries are very reluctant to release any microdata. Out of the 21 responding countries, 14 do not release microdata on

[13] http://www.entrust.com/

natural persons and 11 do not release microdata on enterprises' (UNECE 2001b, p.9; see Chapter 2 of this volume for further detail).

The Netherlands Central Bureau of Statistics has been pivotal in the development of freeware for disclosure control through the ARGUS software suites. Willenborg (1996) describes the early versions τ-ARGUS, and fuller details are in the *Tau-Argus Version 2.0 User Manual* (Hundepool *et al.* 1998). Nordholt (2001) reviews the adoption of disclosure control software in European transition countries. While software such as ARGUS is free, there are real overhead costs in training, human capital, and associated IT infrastructure. For countries such as Lithuania, the situation seems to exceed the capabilities of software. 'As a small country they have an additional dominance problem: often just one or two monopolistic enterprises dominate in some economic activities. Suppression of such a cell would imply that the statistical tables published would not meet its goals. Maybe individual consents of some of these enterprises could solve this problem' (Nordholt 2001, p.2).

Costs therefore involve collaboration and capacity-building by statistical agencies through the sharing of expertise and through direct assistance in the development of statistical series and dissemination strategies. This also puts further responsibility on the disclosure control strategies of agencies in the developed world, which not only have to assure their own citizens of confidentiality, but also have to provide convincing role models to developing countries.

5. The Context Into 2001

Privacy, confidentiality, data access, and disclosure are issues entwined in the context of rapidly developing technological innovation and threat. Like the laws of physics, for every positive there is a negative to counterbalance it. More extensive and detailed databases present public policy opportunities but, as Duncan advises, 'statistical agencies and researchers will (must) increasingly assume the role of data stewards. As in the biblical parable, the best steward is one who ensures effective use of the data, not the one who protects it against any risk by hiding it, unused' (Duncan and Pearson 1991, p.230).

The interface between the public and the private aspects of privacy are investigated by Shapiro, who is critical of the attempts to draw absolute lines between privacy and lack of privacy, preferring to view privacy 'in terms of the confluence of various boundaries, both physical and virtual' (Shapiro 1998, p.283). The printed media and the research literature are well populated with concerns about data threats, including the unintended disclosure of the identities of sex offenders (Sullivan 2001); identity theft (FTC 2000a and 2000b; Regan 2000; Travis 2000); the difficult balance between secrecy and confidentiality (Ponting 1990, p.45); the increased use of electronically stored records on individuals, or 'dataveillance' (Clarke 1991; Clarke 1994); the use of data in the retail sector to influence consumer behavior (Gelernter 2000); and government use of data integration for identity cards (Attaran and VanLaar 1999, p.241; Patel 1998).

In an era of national and international information infrastructures and the ability to link data on a global scale, 'the transition to the Information Age calls for a re-examination of the proper balance between the competing values of personal privacy and the free flow of information in a democratic society' (NII 1997). At a federal level in the United States there are concerns that commercial database owners are not taking disclosure and privacy issues seriously enough and that perhaps stronger regulation is needed (Ross 2001). Other commentators argue for a much stronger citizen appreciation of the checks and balances that operate on their behalf. Varney uses the analogy of the village community where his grandparents lived: 'They may have chafed sometimes at their lack of privacy, but it gave them a sense of belonging to a larger community—as long as it was not abused. Common sense, in the end, prevailed' (Varney 2001).

Or perhaps we should acknowledge that the current position with regard to disclosure is just one position along the historical continuum. Things will change in the future, and may change in unpredictable ways. History also provides us with useful analogies and reassurances (Ware 1980, p.15): Disaffection with the census is not new 'due to the public's innate aversion to government prying, amplified by an unsubtle campaign to discredit the Census as too intrusive' ('200 Years and Counting' 2000). Fienberg even argues that, given all the imponderables and complexities of the social, political, cultural, technological, and methodological processes, 'disclosure is an inherently statistical issue, *i.e.,* one cannot eliminate the risk of disclosure, simply reduce it, unless one restricts access to the data' (Fienberg 2000, p.10).

6. Conclusion

The debate about access to confidential or disclosive official statistics risks becoming a visceral confrontation between the absolutes of protection versus dissemination. In a similarly emotive debate about dissemination policy, Jessica Litman argued that 'the debate has seemed to be fuelled by some combination of almost religious faith...with self interest' (Litman 1994), and privacy, disclosure, and confidentiality excite similar fundamentalist views.

Public perceptions and fears about intrusive data collection will continue to influence policies on access to disclosive official statistics. To date the strongly conservative approaches, which seem to fly in the face of commercial practice and technological capability, can appear anachronistic and overcautious. However, statistical agencies sensibly may choose to continue approaching the issue with the goal of achieving the fullest possible public compliance in data provision, a compliance that increasingly is under seige by technological threats in areas of identity theft, general surveillance, and the expansion of commercially owned databases of individual information, such as the huge individual profile holdings emanating from retail loyalty cards.

It has always been technologically feasible to provide remote access to data. Taking paper files physically, posting outputs, faxing material, posting to ftp sites and sending e-mail attachments all are parts of the remote access process. Innovation will continue, as it has done to date, but within defined legislative, technical, ethical, and organizational practices. Certainly, the task facing statistical agencies does not seem to be getting any easier.

References

Note: In the on-line listings below, wherever possible URLs are given with direct links. For websites that have dynamic URLs, the home page address is shown instead. The date of access for each site is shown in parentheses.

Abramson, R. (2000) 'Federal Agencies Caught With Hand in the Cookie Jar', *Industry Standard,* October 23. (Accessed October 25, 2000), http://www.thestandard.com/article/display/0,1151,19600,00.html.

Abreu, E. (2001) '3-Year-Long Cyberattack Reveals Cracks in U.S. Defense', IDG News Group. May 8 (Accessed May 9, 2001), http://www.idg.net/ic_531968_1794_9-10000.html.

Ackerman, W.A. (1998) 'Encryption: A 21st Century National Security Dilemma', *International Review of Law Computers & Technology,* 12(2), pp.371-94.

Ahn, G.-J. (2000) 'Role-Based Access Control in DCOM', *Journal of Systems Architecture,* 46(13), pp.1175-84.

Al, P.G., and J.W. Altena (2000) 'Data Security, Privacy and the SSB', *Netherlands Official Statistics* 15 (Summer), pp.47-51.

Anon. (2000) 'Lumeria, I-Privacy: New Privacy Services Will Allow Consumers to Buy Anonymously Online', *Economist,* December 8. (Accessed December 9, 2000), http://www.ebusinessforum.com.

Attaran, M., and I. VanLaar (1999) 'Privacy and Security on the Internet: How to Secure Your Personal Information and Company Data', *Information Management & Computer Security,* 7(5), pp.241-6.

Australian Bureau of Statistics [ABS] (1999) *International Merchandise Trade Statistics, Australia: Data Confidentiality,* Information Paper 5487.0, December 20. (Accessed November 14, 2000) http://www.abs.gov.au/ausstats/abs@.nsf/Lookup/NT0000E36E.

Bace, R.G. (2000) *Intrusion Detection*, Indianapolis, Ind.: Macmillan Technical Publishing.

Bennett, P. (2000) 'Access Control by Audio-Visual Recognition', *Work Study,* 49(1), pp.23-6.

Bertino, E. (1998) 'Data Security', *Data and Knowledge Engineering,* 25(1), pp.199-216.

Blakemore, M.J. (1998) 'Customer-Driven Solutions for Disseminating Confidential Employer and Labour Market Data', in *Proceedings of the International Symposium on Linked Employer-Employee Data,* Arlington, Va.: U.S. Bureau of the Census.

Camp, L.J. (1999) 'Web Security and Privacy: An American Perspective', *Information Society,* 15, pp.249-56.

Cathie Marsh Centre for Census and Survey Research [CSSR] (2000) 'The Samples of Anonymised Records From the 1991 Census', University of Manchester, October. (Accessed October 23, 2000), http://les1.man.ac.uk/ccsr/.

Census (2000) 'American FactFinder', U.S. Bureau of the Census, November. (Accessed November 20, 2000), http://factfinder.census.gov/servlet/BasicFactsServlet.

Centers for Disease Control and Prevention and Agency for Toxic Substances and Disease Registry [CDC/ATSDR] (1999) *Secure Data Network Standards and Procedures*, HISSB Adopted July 15, p.9.

Chadwick, D.W., P.J. Crook, A.J. Young, D.M. McDowell, T.L. Dorman, and J.P. New (2000) 'Using the Internet to Access Confidential Patient Records: A Case Study', *British Medical Journal,* September 9. (Accessed September 14, 2000), http://www.bmj.com/cgi/content/short/321/7261/612.

Clarke, R. (1994) 'Dataveillance by Governments: The Technique of Computer Matching', *Information Technology & People,* 7(2), pp.46-85.

Clarke, R.A. (1991) 'Information Technology and Dataveillance', in C. Dunlop and R. Kling (eds) *Computerization and Controversy,* Boston: Academic Press, pp.496-522.

Computer Science and Telecommunications Board [CSTB] (2000) 'Summary of a Workshop on Information Technology Research for Federal Statistics', Washington, D.C.: Computer Science and Telecommunications Board and National Academy Press, (Accessed April 20, 2001), http://books.nap.edu/html/itr_federal_stats/.

Dale, A., and M. Elliott (1998) 'A Report on the Disclosure Risk of Proposals for SARs from the 2001 Census', Manchester: CSSR, Faculty of Economics, University of Manchester, on-line paper (Accessed August 22, 2001), http://les1.man.ac.uk/CCSR/publications/working/risk.htm.

Delio, M. (2001) 'How Secure Is Digital Hospital?', *Wired.com,* March 28. (Accessed March 31, 2001), http://www.wired.com/news/technology/0,1282,42656,00.html.

Denning, D.E. (1980) 'Secure Statistical Databases with Random Sample Queries', *ACM Transactions on Database Systems,* 5(3), pp.291-315.

Dinnie, G. (1999) 'The Second Annual Global Information Security Survey', *Information Management & Computer Security,* 7(3), pp.112-20.

Domingo-Ferrer, J. (2001), *Impact of New Technological Developments in Software, Communications and Computing on SDC*, UNECE Statistical Division, Work Session on Statistical Data Confidentiality, Skopje. (Accessed May 19, 2001), http://www.unece.org/stats/documents/2001.3.confidentiality.htm.

Donaldson, L., and R. May (1999) *Health Implications of Genetically Modified Foods,* London: Department of Health, p.26.

Duke-Williams, O., and P. Rees (1998) 'Can Census Offices Publish Statistics for More Than One Small Area Geography? An Analysis of the Differencing Problem in Statistical Disclosure', *International Journal of Geographical Information Science,* 12(6), pp.579-605.

Duncan, G.T. (1998), *Managing Information Privacy and Information Access in the Public Sector,* Pittsburgh, Pa.: Carnegie Mellon University.

Duncan, G.T., and R.W. Pearson (1991) 'Enhancing Access to Microdata While Protecting Confidentiality: Prospects for the Future. *Statistical Science* 6(3), pp.219-39.

Eaton, J. (1999) '"Open, Sesame?"—The Problems of Digital Identity and Secure Access to Information in the Internet Era: Issues for the Information Industry', *Business Information Review* 16(4), pp.184-91.

Edinburgh Data and Information Access [EDINA] (2001) 'DIGIMAP', January 11. (Accessed January 11, 2001), http://www.edina.ac.uk/digimap.

Environmental Systems Research Institute [ESRI] (2000) 'Census American FactFinder Receives Vice President Gore's Hammer Award', ESRI Inc., January 28. (Accessed November 10, 2000), http://www.esri.com/news/releases/00_1qtr/factfinder.html.

ESRC Data Archive (2000) 'The Labour Force Survey', University of Essex, September. (Accessed September 15, 2000), http://www.data-archive.ac.uk/index.asp.

Felligi, I.P. (1972) 'On the Question of Statistical Confidentiality', *Journal of the American Statistical Association,* 67(337), pp.7-18.

Fienberg, S. (1997) 'A Glimpse at the Future of Social Science. Statistical Data: New Forms of Data Analysis, New Types of Access, and New Issues for Data Providers', *IASSIST Quarterly,* 21(2), pp.8-11.

Fienberg, S.E. (2000). 'Confidentiality and Data Protection Through Disclosure Limitation: Evolving Principles and Technical Advances', IAOS/ISI International Conference on Statistics, Development, and Human Rights, September 4-9. (Accessed December 10, 2000), http://www.statistik.admin.ch/about/international/fienberg_final_paper .doc.

Flaherty, D.H. (1979) *Privacy and Government Data Banks: An International Perspective.* London: Mansell.

Fox, S., and O. Lewis (2001). 'Fear of Online Crime: Americans Support FBI Interception of Criminal Suspects' Email and New Laws to Protect Online Privacy', Washington, D.C.: Pew Internet & American Life Project, p. 10. (Accessed April 10, 2001) http:// www.pewinternet.org.

Freedman, W. (1987) *The Right of Privacy in the Computer Age.* New York: Quorum Books.

Federal Trade Commission [FTC] (2000a) 'ID Theft: When Bad Things Happen to Your Good Name', August. (Accessed: September 1, 2000), http://www.ftc.gov/bcp/conline/pubs/ credit/idtheft.htm.

— (1987b), 'A Constructive Procedure for Unbiased Controlled Rounding', *Journal of the American Statistical Association,* 82, pp.520–4.

(2000b). 'Identity Theft Complaints Triple in Last Six Months: FTC Victim Assistance Workshop To Be Convened October 23-24', August 30. (Accessed September 1, 2000), http:/ /www.ftc.gov/opa/2000/08/caidttest.htm.

Furnell, S.M., and P.S. Dowland (2000) 'A Conceptual Architecture for Real-Time Intrusion Monitoring', *Information Management & Computer Security* 8(2), pp.65-75.

Garrison, W.V., and G.A. McClellan (1997) 'TAO of Gateway: Providing Internet Access to Licensed Databases', *Library Hi Tech,* 15(1-2), pp.39-54.

Gates, G.W. (2001) 'A Holistic Approach to Confidentiality Assurance in Statistical Data', UNECE Statistical Division, Work Session on Statistical Data Confidentiality, Skopje. (Accessed May 19, 2001), http://www.unece.org/stats/documents/2001.3.confidentiality.htm.

Gelernter, D. (2000) 'Visions of the 21st Century: Will We Have Any Privacy Left?', *Time Magazine,* February 21. (Accessed April 27, 2000), http://www.time.com/time/reports/v21/ live/privacy_mag.html.

Griew, A., E. Briscoe, G. Gold, and S. Groves-Phillips (1999). 'Need to Know; Allowed to Know—The Health Care Professional and Electronic Confidentiality', *Information Technology & People,* 12(3), pp.276-86.

General Register Office [GRO] (2001) '2001 Census of Population. Report of the Reviews on Security and Confidentiality', March. (Accessed April 19, 2001), http://wood.ccta.gov.uk/grosweb/grosweb.nsf/pages/censc.

Hancock, B. (2001) 'The Chief Security Officer's Top Ten List for 2001', *Computers and Security* 20(1), pp.10-14.

Hawala, S. (2001) 'American FactFinder: U.S. Bureau of the Census Works Towards Meeting the Needs of Users While Protecting Confidentiality', UNECE Statistical Division, Work Session on Statistical Data Confidentiality, Skopje. (Accessed May 19, 2001), http://www.unece.org/stats/documents/2001.3.confidentiality.htm.

Hawkins, S., D.C. Yen, and D.C. Chou (2000) 'Awareness and Challenges of Internet Security', *Information Management & Computer Security* 8(3), pp.131-43.

Hencke, D. (2000) 'Privacy on the Net: Special Report', *Guardian* (London), July 13. (Accessed July 13, 2000), http://www.guardianunlimited.co.uk/uk_news/story/0,3604,342627,00.html.

Henderson, S.C., and C.A. Snyder (1999) 'Personal Information Privacy: Implications for MIS Managers', *Information and Management,* 36(4), pp.213-20.

Hundepool, A., L. Willenborg, L. van Gemerden, A. Wessels, M. Fischetti, J-J. Salazar, and A. Caprara (1998) *Tau-Argus Version 2.0 User Manual.* Voorburg: Statistics Netherlands, p.28.

Hurt, E. (2001) 'Government Shirks Privacy Standards', *Business2.0 Online*, April 17. (Accessed April 28, 2001), http://www.business2.com/ebusiness/2001/04/scient.htm.

IPlanet (2001) 'iPlanet Portal Platform Selected by the US Department of Defense for a Scalable, Reliable and More Secure E-Government Portal Solution', *iPlanet E-Commerce Solutions*, March 22. (Accessed March 30, 2001), http://www.iplanet.com/about_us/press_release/portal_dod_5_1_1aj.html.

Keller, W.J. (1999) 'Preparing for a New Era in Statistical Processing: How New Technologies and Methodologies Will Affect Statistical Processes and Their Organisation', in *Proceedings of the Strategic Reflection Colloquium on IT Issues for Statistics,* Luxembourg: Eurostat, pp.17-31.

Keller-McNulty, S. (1997) 'White Paper for the Workshop on Research and Development Opportunities in Federal Information Services: Data Access and Data Confidentiality in a Decentralized Federal Statistical System', International Statistical Institute. (Accessed May 2, 2001), http://www.isi.edu/nsf/papers/sallie.htm.

Kingsbury, N.R. (2001) *Record Linkage and Privacy: Issues in Creating New Federal Research and Statistical Information,* Washington D.C.: General Accounting Office.

Kruskal, W. (1981) 'Statistics in Society: Problems Unsolved and Unformulated', *Journal of the American Statistical Association*, 76(375), pp.505-15.

Laskowski, M.S. (2000) 'The Impact of Electronic Access to Government Information: What Users and Documents Specialists Think', *Journal of Government Information,* 27(2), pp.173-85.

Lemos, R. (2001) 'Lawyers Slam FBI "Hack"', *ZDNet News,* May 1. (Accessed May 3, 2001), http://www.zdnet.com/zdnn/stories/news/0,4586,5082126,00.html.

Lin, A., and R. Brown (2000) 'The Application of Security Policy to Role-Based Access Control and the Common Data Security Architecture', *Computer Communications,* 23(17), pp.1584-93.

Luxembourg Income Study [LIS] (1997) 'LIS Project. Rules of Organization, Structure, and Governance. Luxembourg: Author.

—— (2000) *Luxembourg Income Study* November 20. (Accessed November 25, 2000), http://www.lis.ceps.lu.

Litman, J. (1994) 'Rights in Government-Generated Data', in *Proceedings of the Conference on Law and Information Policy for Spatial Databases,* October 28-29. (Accessed April 13, 2000), http://www.spatial.maine.edu/tempe/litman.html.

Lynch, C.A. (1997) 'The Changing Role in a Networked Information Environment', *Library Hi Tech*, 15(1-2), pp.30-8.

Mabuchi, K., H. Fujie, G. Dhillon, and J. Backhouse (1996). 'Risks in the Use of Information Technology Within Organizations', *International Journal of Information Management*, 16(1), pp.65-74.

Malaguerra, C. (2000) 'Discussion', in Eurostat (ed) *Proceedings of the 86th DGINS Conference,* September 4-8, Porto, Portugal, pp.87-91.

Marx, G.T. (1998) 'Ethics for the New Surveillance', *The Information Society,* 14(3), pp.171-85.

McGuire, D. (2001) 'Commerce Department Creates "Privacy Advisor" Position', *Newsbytes.com*, May 2. (Accessed May 2, 2001), http://www.computeruser.com/news/01/05/02/news2.html.

McNulty, J. (1979) 'The Information Explosion', in P. Hewitt (ed) *Computers, Records and the Right to Privacy,* Purley, Surrey: Input Two-Nine, pp.95-103.

Mitchell, R.C., R. Marcella, and G. Baxter (1999) 'Corporate Information Security Management', *New Library World*, 100(5), pp.213-27.

National Center for Health Statistics [NCHS] (2000) 'NCHS Research Data Center Opens', November 6 (Accessed November 20, 2000), http://www.cdc.gov/nchs/r&d/rdc.htm.

—— (2001) 'How NCHS Protects Your Privacy', (Accessed May 22, 2001), http://www.cdc.gov/nchs/about/policy/confiden.htm.

National Information Infrastructure Task Force [NII] (1997) 'Options for Promoting Privacy on the National Information Infrastructure', April. (Accessed October 17, 2000), http://www.iitf.nist.gov/ipc/privacy.htm.

National Infrastructure Protection Center [NIPC] (2000) November 1. (Accessed November 1, 2000), http://www.nipc.gov.

Nordholt, E.S. (2001) 'Progress in the Implementation of SDC Methods and Techniques in Central and Eastern Europe', UNECE Statistical Division, Work Session on Statistical Data Confidentiality, Skopje. (Accessed May 19, 2001), http://www.unece.org/stats/documents/2001.3.confidentiality.htm.

Øhrn, A., and L. Ohno-Machado (1999) 'Using Boolean Reasoning to Anonymize Databases', *Artificial Intelligence in Medicine*, 15(3), pp.235-54.

Office of Management and Budget [OMB] (1997) 'Order Providing for the Confidentiality of Statistical Information and Extending the Coverage of Energy Statistical Programs Under the Federal Statistical Confidentiality Order', Notice, Washington, D.C.: Author, pp.35,043-9.

—— (1999) *Statistical Programs of the United States Government—Fiscal Year 2000,* Washington, D.C.: Author, p. 79.

Openshaw, S., O. Duke-Williams, and P. Rees (1997). 'Measuring Confidentiality Risks in Census Data', Leeds: University of Leeds, Department of Geography.

Openshaw, S., and L. Rao (1995). 'Re-engineering 1991 Census Geography: Serial and Parallel Algorithms for Unconstrained Zone Design' University of Leeds, Department of Geography. (Accessed February 20, 2001), http://www.leeds.ac.uk/research/papers/95-3.

Oppliger, R. (2000) *Security Technologies for the World Wide Web,* Boston: Artech House.

Patel, L. (1998) 'The Government Has Our Numbers', *Wired.com*, July 3. (Accessed January 10, 2001), http://www.wired.com/news/politics/0,1283,13441,00.html.

Ponting, C. (1990) *Secrecy in Britain,* Oxford: Blackwell.

Posner, R.A. (1984) 'An Economic Theory of Privacy', in F. Schoeman (ed) *Philosophical Dimensions of Privacy: An Anthology,* Cambridge: Cambridge University Press.

Quantina, C., H. Bouzelata, F.A.A. Allaerb, A.M. Benhamichec, J. Faivrec, and L. Dusserrea (1998) 'How to Ensure Data Security of an Epidemiological Follow-Up: Quality Assessment of an Anonymous Record Linkage Procedure', *International Journal of Medical Informatics,* 49(1), pp.117-22.

Rathmell, A. (2001) 'Protecting Critical Information Infrastructures', *Computers & Security,* 20(1), pp.43-52.

Regan, K. (2000) 'Net Blamed for Identity Theft Spike', *E-Commerce Times,* August 31. (Accessed September 1, 2000), http://www.ecommercetimes.com/news/articles2000/000831-4.shtml.

Ross, P. (2001) 'Momentum Grows for Federal Online Privacy Laws', *CNET News.com,* October 11. (Accessed January 10, 2000), http://news.cnet.com/news/0-1005-200-3165244.html.

Rowland, S., and L. Zayatz (2001) 'Automating Access with Confidentiality Protection: The American FactFinder', in *Proceedings of the Social Statistics Section, American Statistical Association.*

Shapiro, S. (1998) 'Places and Spaces: The Historical Interaction of Technology, Home, and Privacy', *Information Society,* 14, pp.275-84.

Siponen, M.T. (2000) 'A Conceptual Foundation for Organizational Information Security Awareness', *Information Management & Computer Security* 8(1), pp.31-41.

Spencer, H. (1997) 'Age of Uncontrolled Information Flow', *The Information Society,* 13, pp.163-70.

Spieker, F. (1999) 'Access to Confidential Data in an Integrated Statistical System', in Eurostat (ed) *Statistical Data Confidentiality,* Luxembourg: European Communities, pp.209-19.

Statcan (2000a) 'Statistical Research Data Centres', Statistics Canada, December 19. (Accessed March 20, 2001), http://www.statcan.ca/Daily/English/001219/d001219k.htm.

— (2000b) 'Statistics Canada: Support Activities', Statistics Canada, November. (Accessed November 10, 2000), http://stcwww.statcan.ca/english/concepts/methodology/2000/supp.htm.

— (2000c) 'Terms of Reference: Expert Panel on Access to Historical Census Records', Statistics Canada, April 17. (Accessed April 30, 2001), http://www.statcan.ca/english/census96/terms.htm.

Statistics (2000a) 'Office for National Statistics: Risk Management. National Statistics', September. (Accessed November 2, 2000), http://www.statistics.gov.uk/nsbase/methods_quality/risk_contents.asp.

— (2000b) 'Social Survey Methodology Unit. National Statistics', June 2. (Accessed September 14, 2000), http://www.statistics.gov.uk/methods_quality/ssmustatistical.asp.

Stratopoulos, T., and B. Dehning (2000) 'Does Successful Investment in Information Technology Solve the Productivity Paradox?' *Information and Management,* 38, pp.103-17.

Sullivan, B. (2001) 'Sex Offender Web Sites Are Insecure', *MSNBC,* January 12. (Accessed January 12, 2001), http://www.zdnet.com/zdnn/stories/news/0,4586,2673945,00.html.

SUN (1999) 'Sun Launches i-Planet: Revolutionary Internet Software Delivers Innovative New Way to Work', Sun Microsystems, April 27. (Accessed March 23, 2001), http://www.sun.com/smi/Press/sunflash/9904/sunflash.990427.4.html.

— (2001) 'iPlanet.com: Creating a Twenty-First Century Web Site in Six Months', Sun Microsystems, March 5. (Accessed March 20, 2001), http://dcb.sun.com/practices/casestudies/iplanet_part1.jsp.

Survey (1998) 'National Interest Mapping Service Agreement (NIMSA) and Trading Fund Status', Ordnance Survey, Information Paper, February 2. (Accessed December 11, 2000), http://www.ordsvy.gov.uk/.

Tambay, J.-L. (2001) 'Providing Greater Accessibility to Survey Data for Analysis', UNECE Statistical Division, Work Session on Statistical Data Confidentiality, Skopje. (Accessed May 19, 2001), http://www.unece.org/stats/documents/2001.3.confidentiality.htm.

Thorogood, D. (1999). 'Better Solutions to Disclosure Control: Research Under the Fifth Framework Programme', in Eurostat (ed) *Statistical Data Confidentiality,* Luxembourg: European Communities, pp.99-103.

Travis, A. (2000) 'Welcome to the Crimes of the Future—Official Report Predicts Surge in Hi-Tech Theft as Criminals Target Personal IDs', *Guardian* (London), March 25, 2000. (Accessed March 30, 2000), http://www.newsunlimited.co.uk.

Trivellato, U. (2000) 'Data Access Versus Data Privacy: An Analytical User's Perspective', in Eurostat (ed) *Proceedings Innovations in Provision and Production of Statistics: The Importance of New Technologies,* Helsinki: Eurostat, pp.92-107.

Turega, M. (2000) 'Issues With Information Dissemination on Global Networks', *Information Management & Computer Security,* 8(5), pp.244-8.

'200 Years and Counting: Census Nosiness Isn't New' (2000) *USA Today,* April 5. (Accessed May 25, 2000), http://www.usatoday.com.

United Nations Economic Commission for Europe [UNECE] (2001a) 'Report of the March 2001 Work Session on Statistical Data Confidentiality', UNECE Statistical Division, Work Session on Statistical Data Confidentiality, Skopje. (Accessed May 19, 2001), http://www.unece.org/stats/documents/2001.3.confidentiality.htm.

— (2001b) 'Statistical Data Confidentiality in the Transition Countries: 2000/2001 Winter Survey', UNECE Statistical Division, Work Session on Statistical Data Confidentiality, Skopje. (Accessed May 19, 2001), http://www.unece.org/stats/documents/2001.3.confidentiality.htm.

Varney, C. (2001) 'Cyberspace Security: The Death of Privacy?', *Newsweek,* January. Special Edition, December 2000-February 2001, pp. 78-79. (Accessed December 29, 2000), http://www.msnbc.com/news/508910.asp?cp1=1.

Ware, W.H. (1980) 'Privacy and Information Technology—the Years Ahead', in L.J. Hoffman (ed) *Computers and Privacy in the Next Decade,* New York: Academic Press, pp.9-22.

Waterhouse, R. (2001) 'Big Brother Is Asking You', *Sunday Times* (London), April 29. (Accessed May 1, 2001), http://www.sunday-times.co.uk/.

Willenborg, L. (1996) 'SDC Project: First Progress Report', Netherlands Central Bureau of Statistics, September. (Accessed May 16, 2001), http://www.cbs.nl/sdc/progres1.htm.

Willson, J., and T. Oulton (2000) 'Controlling Access to the Internet in UK Public Libraries', *OCLC Systems and Services,* 16(4), pp.194-201.

— (2000b) 'Protocols for Official Statistics', Statistics New Zealand, September. (Accessed September 14, 2000), http://www.stats.govt.nz.

Confidentiality, Disclosure, and Data Access: Theory and Practical Application for Statistical Agencies
Pat Doyle, Julia I. Lane, Jules J.M. Theeuwes and Laura M. Zayatz (Eds)

Chapter 14

Public Perceptions of Confidentiality and Attitudes Toward Data Sharing by Federal Agencies*

Eleanor Singer
University of Michigan

1. Introduction

This chapter differs from most of the others in this book in that it focuses not on actual risks to respondent confidentiality or on methods to reduce those risks, but rather on what the public believes about the confidentiality of data gathered by the U.S. Bureau of the Census and how it regards the prospect of data sharing among federal agencies. Furthermore, it examines changes in these beliefs and attitudes over time. It looks primarily at changes over a span of five years during which four identical surveys of public opinion were carried out, three of them under contract with the Census Bureau. It also takes into account research done shortly after the 1990 census in order to explain the unexpectedly large decline in return rates to that census. Thus, the concerns that gave rise to the research reported here have less to do with statistical disclosure and disclosure limitation, which is the focus of most of the chapters in this volume, than with declining response rates to surveys, including the U.S. census, and the hypothesis that public concerns about privacy and confidentiality might be contributing to that decline.

* This chapter is based in part on work done under Contract 50YABC-7-66019, Task Order 46-YABC-9-00002, entitled "Study of Privacy Attitudes in 2000," between the U.S. Census Bureau and the University of Michigan, as well as on a 1996 survey of the public carried out by Westat for the Census Bureau and on a 1995 survey carried out in collaboration with Stanley Presser at the University of Maryland with consultation by the Census Bureau. I would like to acknowledge the help of my colleagues at the Census Bureau, in particular Randall J. Neugebauer, who managed the most recent study, as well as my colleagues at The Gallup Organization who were responsible for carrying it out, in particular, Roger Tourangeau (now at the University of Michigan) and Darby Miller Steiger. I would also like to express my appreciation to Jeff Kerwin of Westat, who conducted the 1996 survey, in which I participated as a co-investigator. Stanley Presser's collaboration and advice throughout this entire period have been invaluable, as has John Van Hoewyk's assistance with the analyses on which this chapter is based. I would also like to thank Stanley Presser and anonymous reviewers at the Census Bureau for comments on an earlier draft. Final responsibility for the content is, of course, my own.

2. Early Research on Confidentiality Concerns

The mail return to the 1990 U.S. census averaged 64.6 percent, some 10 percentage points less than in 1980 and 5 percentage points less than had been anticipated by the Census Bureau (U.S. General Accounting Office 1992). One hypothesis put forward to explain the reduced return rate was increased public concern about privacy, documented in a series of surveys by the Harris Organization (Westin 1994), and about confidentiality[1]. Although the Outreach Evaluation Study, carried out by the Census Bureau in 1990, found that the large majority of respondents believed that census data are kept confidential (Fay *et al.* 1991, p.18), and that such beliefs had not declined since the last decennial census, it also documented a significant change in the relationship between trust in the Census Bureau's assurance of confidentiality and self-reported census return rate. Whereas trust was not predictive of self-reported returns in 1980, it was predictive of such returns in 1990 (Fay *et al.* 1991, p.18 and Table 4), with some 13 percentage points separating the self-reported return rates of those with a high and a low degree of trust[2].

In an analysis of actual census mail return rates and attitudes toward privacy and confidentiality (as measured in the Survey of Census Participation, carried out by the National Opinion Research Center in the summer of 1990), Singer *et al.* (1993) found that both attitudes were predictive of actual returns, with concerns about confidentiality, measured by a series of items all pertaining to the census, the stronger predictor of the two[3]. In a subsequent analysis that pitted concerns about privacy and confidentiality against other attitudes, demographic characteristics, and various measures of competing demands as well as access and capacity, Couper *et al.* (1998) demonstrated that confidentiality concerns (but not concerns about privacy) remained a significant predictor of mail returns to the 1990 census.

Adding to the Census Bureau's unease was a National Academy of Sciences panel recommendation that it consider using administrative records to improve the accuracy of the 2000 count (Steffey and Bradburn 1994). It was hypothesized that such data sharing among federal agencies, if it became public knowledge, might increase confidentiality concerns, as might a request for the respondent's Social Security number to facilitate the merging of information.

As a result of these various developments, the Census Bureau in the early 1990s embarked on a program of privacy-related research, including focus groups, large-

---

[1] By concern about confidentiality, I refer to a desire to keep information already given to one agent out of the hands of others; by concern about privacy, I refer to a desire to keep information out of the hands of others altogether (Singer *et al.* 1993). Although there is some evidence that the public may be blurring the distinction between these concepts (Martin 2000), the distinction appears to be meaningful in much of the research reported here.

[2] In 1999 and 2000, the relationship between trust in the Census Bureau's promise of confidentiality and self-reported return of the census form was smaller but still statistically significant.

[3] The privacy index consisted of eight items, only two of which dealt explicitly with the Census Bureau or the census.

scale experiments, and commitment to support a series of cross-sectional surveys that would track attitudes about privacy and confidentiality, especially as they related to the decennial census and the proposal to supplement the traditional count by use of administrative records. This chapter is based primarily on the four Surveys of Privacy Attitudes conducted under this program, with other research brought in as appropriate[4].

The rest of this chapter is divided into five major sections: a description of the methods used; trends in beliefs about confidentiality; trends in attitudes toward privacy; trends in attitudes toward data sharing; and an examination of the relationship between attitudes and behavior. A concluding section summarizes the findings and draws some implications from them.

3. Methods

Four surveys of the U.S. telephone population 18 and over have measured privacy, confidentiality, and data sharing attitudes between 1995 and 2000. The first was developed in consultation with the Census Bureau as part of the University of Maryland's 1995 Joint Program in Survey Methodology practicum. The second, which used a questionnaire virtually identical to that used in 1995, was carried out by Westat in 1996 under contract to the Census Bureau. The third and fourth, done in July through October 1999, just before the start of the public relations campaign and nationwide field recruiting for Census 2000, and from April to July 2000, after delivery of census forms to U.S. households, were conducted by the University of Michigan under contract with the Census Bureau, with data collected by the Gallup Organization.

All four surveys used virtually identical methods and achieved very similar response rates. All were random digit dialed surveys with one member of the household age 18 or over randomly selected after household listing by the interviewer. The response rates (interviews divided by the total sample less businesses, nonworking numbers, and [in 1995] numbers that were never answered after a minimum of 20 calls or [in 1996, 1999, and 2000] the estimated number of ineligibles among the noncontacts) were 61 percent, 60 percent, 62 percent, and 61 percent. Sample sizes were 1,443 for 1995, 1,215 for 1996, 1,677 for 1999, and 1,978 for 2000. All analyses in this chapter are weighted to correct for unequal probabilities of selection (due to households containing different numbers of adults and phone lines) and poststratified to Current Population Survey distributions on region, sex, race, age, and education. When multiple linear or logistic regression results are

---

4 Excluded from this chapter is an examination of public opinion about the confidentiality of health care data, summarized in Singer, Shapiro, and Jacobs (1997, and see the references cited there).

reported, the analyses also impute for item-missing data using the multiple imputation strategy described by Raghunathan and colleagues (in press)[5].

Because we believed that questions about data sharing and about the confidentiality of census information would be more meaningful if respondents had some idea of the kind of information involved, each interview began with the census short-form questions: sex, age, race, ethnicity, and (in 1995 and 1996) marital status (which was not asked on the 2000 census). Subsequent questions about confidentiality and data sharing referred back to the content of these five items. As a result, it was hoped that responses to questions probing attitudes toward agencies' sharing of data with the Census Bureau and beliefs about the Census Bureau's safeguarding of identified data would be anchored in the specific content of this information.

Besides anchoring questions about data sharing in the specifics of the census short form, we also informed (or reminded) respondents that the 1990 census had failed to count a significant number of people, and that as a result the communities in which these people lived were deprived of full political representation and economic benefits. Thus, the context for the questions about data sharing was the undercount and its consequences. Later, we also introduced the burden of answering the long form, and asked whether data sharing would be acceptable in order to reduce this burden.

A number of criticisms have been leveled against public opinion surveys about privacy, confidentiality, and data sharing in general. Among these criticisms are the relatively low response rates achieved by many of them, including the four above, and the potential bias this low rate introduces; the relative ignorance of the public about data confidentiality and how the Census Bureau protects it; and the relatively unformed nature of public opinion on these issues, which means that the opinions expressed may be quite unstable. Before we discuss the survey findings, it is important to address these criticisms.

Low Response Rates

Because it is the potential bias introduced by nonresponse, not the proportion of nonrespondents, that is of concern, it is worth noting that several recent studies have failed to demonstrate nonresponse bias even with rather drastic reductions in response rates (Curtin *et al.* 2000; Keeter *et al.* 2000). These studies, however, lack any theoretical basis for their findings, so that it is impossible to know how widely generalizable the findings are. Groves, *et al.* (2000) have recently advanced such a theory, which suggests that survey introductions have the potential for inducing

[5] The imputations were created through a sequence of univariate regressions with the covariates including all other variables observed or imputed for the individual. The type of regression used (*i.e.*, linear versus logistic) depended on the variable to be imputed. The sequence of imputing missing values was continued in a cyclic manner, each time overwriting the previously imputed values to build more interdependence and exploit the correlational structure of the data.

some respondents to participate and for alienating others, for example by making certain topics and sponsors salient. A study by Groves, Presser, and Dipko (2001) supports this hypothesis with respect to survey topic. Thus, mention of the Census Bureau and of confidentiality in the introduction to the four Census Bureau surveys might a) induce people who are favorable toward the Census Bureau and concerned about confidentiality issues to respond, while discouraging those who are indifferent to such issues or hostile to the Census Bureau; and/or b) discourage people who are very concerned about privacy and confidentiality from responding, because the survey itself is viewed as an invasion of privacy.

There has never been a large enough budget or sample for the privacy surveys to permit a true nonresponse study. But a detailed examination of the possible response bias due to nonresponse in the 1996 survey (Brick *et al.* 1997, Chapter 4), based on a comparison of those who ever refused with those who never refused, led to two conclusions. First, there was some reason to believe that, had more refusals been converted, opposition to data sharing would have been somewhat higher than measured. But second, the size of this difference would likely have been very small.

Ignorance About Issues and Instability of Response

It is unlikely that most respondents to these surveys had ever considered the issues they were questioned about before the survey itself. That being the case, what importance should be attached to their answers (*cf.* Fishkin 1995; Yankelovich 1991)?

To forestall this criticism, we tried to incorporate relevant information into survey questions whenever possible, even at the risk of making the questions longer than is customary. For example, we informed respondents about the nature and size of the undercount before we asked whether they would be willing to have the Census Bureau use records from other government agencies to reduce it; and we told people about the kind of information respondents to the short form had to provide before we asked how much they would be bothered by a breach of confidentiality. The issue of 'nonattitudes' is considered further below, under the section on attitudes toward data sharing. In brief, the argument is that respondents answer questions about novel phenomena not in random fashion, but rather in light of their attitudes toward related, known stimuli.

Still another criterion for judging the usefulness of the surveys is how well the attitudes they measure predict behavior, a relationship that is examined later in this chapter.

Changes in Methods

The four surveys whose findings are reported in this chapter were done by three different survey organizations, and several studies have indicated that so-called house effects can complicate comparisons (*e.g.,* Smith 1982; Turner 1981; Turner and Krauss 1978). However, these effects are ordinarily attributable to variations in question order or differences in survey organizations' probing of Don't Know responses. Both were well controlled in the four surveys under discussion here. On most questions, for example, Don't Know and Not Sure rates are very similar across the four years. Further, examination of changes in responses over time indicates that they vary from question to question, thus making it unlikely that there is a consistent 'house effect' in these data.

4. Trends in Beliefs About Confidentiality

Trends in beliefs about the Census Bureau's treatment of personal information were measured in several different ways. Early in the interview, respondents were asked for their beliefs about Census Bureau practices. Later questions probed their knowledge of the laws governing confidentiality practices, and then those knowledgeable about the relevant laws were asked whether they trusted the Bureau to obey them. Finally, at the very end of the interview, respondents were asked several questions about potential misuses of census data involving breaches of confidentiality. Most questions were asked in all four years, but some were asked in only three of the survey.

The first question designed to probe beliefs about actual practices asked, 'Do you believe other agencies, outside the Census Bureau, can or cannot get people's names and addresses along with their answers to the census, or are you not sure?' The introduction to the question referred back to the demographic questions asked on the short form and informed people that 'the person in the household who fills out the form must list the full name of everyone who lives there along with each person's age, sex, race, [and marital status.]'. The second question, asked for the first time in 1996 to assess whether use of the term 'confidentiality' would change the pattern of responses, was, 'Do you think the Census Bureau does or does not protect the confidentiality of this information, or are you not sure?', with an introduction identical to that already quoted. Respondents in 1996 were randomly assigned to one question or the other. Finally, in 1999 and 2000, to try to clarify earlier inconsistencies, one-third of the sample was asked both these questions (with the order of questions randomized), followed by an open-ended question about the meaning of confidentiality to the respondent.

Responses to the two questions inquiring into beliefs about Census Bureau practices are shown in Tables 1 and 2 for all four years[6]. For both questions, there

[6] To examine trends in answers to these two questions, we looked at three versions of the 1999-2000 questions. First, we looked at respondents who were asked only one question or the other. Second, we looked at respondents who were asked only one question plus those who were asked the question first in the sequence. Finally, we looked at all respondents who answered the question, regardless of the order in which it was asked. Regardless of order, the results were essentially the same as those shown in Tables 1 and 2, which combine the responses of those who answered only one question and those who answered the question first.

is a significant increase between 1996 and 2000 in the proportion giving the correct response (that other agencies cannot get the data, and that the Census Bureau protects confidentiality)—from 6.1 percent to 17.3 percent in the case of 'can get', and from 12.9 percent to 25.1 percent in the case of confidentiality. Unlike later questions discussed in this section, these questions offered an explicit Not Sure category to respondents. The very large proportion of Not Sure answers, which is perhaps the most striking feature of both tables, is therefore a function of both the public's lack of information and the response options offered by the question (*cf.* Schuman and Presser 1981). In 1996, for example, when the questions were asked both with an explicit Not Sure option and, in split-ballot form, without such an option, the Not Sure rate shown in Table 1 dropped from 46.8 percent to 7.7 percent; however, the ratio of correct to incorrect responses did not change[7].

Table 1. Beliefs About Sharing of Census Responses: By Year

Do you think other government agencies, outside the Census Bureau, can or cannot get people's names and addresses along with their answers to the census, or are you not sure?

Response	1995	1996	1999	2000
Other Agencies Can Get Names	50.1%	47.1%	43.9%	42.0%
Other Agencies Cannot Get Names	9.2%	6.1%	12.2%	17.3%
Not Sure	40.7%	46.8%	44.0%	40.7%
N (weighted)	1,443	317	830	989

Source: 1995, Question 7; 1996, Question 7_1; 1999-2000, Question 7a1 or 7a3.

Table 2. Beliefs About Protection of Confidentiality: By Year

Do you think the Census Bureau does or does not protect the confidentiality of this information, or are you not sure?

Response	1996	1999	2000
Census Protects	12.9%	22.8%	25.1%
Census Does Not Protect	9.6%	11.5%	9.4%
Not Sure	77.5%	65.7%	65.5%
N (weighted)	289	827	975

Source: 1996, Question 7_3; 1999-2000, Question 7a2 or 7a4.

[7] When respondents who answered Not Sure were asked to guess, the percentage giving the correct response increased to 52.8 percent in 1999 and 60.5 percent in 1999 and 2000 for the question about confidentiality, and to 20.8 percent and 24.2 percent, respectively, for the question about other agencies (calculated from Table 1 in Tourangeau *et al.* 2001).

Data comparable to those in Table 1 are also available from a National Research Council (1979) study inquiring into confidentiality concerns as factors in survey response, which asked an almost identical question. Reanalyzing the responses to this question, Brick *et al.* (1997) report that 39 percent believed Census records were available to other agencies, 9 percent believed they were not, and 51 percent said they did not know. These figures are quite similar to those obtained in the later surveys, although a larger percentage answered Don't Know and a smaller percentage offered the incorrect response in 1979.

Near the end of the 1996 interview, respondents were asked for the first time whether the Census Bureau was forbidden by law from sharing data with other agencies, or (in a split-ballot version) whether the Census Bureau was required by law to keep the data confidential. These questions were repeated in 1999 and 2000. Trends in responses to the 'forbidden by law' question are presented in Table 3, which shows a large increase in the proportion giving the correct response ('forbidden by law') between 1996 and 1999, and a further proportional increase between 1999 and 2000, although even in 2000 less than half the sample gave the correct response. Incorrect responses also increased between 1996 and 1999, but this trend was dramatically reversed in 2000, perhaps as a result of the Census Bureau's public relations campaign in connection with the decennial census. Responses to the 'required to keep confidential' question show a similar trend (data not shown)[8], although in every year the proportion believing that there is a law requiring confidentiality is much larger than the proportion believing that there is a law forbidding data sharing with other agencies, rising to 76 percent in 2000.

Table 3. Is Census Bureau Forbidden by Law from Sharing Information: By Year

Response	1996	1999	2000
Yes	28.3%	43.3%	48.9%
No	17.1%	29.7%	19.0%
Don't know	54.6%	27.0%	32.1%
N (weighted)	579	762	973

Source: 1996, Question 22a; 1999-2000, Question 24a.

[8] Data not shown in this chapter are included in a final report on these surveys to the Census Bureau. See Singer *et al.* (2001).

However, just as the percentage of those correctly perceiving the Census Bureau's protection of confidentiality increased between 1995 and 2000, so did the percentage of those saying it would bother them 'a lot' if another government agency got their answers to the census, along with their name and address, or if their answers to the census were not kept confidential. The proportion responding 'a lot' to the former question increased significantly, from 36.8 percent to 45.6 percent, between 1995 and 2000; corresponding responses to the latter question increased from 36.6 percent to 49.6 percent between 1996, the first time the question was asked, and 2000. In both cases, the largest increase occurred between 1996 and 1999, with the further change between 1999 and 2000 not statistically significant.

In all three years, respondents who indicated that there were laws forbidding data sharing or requiring confidentiality were asked whether they trusted the Census Bureau to obey these laws. Table 4 shows trends in responses to this question (because responses did not differ depending on which version of the preceding question respondents received, they have been combined in this table). The small fluctuations over time in the percentage saying they would trust the Bureau are not statistically significant. But, coupled with the increased awareness of the relevant laws, this means that a larger number of people trusted the Census Bureau in 2000 than did so in 1996.

Table 4. Trust Census Bureau to Keep Information Confidential (Those Who Know the Law Only): By Year

Response	1996	1999	2000
Yes	66.7%	69.3%	67.8%
No	33.3%	30.7%	32.2%
N (weighted)	464	957	1,197

Source: 1996, Question 22a1; 1999-2000, Question 24a1.

Almost at the end of the questionnaire, respondents were asked three questions designed to measure the prevalence of suspicions sometimes voiced about the misuse of census data for law enforcement purposes. The first of these questions (asked in 1995, 1999, and 2000) was, 'Do you believe the police and the FBI use the census to keep track of troublemakers?' The proportion of those giving the correct response (that it is not used for that purpose) increased slightly, from 49.0 percent to 52.1 percent, between 1995 and 1999, and then substantially, to 63.5 percent, between 1999 and 2000 (the overall change is statistically significant at 0.001). The second question, asked only in 1999 and 2000, was, 'How about to locate illegal aliens? Do you believe the census is used for that?' The proportion voicing this belief declined significantly, from 50.3 percent in 1999 to 42.1 percent in 2000. Finally, respondents in 1999 and 2000 were asked, 'Do you agree or disagree that people's answers to the census can be used against them?' The proportion agreeing declined from 39.2 percent to 37.3 percent, but this change was not statistically significant[9].

5. Trends in Attitudes Toward Privacy

So far, we have considered trends in beliefs about confidentiality. We also asked questions about privacy, as distinct from confidentiality. One question asked specifically whether the respondent regarded the Census Bureau's asking about age, race, and sex, along with name and address, as an invasion of privacy; other questions were more general[10]. Some of these questions were asked in all four years; most were asked only in 1995, and then again in 1999 and 2000.

There was a small but significant decline between 1995 and 2000 in the proportion of the sample who regarded the questions asked on the census short form as an invasion of privacy; 23.5 percent regarded it as an invasion in 1995, and 20.9 percent did so in 2000.

Scores on the Privacy Index, consisting of answers to the five more general privacy questions, are shown in Table 5 for each of the three years in which they were asked. The overall change is significant ($F = 4.75$, $df = 5097$, $p < 0.01$), as is the change from 1999 to 2000. Thus, general concerns about privacy declined slightly but significantly between 1995 and 2000.

---

[9] Martin (2000) reports that agreement that people's answers can be used against them increased significantly during the period following mailing of the 2000 census forms; this is based on Inter-Survey tracking surveys with independent samples. We also found an increase in agreement with this statement by interview date (logged) during the 2000 survey ($p = 0.2$), but ours were not independent samples, simply people interviewed later and earlier in the interviewing period. However, the analysis controlled for a series of demographic characteristics and self-reported exposure to positive and negative publicity about the census, as well as attitudes toward the census and privacy and confidentiality issues.

[10] For the wording of the questions and the way they were scored, see the Appendix to this chapter. Question numbers refer to the 1999 and 2000 surveys; questions were asked in the same relative order in 1995.

Table 5. Scores on Privacy Index* in 1995, 1999, and 2000

Year	Mean	Standard Deviation	N
1995	15.202	2.722	1443
1999	15.129	2.819	1677
2000	14.918	2.944	1978

*For question wording and scoring, see Appendix to this chapter. Imputed data were used in calculating the Privacy Index.

Respondents were also asked to weigh possible gains in efficiency from the use of administrative records against possible loss of privacy. Specifically, they were asked, 'Sharing information between different government agencies saves time and money, but it also means some loss of privacy for the individual. Do you think the benefits of saving time and money outweigh the loss of privacy?'[11] The proportion saying the benefits of saving time and money outweighed possible privacy losses dropped from 44.9 percent in 1996, the first time this question was asked, to 40.0 percent in 1999, and then rose slightly to 41.1 percent in 2000. These distributions did not differ significantly (chi-square = 4.03, $df = 2$, $p = 0.13$).

6. Trends in Attitudes Toward Data Sharing

It seems likely that public beliefs and feelings about data sharing among federal statistical agencies do not conform to the classical definition of attitudes, which carries with it the implication of an *enduring* predisposition (Allport 1935). On the contrary, the survey itself may be the first time those questioned have encountered the topic, and as a result the questions themselves, as well as the way they are framed, are likely to have stronger effects on their answers than would be the case if the attitudes were firmly held and buttressed by other belief and value systems.

In fact, attitudes about data sharing are very likely to fall into the category of 'nonattitudes', or 'pseudo-attitudes' (Bishop *et al.* 1980). Schuman and Presser (1981, p.153), however, point out that, when people are confronted with an issue they have not encountered before, they tend to assimilate it to information and attitudes which they already possess. Drawing on their arguments, Singer, Schaeffer, and Raghunathan (1997) showed that opinions about data sharing are related in predictable ways to trust in government, to confidence in the Census Bureau's promise of confidentiality, to feelings of political effectiveness, and to a more general inclination to share or withhold personal information, even though 55 percent

[11] It might be argued that what is lost is not privacy, but the confidentiality of information given to one agency and now shared with another.

of the sample said, in response to a question, that they had heard or read little or nothing at all about data sharing among federal agencies. Although such opinions may shift in response to media attention to the issue (Kerwin and Forsyth 1998, p.19), they can usefully be regarded as reflecting these general predispositions.

The series of surveys under discussion here explore trends in attitudes toward three different issues: the use of administrative records to reduce the undercount; the use of such records to replace the conventional census; and the use of administrative records to provide the information currently collected by means of the census long form[12]. Questions about reducing the undercount were asked in terms of data sharing by the Social Security Administration, the Internal Revenue Service (IRS), and one additional agency, which varied from year to year[13]; the order in which agencies were asked about was randomly rotated. Questions about a records-only census did not specify any particular agency, and questions about the long form were asked only about the IRS and a second agency, which varied from year to year, as described in footnote 12. In each case, the question about administrative records was preceded by a short description of the problem their use was designed to address. Thus, respondents were first informed about the existence of the undercount, and then asked how they felt about specific federal agencies sharing data with the Census Bureau in order to 'identify people who are missed in the census'. To justify the use of administrative records to replace the conventional count, respondents were told, 'No one would be asked to fill out a [census] form. Instead, the Census Bureau would count the entire population by getting information from other government agencies'. The question about replacing the long form was preceded by a question probing awareness of the existence of the long form, and the question itself contained a fairly lengthy rationale: 'Other government agencies . . . already have some of the information asked on the long form. It has been proposed that they give this information to the Census Bureau. Combining information from agencies would mean that everyone could fill out the short form instead of some people having to fill out the longer form. To make this possible, would you favor or oppose . . .'.

Table 6 shows trends in attitudes toward data sharing by the IRS in order to reduce the undercount; attitudes toward the Social Security Administration are similar but more favorable in all years. The largest drop in approval occurs between 1996 and 1999 for all three agencies. In every year, those strongly opposed outnumber those strongly in favor by almost two to one (data not shown).

---

[12] The findings of earlier studies are reviewed in Blair (1994).

[13] In 1995, it was the Immigration and Naturalization Service; in 1996, it was the Food Stamp Office; and in 1999-2000 it was 'agencies providing public housing assistance'.

Table 6. Opinions on the IRS Sharing Short Form Information With the Census Bureau: By Year

Response	1995	1996	1999	2000
Favor	70.5%	69.3%	54.0%	55.2%
Oppose	29.5%	30.7%	46.0%	44.8%
N (weighted)	1,366	1,167	1,619	1,925

Source: 1995, Question 12a, b, or c, depending on order; 1996–2000, Question 10, 12, or 13, depending on order.

Trends in attitudes toward a records-only census are shown in Table 7. The decline in support for this option is almost linear, with the total drop between 1995 and 2000 amounting to approximately 17 percentage points. Those opposed to a records-only census were then asked whether they would favor it if it led to increased accuracy and (if they were still opposed) if it saved money[14]. The argument about accuracy persuaded more people than the argument about economy. The proportion remaining opposed in the face of both arguments increased from 16 percent in 1996 to 23 percent in 1999 and 24 percent in 2000. Those who remained opposed were asked about the reasons for their opposition. The most frequently cited reasons involved concerns about privacy and confidentiality, given by 22 percent in 1999 and 29 percent in 2000; the second most frequent reason was a belief that such a census would be less accurate (17 percent in 1999 and 19 percent in 2000).

Table 7. Opinions on a Records-Only Census: By Year

Would you favor or oppose the Census Bureau getting everyone's name, address, age, sex, race, [and marital status] from the records of other government agencies so no one would have to fill out a census form?

Response	1995	1996	1999	2000
Favor	59.0%	54.7%	46.5%	42.3%
Oppose	41.0%	45.3%	53.5%	57.7%
N (weighted)	1,338	1,137	1,629	1,915

Source: 1995, Question 13; 1996–2000, Question 14.

[14] The order of asking about accuracy and economy was randomized, with those who continued their opposition to a records-only census being offered the second reason for changing their views.

During the decennial census, one-sixth of the population receives a longer questionnaire (the so-called long form), which asks about such things as jobs and income in addition to the basic questions needed to enumerate the population. During Census 2000 the long form became the object of brief but intense negative publicity[15]. Whether as a result of this negative publicity or for other reasons, preliminary analyses indicated that differences in response rates between the long and the short form increased from 5 percent in 1990 to 11 percent in 2000[16].

Since 1995, the Surveys of Privacy Attitudes have asked whether people were aware of the long form and whether they would be willing to have government agencies share data with the Census Bureau in order to eliminate it. Only about one-fifth of the population were aware of the existence of the long form in 1996, down somewhat from 1995, and the figure had declined to 17 percent in 1999. But by the time of the 2000 survey, which went into the field the week after census forms were delivered to every U.S. household, some 59 percent claimed awareness of the long form. However, increased awareness did not translate into increased approval of having government agencies such as the IRS share data with the Census Bureau in order to eliminate the need for the long form. The proportion favoring data sharing for this purpose declined from 52.2 percent in 1995 to 42.9 percent in 2000, at an average rate of about 2 percentage points per year; and, as in the case of data sharing to reduce the undercount, those strongly opposed to data sharing of long-form information outnumbered those strongly in favor by roughly two to one (data not shown). The relationship between awareness of the long form and approval of government agencies sharing data with the Census Bureau in order to eliminate the long form was significant in only one of the four years (1996), with those more aware significantly more favorable toward sharing[17].

In every year, the public was more reluctant to permit sharing of sensitive data than to permit sharing of the information needed to produce a count of the population. However, the gap between the long and short forms actually declined, over the years, from about 18 percentage points in 1995 to about 12 percentage points in 2000, because reluctance to permit sharing even short-form information by the IRS declined at a greater rate. Not unexpectedly, those who believed the Census Bureau protects data confidentiality were significantly more willing to have other agencies share long-form data with the Census Bureau in all three years (data not shown).

---

[15] See, for example, D'Vera Cohn, 'Census Vote Suggests Census Reply Choices', *Washington Post,* April 8, 2000, p.A02; Haya El Nasser, 'Census Shaken by Grumbling', *USA Today,* April 10, 2000, p.4A; and D'Vera Cohn, 'Census Complaints Hit Home', *Washington Post*, May 4, 2000, p.A09.

[16] Reported in Steven A. Holmes, 'Defying Forecasts, Census Response Ends Declining Trend', *New York Times,* September 20, 2000. The final census calculations of these rates were not available when this chapter was written.

[17] In 1999 and 2000 this relationship was reversed, with those more aware less favorable toward sharing, but the relationship was not significant.

Willingness to Provide Social Security Number

One question of particular interest to the Census Bureau is the extent to which people would be willing to provide their Social Security number (SSN) to the Census Bureau in order to permit more precise matching of administrative and census records. Evidence from earlier Census Bureau research is conflicting in this regard. On the one hand, respondents in four out of five focus groups overwhelmingly opposed this practice when they were asked about it in 1992 (Singer and Miller 1992). On the other hand, individuals in a field experiment in 1992 were only 3.4 percentage points less likely to return a census form when it requested their SSN than when it did not; an additional 13.9 percent returned the form but did not provide an SSN (Singer, Bates, and Miller 1992). An experiment carried out by the Census Bureau during Census 2000 indicates that a request for SSN decreased mail returns as well as the number of forms returned with some missing data; but the effect was not nearly as strong as is suggested by the survey data reported below (Guarino *et al.* 2001).

A question about SSN was included for the first time in the 1996 survey. The question read as follows:

> The Census Bureau is considering ways to combine information from federal, state, and local agencies to reduce the costs of trying to count every person in this country. Access to Social Security numbers makes it easier to do this. If the census form asked for your Social Security number, would you be willing to provide it?

Responses to this question parallel responses to the data-sharing questions, with willingness to provide the SSN declining significantly from 68 percent in 1996, the first time the question was asked, to 55 percent in 1999 and 56 percent in 2000.

Analyses not shown indicate that two kinds of reasons are associated with expressed reluctance to provide one's SSN to the Census Bureau. First, there are reasons related to beliefs about the census: People who were less aware of the census, who considered it less important, and who were less favorable toward the idea of data sharing were significantly less willing to provide their SSN. Low levels of education are also associated with these characteristics. Second is a set of beliefs and attitudes concerning privacy, confidentiality, and trust: People who were more concerned about privacy, who had less trust in the Census Bureau's maintenance of confidentiality, and who were less trusting of government in general were much less likely to say they would provide their SSN to the Census Bureau. Women are in general more concerned about privacy than men, and they were also less willing to say they would provide their SSN.

7. Attitudes and Behavior

So far, we have considered trends in attitudes toward privacy, confidentiality, and data sharing. The question we now turn to is the extent to which these attitudes are indicative of behavior.

We have already seen that although requesting an SSN reduces both unit and item response, it does not reduce it nearly as much as suggested by responses to these surveys. We have also seen, in the introduction to this chapter, that concerns about privacy and confidentiality predicted mail return to the 1990 census, although their contribution to the explained variance was relatively small (1.3 percent). Because the Census Bureau matched 1999 and 2000 survey responses to its Master Address File, it was possible to examine the relationship between attitudes and behavior—*i.e.*, returning the 2000 census form—which previous surveys, undertaken in non-census years, had not been able to do.

At the conclusion of the interview, all respondents to the 1999 and 2000 surveys were asked by Gallup interviewers for their address. (If the address had already been obtained before the survey, the interviewer merely verified it with the respondent. As justification for the request, interviewers said, 'We need your name and address in case the Census Bureau wants to do any follow-up research'.) In 1999, Gallup interviewers obtained 1,399 addresses from 1,677 respondents, or 83.4 percent; in 2000, they obtained 1,682 addresses from 1,978 respondents, or 85 percent. Taking both years together, some 574 (15.7 percent) respondents did not supply addresses at all, and an additional 175 (4.8 percent) did not provide enough information to permit the Census Bureau to attempt to match the address. Of the remaining 2,906 addresses, the Census Bureau matched 2,182, or 75 percent, at the household level. Thus, the analyses in this section are based on the 2,182 of 3,655 respondents (59.7 percent) who provided an address that was matched by the Census Bureau. Because this is a very low percentage[18], we also consider, later in the chapter, the extent to which respondents who either did not supply an address or whose address could not be matched differ on attitudinal and demographic characteristics from those whose address was successfully matched by the Census Bureau.

[18] For example, for their analysis of privacy and confidentiality as factors in response to the 1990 census, Singer *et al.* (1993) used respondents to the face-to-face Survey of Census Participation, carried out in the summer of 1990 by the National Opinion Research Center with a response rate of 89.8 percent. Respondents to this survey had been linked to decennial census information as part of a larger project on survey participation (see Groves and Couper 1992); 97.6 percent of the addresses were successfully matched at the household level. For details of the match operation, see Couper and Groves (1992).

Mail Return Rates Among Respondents

All but four respondents in 1999, and all but four in 2000, were designated to return their census form by mail. These eight respondents are excluded from the analyses that follow. Among respondents interviewed in 1999 and designated for mail return, 85.6 percent of those whose addresses were matched by the Census Bureau returned their census form by mail; in 2000, the percentage was 86.2[19]. This was considerably higher than the rate of 76.1 percent reported in Singer *et al.* (1993)—an indication that those who provided their addresses in 1999 and 2000 were generally more cooperative respondents.

Of those interviewed in 1999, 16.6 percent said they had received the long form, compared with 19.6 percent of those interviewed in 2000. In both years, the return rate varied according to which form had been received. For those interviewed in 1999, it was 87.1 percent for the short form vs. 78.3 percent for the long form, a difference of 8.8 percentage points; for those interviewed in 2000, it was 87.8 percent vs. 80.9 percent, a difference of 6.9 percentage points[20].

Predictors of Mail Return

To determine the effect of attitudes on behavior, we estimated a logistic regression equation with probability of return as the dependent variable and form type, six demographic variables, and eleven attitudinal variables as predictors, separately for 1999 and 2000[21]. Another behavioral indicator, refusal or inability to provide income on the survey, was also included as a predictor.

Results are shown in Tables 8 and 9[22]. The effect of form type is highly significant in both years; with other variables controlled, those receiving the long form were only about half as likely to return the form by mail as those receiving the short form. Age and education were also significant in both years, with older and better-educated respondents more likely to return their census form by mail. In the 1999 but not the 2000 sample, women were significantly more likely to return their

[19] An alternative way of computing a return rate, based on the Census Bureau's Nonresponse Follow-up Universe variable (NRU), yields a slightly lower return rate: 80.4 percent in 1999 and 81.7 percent in 2000. NRU takes into account the date of return, classifying all census forms received after April 19 as non-returns. We have included all returns, regardless of the date, and have even included one telephone and two Be Counted returns in the analyses reported in this section, but we ran all analyses with both dependent variables and comment on differences as appropriate.

[20] The differences between long and short forms are larger if the NRU variable is used: 15.2 percent in 1999 and 12.7 percent in 2000. As noted above, preliminary Census Bureau analyses indicate a difference of 11 percentage points in the population as a whole between the long and the short forms (Holmes; see note 16).

[21] For a definition of the attitudinal variables, see Appendix A.

[22] The effect of region, estimated from four dummy variables, was not significant, and region was therefore removed from the regressions shown in Tables 8 and 9.

census form. In the 2000 but not the 1999 sample, nonwhites were significantly less likely to do so.

Table 8. Demographic and Attitudinal Predictors of Census Mail Returns, 1999

Variable	Parameter Estimate	Standard Error
Intercept	–1.453	1.424
Form type (long)	–0.673***	0.224
Female	0.387*	0.195
Age (logged)	0.934***	0.275
Nonwhite	0.164	0.292
Hispanic	–0.034	0.395
Income	0.051	0.079
Education	0.211**	0.098
Privacy Index	–0.054	0.039
Invasion of Privacy	0.150	0.289
Knowledge About Census	0.020	0.045
Importance	0.027	0.083
Census Misused	0.091	0.093
Share to Reduce Undercount	–0.035	0.226
Share to Eliminate Census	0.103	0.201
Share to Eliminate Long Form	0.001	0.207
Willing to Give SSN	0.246	0.207
Trust Government	–0.063	0.092
Obligation to Cooperate with Census	–0.08	0.205
Income Imputed	0.401	0.351

* $p < 0.10$

** $p < 0.05$

*** $p < 0.001$

Table 9. Demographic and Attitudinal Predictors of Census Mail Returns, 2000

Variable	Parameter Estimate	Standard Error
Intercept	–1.743	1.308
Form type (long)	–0.506**	0.203
Female	–0.048	0.184
Age (logged)	1.230***	0.252
Nonwhite	–0.543**	0.213
Hispanic	0.040	0.326
Income	–0.026	0.072
Education	0.167*	0.087
Privacy Index	–0.079**	0.036
Invasion of Privacy	0.125	0.249
Knowledge About Census	0.030	0.042
Importance	0.031	0.080
Census Misused	–0.172**	0.084
Share to Reduce Undercount	–0.058	0.211
Share to Eliminate Census	–0.124	0.184
Share to Eliminate Long Form	–0.107	0.194
Willing to Give SSN	–0.115	0.197
Trust Government	0.100	0.081
Obligation to Cooperate with Census	0.267	0.199
Income Imputed	0.134	0.300

* $p < 0.10$
** $p < 0.05$
*** $p < 0.001$

None of the attitudinal variables predicted census return among 1999 respondents, and only two did so among 2000 respondents: privacy concerns and the belief that the census is misused. Those who were more concerned about privacy—that is, scored higher on the Privacy Index—were significantly less likely than those less concerned to return their census form, although the effect is relatively small (odds ratio of 0.924). Likewise, those who believed that the confidentiality of census data might be breached to serve law enforcement purposes were also less likely to return their census form. Thus, concerns about privacy and the possibility of census information being misused appear to be predictive of cooperation when they are measured in the context of the actual census; they do not predict mail returns when they are measured a year earlier. This parallels the finding, noted earlier, that privacy and confidentiality concerns, as measured in 1990, significantly predicted mail returns to the 1990 census (Couper *et al.* 1998; Singer *et al.* 1993)[23].

The 1999 survey contained a measure of the household's intention to return the census form next year, and the 2000 survey asked whether or not the household had returned it. The 'intention' variable in 1999 was not significant in predicting actual returns, but the report on past behavior in 2000 was highly significant. Respondents who said they had returned their census form were 4.7 times as likely to have returned it as those who said they had not, or were not sure, controlling for the effect of form type (long vs. short). The interaction between form type and self-reported return (or intention to return) was not significant.

Matched and Unmatched Respondents

Tables 10 and 11 display, for 1999 and 2000, respectively, the characteristics that differentiate those households whose survey responses we were able to match to census information from those for whom no match was made. Among 1999 respondents, three demographic characteristics significantly predict a match (at the 0.10 level). Older people were significantly more likely to be matched, whereas nonwhites were significantly less likely to be matched. There were regional variations as well: Respondents from the Midwest and South were significantly more likely to be matched than those from the West (the reference category).

Among respondents in 2000, again two demographic characteristics were significant predictors of matchability. Hispanics were less likely to be matched, and those with higher incomes were significantly more likely to be matched. Regional variations were again significant in 2000, with respondents from the Northeast significantly less likely to be matched than those from the West.

[23] Because of the split ballot experiments described earlier, confidentiality concerns were not measured for the sample as a whole in 1999 or 2000. As a result, we cannot measure the effect of confidentiality concerns relative to privacy concerns in these years.

Table 10. Demographic and Attitudinal Predictors of Match Between Survey and Census Records, 1999

Variable	Parameter Estimate	Standard Error
Intercept	–1.764**	0.775
Female	0.149	0.108
Age (logged)	0.438***	0.155
Nonwhite	–0.259*	0.150
Hispanic	–0.011*	0.223
Income	0.042	0.044
Education	–0.025	0.052
Northeast	–0.038	0.164
Midwest	0.338**	0.159
South	0.262*	0.146
Privacy Index	0.013	0.021
Invasion of Privacy	–0.675***	0.134
Knowledge About Census	0.015	0.024
Importance	–0.049	0.046
Census Misused	–0.087	0.050
Share to Reduce Undercount	0.228*	0.121
Share to Eliminate Census	–0.036	0.113
Share to Eliminate Long Form	0.268**	0.115
Willing to Give SSN	0.337***	0.116
Trust Government	0.023	0.049
Obligation to Cooperate with Census	0.018	0.114
Income Imputed	–0.543***	0.153

* $p < 0.10$
** $p < 0.05$
*** $p < 0.001$

Table 11. Demographic and Attitudinal Predictors of Match Between Survey and Census Records, 2000

Variable	Parameter Estimate	Standard Error
Intercept	0.189	0.739
Female	–0.133	0.102
Age (logged)	0.094	0.142
Nonwhite	–0.122	0.129
Hispanic	–0.376***	0.182
Income	0.108**	0.040
Education	–0.065	0.046
Northeast	– 0.509***	0.157
Midwest	0.014	0.153
South	–0.174	0.139
Privacy Index	–0.032*	0.019
Invasion of Privacy	– 0.279**	0.129
Knowledge About Census	0.020	0.023
Importance	–0.003	0.045
Census Misused	–0.069	0.047
Share to Reduce Undercount	0.294***	0.111
Share to Eliminate Census	–0.024	0.104
Share to Eliminate Long Form	0.180*	0.108
Willing to Give SSN	0.223**	0.105
Trust Government	0.052	0.044
Obligation to Cooperate with Census	0.143	0.113
Income Imputed	–0.612***	0.133

* $p < 0.10$
** $p < 0.05$
*** $p < 0.01$

In each of the two years, several attitudinal variables significantly differentiated respondents for whom a match could be made from those for whom it could not. In both years, those who considered the census an invasion of privacy and those for whom income had to be imputed were significantly less likely to be matched, and those who were willing to provide their SSN and who approved of using administrative records to reduce the undercount and to eliminate the long form were significantly more likely to be matched. In 1999, those who believed the census might be misused were also significantly less likely to be matched; and in 2000, this was true of those who scored higher on the Privacy Index.

This profile of demographic and attitudinal characteristics generally reinforces a perception that respondents providing matchable addresses were less concerned about privacy issues and, perhaps, more favorable toward the Census Bureau than those whose addresses could *not* be matched. In all likelihood, then, the inability to include some 40 percent of the sample in the analysis of the relationship between attitudes and behavior serves to understate the extent to which concerns about privacy negatively affect willingness to cooperate with the decennial census. Even with the large sample loss, however, the negative impact of privacy concerns and of the perception that census data are misused for law enforcement purposes is clearly significant, and remains so despite a variety of demographic controls on the relationship.

8. Summary and Conclusions

One of the striking findings of a comparison between the 1995 and 1996 surveys was the absence of significant change in most of the measures directly related to the census and the Census Bureau (Singer, Presser, and Van Hoewyk 1997). Furthermore, there was no particular pattern to those changes (5 of 22 questions about the Census Bureau were significant at the 0.10 level) that did occur. At the same time, there were significant changes over the course of a year in attitudes of trust in government, concern about privacy, and feelings of political efficacy, all in the direction of less trust, less efficacy, and greater concern about privacy.

But from the perspective of five years, many of the small changes that failed to register as statistically significant over the space of a year turn out to be significant secular trends, whereas some of what appeared as short-term change appears, in retrospect, to have been merely fluctuation. In this section, several different patterns of attitude change are distinguished.

One distinct pattern of change is apparent with respect to knowledge and awareness of the census, measured by questions that asked how important it was to count the population, whether people had heard of the undercount, whether they were aware of the uses of the census and of the census long form, and how important it was to cooperate with the census. These questions show small fluctuations between 1995 and 1999, and then large changes between 1999 and 2000, all in the direction of greater knowledge and awareness. Undoubtedly, this pattern is attributable to what has been referred to as the 'census climate'—the huge amount of media attention generated by the census in the decennial year. Other things being equal, these responses can be expected to return to 'normal' by the middle of the decade, and to resemble those in 1995.

Another pattern of responses characterizes questions tapping knowledge specifically about Census Bureau confidentiality practices—questions that inquire into knowledge of laws or beliefs about practices. All these questions show small but significant trends in the direction of greater accuracy. With two exceptions, most of these trends are rather evenly spread over the five years and do not appear to be attributable to the Census Bureau public relations campaign. The exceptions are correct responses to the question whether other agencies can get identified census data, which increased from 12.2 percent to 17.3 percent between 1999 and 2000, and a decline in incorrect responses to the question whether the Census Bureau is required by law to keep information confidential.

Paralleling this pattern of a secular increase in knowledge about confidentiality, however, is a significant increase over time in the percentage saying they would be bothered 'a lot' if their census data were shared with anyone outside the Census Bureau, as well as a decline in approval of data sharing for all three of the purposes asked about. With the exception of support for data sharing to reduce the undercount, which stabilized between 1999 and 2000, these trends appear to be linear. Expressed willingness to provide one's Social Security number also declined between 1996 and 1999, with no further change in 2000.

Interestingly enough, these changes are *not* paralleled either by increasing distrust of the uses to which census data might be put, by increasing concerns about privacy in general, or by declining trust in government. The three questions about possible misuse of census data all show a decline in distrust between 1999 and 2000, with two of the three statistically significant. The question asking whether people trust the Census Bureau to keep data confidential (if they correctly perceived that there were laws governing confidentiality) shows no significant change. The question asking whether the census short form is an invasion of privacy shows a small but significant decline between 1995 and 2000, and an index of general privacy concern also shows a small but significant decline between 1995 and 2000. Finally, people's trust in 'the government in Washington' shows an increase between 1996 and 2000 after declining from 1995 to 1996.

What is the significance of these findings for respondents' behavior?

There is, clearly, no one-for-one relationship between what people say and what they actually do. Many more people, for example, say they would not be willing to provide their Social Security number to the Census Bureau than actually fail to provide it when it is asked for on their census form. This is not unlike the differences between the laboratory and the field. Findings resulting from a laboratory experiment are ordinarily sharper and more clear-cut than those from a field experiment, because in the lab the experimenter can heighten the effect of the experimental stimulus and exercise more careful control over extraneous stimuli. Similarly, the survey interviewer focuses the respondent's attention on one stimulus at a time—the question about providing a Social Security number, for example—whereas the actual request for the number comes in the context of a series of other questions on an official printed government form, accompanied by materials urging cooperation.

At the same time, as we have seen, concerns about privacy do have a significant effect on whether respondents return their census form. Not only that, but feelings that the census is an invasion of privacy, expressed willingness to provide Social

Security number, and approved of having other agencies share data with the Census Bureau also significantly predict another behavior—whether respondents will provide the address needed to match their survey answers with census files.

This series of surveys leads to another conclusion as well. Feelings, as opposed to beliefs and knowledge, are remarkably resistant to change. Whereas knowledge about the uses of the census, beliefs about its importance, information about the undercount, and even beliefs about the Census Bureau's confidentiality practices all show significant and large effects of the so-called census climate, trust in the Census Bureau and the government in general, as well as concerns about privacy, are relatively impervious to change. And yet it is feelings, more strongly than knowledge or beliefs, that predict not only expressed willingness to cooperate with the census (*cf.* Presser *et al.* 2000), but actual cooperation as well.

References

Allport, G. (1935) 'Attitudes', in C. Murchison (ed) *Handbook of Social Psychology,* Worcester, Mass.: Clark University Press.

Bishop, G. F., R.W. Oldendick, A.J. Tuchfarber, and S.E. Bennett (1980) 'Pseudo-Opinions on Public Affairs', *Public Opinion Quarterly,* 44, pp.198-209.

Blair, J. (1994) 'Ancillary Uses of Government Administrative Data on Individuals: Perceptions and Attitudes', unpublished paper prepared for the National Academy of Sciences Panel on Census Methods.

Brick, J.M., D. Cantor, J. Kerwin, G. Pamela, and E. Singer (1997) *Review of Methods to Increase Coverage and Response Rates on the Study of Privacy Attitudes*, report to the U.S. Census Bureau under Contract 50-YABC-2-66025 (Deborah Bolton, contract manager).

Couper, M.P., and R.M. Groves (1992) 'Survey-Census Match Procedures', unpublished memorandum, Washington, D.C.: U.S. Bureau of the Census.

Couper, M.P., E. Singer, and R.A. Kulka (1998) 'Participation in the 1990 Decennial Census', *American Politics Quarterly,* 26, pp.59-80.

Curtin, R., S. Presser, and E. Singer (2000) 'The Effects of Changes in Response Rate on the Index of Consumer Confidence', *Public Opinion Quarterly,* 64, pp.413-28.

Fay, R.L., N. Bates, and J. Moore (1991) 'Lower Mail Response in the 1990 Census: A Preliminary Interpretation', in *Proceedings of the Annual Research Conference,* Washington, D.C.: U.S. Bureau of the Census.

Fishkin, J.S. (1995) *The Voice of the People: Public Opinion and Democracy,* New Haven, Conn.: Yale University Press.

Groves, R.M., and M.P. Couper (1992) 'Correlates of Nonresponse in Personal Visit Surveys', in *Proceedings of the Section on Survey Research Methods,* Alexandria, Va.: American Statistical Association.

Groves, R.M., S. Presser, and S. Dipko (2001) 'The Role of Topic Salience in Survey Participation Decisions', unpublished paper, College Park, Md.: University of Maryland.

Groves, R.M., E. Singer, and A. Corning (2000) 'Leverage-Saliency Theory of Survey Participation', *Public Opinion Quarterly,* 64, pp.299-308.

Guarino, J., J. Hill, and H. Woltman (2001) 'The Effect of SSN Requests and Notification of Administrative Record Use on Response Behavior in Census 2000', paper presented at the annual meeting of the American Association for Public Opinion Research, Montreal, May 19.

Keeter, S., C. Miller, A. Kohut, R.M. Groves, and S. Presser (2000) 'Consequences of Reducing Nonresponse in a National Telephone Survey', *Public Opinion Quarterly,* 64, pp.124-48.

Kerwin, J., and B. Forsyth (1998) *Report on Privacy and an Administrative Records Census*, report to the U.S. Census Bureau under Contract 50-YABC-2-66025.

Martin, Elizabeth (2000) 'Changes in Public Opinion During the Census', paper presented at the Census Advisory Committee of Professional Associations, Washington D.C., October 19.

National Research Council (1979) *Privacy and Confidentiality as Factors in Survey Response,* Washington, D.C.: National Academy of Sciences.

Raghunathan, T. E., J.M. Lepkowski, J. Van Hoewyk, and P. Solenberger (in press) 'A Multivariate Technique for Multiply Imputing Missing Values Using a Sequence of Regression Models', *Survey Methodology.*

Schuman, H., and S. Presser (1981) *Questions and Answers in Attitude Surveys,* New York: Academic Press.

Singer, E., S. Presser, and J.Van Hoewyk (2000) 'Knowing v. Feeling as Factors in Willingness to Provide Information to the Census', *Social Science Research,* 29, pp.140-7.

Singer, E., and E.R. Miller (1992) *Report on Focus Groups* Washington, D.C.: U.S. Bureau of the Census, Center for Survey Methods Research.

Singer, E., N. Bates, and E.R. Miller (1992) 'Memorandum for Susan Miskura', Washington, D.C.: U.S. Bureau of the Census, July 15.

Singer, E., N.A. Mathiowetz, and M.P. Couper (1993) 'The Impact of Privacy and Confidentiality Concerns on Survey Participation: The Case of the 1990 U.S. Census', *Public Opinion Quarterly,* 57, pp.465-82.

Singer, E., S. Presser, and J. Van Hoewyk (1997) *Report to the Census Bureau on Findings from the 1996 Survey of Public Opinion about the Use of Administrative Records in the Census*, report to the U.S. Census Bureau under Contract 50-YABC-2-66025 (Deborah Bolton, contract manager).

Singer, E., N.C. Schaeffer, and T.E. Raghunathan (1997) 'Public Attitudes Toward Data Sharing Among Federal Agencies', *International Journal of Public Opinion Research,* 9, pp.277-84.

Singer, E., R.Y. Shapiro, and L.R. Jacobs (1997) 'Privacy of Health Care Data: What Does the Public Know? How Much Do They Care?', in A.R. Chapman (ed) *Health Care and Information Ethics,* Kansas City: Ward Sheed.

Singer, E., J. Van Hoewyk, R. Tourangeau, D.M. Steiger, M. Montgomery, and R. Montgomery (2001) *Final Report on the 1999-2000 Surveys of Privacy Attitudes*, report to the U.S. Census Bureau.

Smith, T.W. (1982) 'House Effects and the Reproducibility of Survey Measurements', *Public Opinion Quarterly,* 46, pp.54-68.

Steffey, D., and N. Bradburn (1994) *Counting People in the Information Age.* Washington, D.C.: National Academy Press.

Tourangeau, R., E. Singer, and S. Presser (2001) 'Context Effects in the 1999 and 2000 Surveys of Privacy Attitudes', paper presented at the Annual Meeting of the American Association for Public Opinion Research, Montreal, May 20.

Turner, C.F. (1981) 'Surveys of Subjective Phenomena', in D. F. Johnson (ed) *The Measure of Subjective Phenomena,* Washington, D.C.: U.S. Bureau of the Census.

Turner, C.F., and E. Krauss (1978) 'Fallible Indicators of the Subjective State of the Nation', *American Psychologist,* 33, pp.456-70.

U.S. General Accounting Office (1992) *Decennial Census: 1990 Results Show Need for Fundamental Reform,* GAO/GGD-92-94. Washington, D.C.: Author.

Westin, A.F. (1994) "Interpretive Essay." In *Consumer Privacy Survey*, ed. Equifax Harris, New York: Lewis Harris & Associates.

Yankelovich, D. (1991) *Coming to Public Judgment: Making Democracy Work in a Complex World,* Syracuse, N.Y.: Syracuse University Press.

Appendix A

The following questions were used in the indexes shown in Tables 8–11:

Privacy index. This index combines responses to five questions tapping general (*i.e.*, not specifically census-related) concerns. The index consists of the sum of Q. 26: 'In general, how worried would you say you are about your personal privacy: very worried, somewhat worried, not very worried, or not worried at all'; Q. 29c: 'Please tell me if you strongly agree, somewhat agree, somewhat disagree, or strongly disagree. People's rights to privacy are well protected'; Q. 29d: 'Please tell me if you strongly agree . . . People have lost all control over how personal information about them is used'; Q. 29e: 'Please tell me if you strongly agree . . . If privacy is to be preserved, the use of computers must be strictly regulated'; and Q. 29f: 'The government knows more about me than it needs to'. Scores were reversed for Q. 29c. High scores indicate high concern about privacy.

The mean score on the Privacy Index was 15.129 (SD = 2.819) in 1999 and 14.918 (SD = 2.944) in 2000; $t = 2.20$, $df = 3653$, $p < 0.05$. Thus, general concern about privacy declined slightly but significantly between 1999 and 2000.

Invasion of privacy. In addition to the general questions above, Q16 asked: 'Do you feel it is an invasion of your privacy for the Census Bureau to ask your age, race, and sex along with your name and address?'

In 1999, 22.8 percent of respondents said that they considered the questions an invasion of privacy; in 2000, this response was chosen by 21.1 percent, a difference that was not statistically significant.

Knowledge about census. To measure knowledge about the census, we asked four questions. First, in Q.8, we asked, 'The census is used in many different ways. It is used to decide how many representatives each state has in Congress. The census is also used to decide how much money communities get from the government. Have you heard about either of these uses of the census?' If respondents answered Yes to Q. 8, they were asked Q. 8a: 'How much would you say you know about how the census is used—a lot, something, a little, or almost nothing?' Q. 9, administered in split-ballot form, asked, 'In the 1990 census about 5 million people were not counted. Some communities/big cities and cities with large minority populations were more likely to be undercounted than others. As a result, undercounted communities got fewer political representatives and less money from the government than they should. Have you heard about some communities/big cities and cities with large minority populations getting fewer political representatives or less money BECAUSE they were undercounted?'[24] If respondents answered Yes to this question, they were asked Q9(1): 'How much would you say you know about the

---

[24] Response distributions to the two versions of the question diverged somewhat in 2000. In 1999, 41 percent said they had heard about the undercount in 'some communities' and 44 percent said they had heard about it 'in big cities'; in 2000, the corresponding percentages were 49 percent and 57 percent.

census undercount—a lot, something, a little, or almost nothing?' An index running from 2 to 10 was constructed from the responses to these questions, with 10 indicating greater knowledge.

The mean score on the Knowledge Index was 4.578 (SD = 2.465) in 1999 and 5.241 (SD = 2.334) in 2000; $t = -8.301$, $df = 3{,}485$, $p < .001$.

Importance. This index consists of the sum of responses to Q.1, 'Every year the Census Bureau counts the people in the United States. How important do you think it is to count the people in the United States—extremely important, very important, somewhat important, not too important?' and Q. 23, 'As I said earlier, some communities/big cities with large minority populations were more likely to be undercounted in the census than others. As a result, undercounted communities get fewer political representatives and less money from the government than they should. Do you think this problem is very serious, somewhat serious, not too serious, or not serious at all?' High scores indicate that the respondent attaches greater importance to the census count.

The mean score on this index was 6.317 (SD = 1.236) in 1999 and 6.549 (SD = 1.235) in 2000; $t = 5.65$, $df = 3{,}653$; $p < .001$.

Census misused. Three questions in 1999 and 2000 assessed whether people believed census information was misused. Q. 32 asked, 'Do you believe the police and the FBI use the census to keep track of troublemakers?' Q. 33 asked, 'How about to locate illegal aliens? Do you believe the census is used for that?' Q. 34 asked, 'Do you agree or disagree that people's answers to the census can be used against them?' The index of census misuse consisted of the sum of the Yes /Agree answers to these questions.

In 1999, the mean score was 1.365 (SD = 1.075); in 2000, it was 1.179 (SD = 1.091); $t = 5.17$, $df = 3653$, $p < 0.001$.

Share to fix undercount. For this index, we used the answer to Q. 10: 'Now I will ask you about a proposal to fix the undercount. It involves using records from a number of government agencies to identify people who are missed in the census. One of the agencies is the Social Security Administration/Internal Revenue Service/Agencies providing public housing assistance. People who have a Social Security record/tax return/public housing agency record could then be counted. Would you favor or oppose giving the Census Bureau the name, address, age, sex [and race] of all the people for whom they have information in their records?' A random third of the sample was asked first about each of the three agencies; subsequent questions probed their attitudes toward the sharing of information by the other two. For this analysis, we used only the information about the first agency asked about; results do not change if we look at the results separately for each agency.

The proportion willing to share data for this purpose in 1999 was 64.0 percent, and in 2000, 64.7 percent, a nonsignificant increase (chi-square = 0.170, $df = 1$).

Share to eliminate census. Q. 14 asked, 'Another proposal is to do away with census forms entirely. No one would be asked to fill out a form. Instead, the

Census Bureau would count the entire population by getting information from other government agencies. Would you favor or oppose the Census Bureau getting everyone's name, address, sex, age, and race from other government agencies, so no one would have to fill out a census form?' The proportion favoring data sharing in order to eliminate the short census form was 46.5 percent in 1999 and 42.3 percent in 2000 (chi-square = 6.22, $df = 1$, $p < 0.01$).

Share to eliminate long form. Finally, we asked (Q. 19/20), 'Other government agencies such as agencies providing public housing assistance/the IRS already have some of the information asked on the long form. It has been proposed that they give this information to the Census Bureau. Combining information from agencies would mean that everyone could fill out the short form instead of some people having to fill out the longer form. To make this possible, would you favor or oppose the agencies providing public housing assistance/IRS giving the Census Bureau information on things like people's jobs and income, along with their name and address?' In 1999, 49.5 percent expressed a willingness to have government share data under these conditions; in 2000, 47.5 percent expressed such willingness, a nonsignificant decline (chi-square = 1.48; $df = 1$).

Willing to give Social Security number. Following the questions about willingness to have agencies share data in order to facilitate the census count, people were asked in Q. 21, 'The Census Bureau is considering ways to combine information from federal, state, and local agencies to reduce the costs of trying to count every person in this country. Access to Social Security numbers makes it easier to do this. If the census form asked for your Social Security number, would you be willing to provide it?'

The increase from 55.1 percent to 55.9 percent between 1999 and 2000 was not statistically significant (chi-square = 0.27, $df = 1$).

Trust in government. Trust in the federal government was measured by the sum of responses to two questions. Q. 30 asked, 'How much do you trust the government in Washington to do what is right? Just about always, most of the time, some of the time, or almost never?' Q. 31 asked, 'How about the people running the government—would you say you have a great deal of confidence, only some confidence, or hardly any confidence at all in the people running the government?'

The mean of this index was 3.962 (SD = 1.148) in 1999 and 3.987 (SD = 1.197) in 2000; $t = -0.657$, $df=3.598$, *ns.*

Obligation to cooperate. Q. 29 asked respondents to indicate the extent of their agreement or disagreement with the statement, 'Everyone has an obligation to cooperate with the census'. In 2000, 66 percent agreed strongly that people should cooperate, a highly significant change from 1999, when only 50 percent endorsed this response option.

Confidentiality, Disclosure, and Data Access: Theory and Practical Application for Statistical Agencies
Pat Doyle, Julia I. Lane, Jules J.M. Theeuwes and Laura M. Zayatz (Eds)

Chapter 15

The Privacy Context of Survey Response: An Ethnographic Account*

Eleanor R. Gerber
U.S. Bureau of the Census

1. Introduction

Many factors affect the public's response to requests for information in government surveys, including personal experience, cultural value systems, and self-interested or self-protective responses to social circumstances. Expectations are formed through experiences with all data collectors and all modes: school forms, job applications, magazine surveys, phone calls from marketers, and so on. The public absorbs potent images of privacy at risk in fictional accounts and news stories.

In undertaking this research, the goal has been to create a preliminary sketch of this wider context, and to locate respondents' reactions to government surveys within it. We found it useful to focus on the decision to provide (or to refuse to provide) information; how this decision is constructed, what factors are taken into account, and what other concerns are evoked in considering this decision. This decision appears to be quite complex. Rather than predefining a set of 'private information', our respondents recreate this decision structure each time a request for information is considered. As a result, privacy must be considered essentially situational; a privacy reaction is driven by the full context in which it occurs. The most important contextual elements are found both in the request for information and in the circumstances of the respondent. In particular, the request for information must come from a legitimate source and be relevant to a legitimate purpose. In addition, there must be a positive balance of benefits and risks for the respondent.

We found relatively unstructured conversations with respondents to be the most appropriate method of eliciting the full web of ideas in the most naturalistic way possible. Therefore, we have adopted exploratory qualitative techniques for this research. Our aim is to portray a spectrum of beliefs and responses to privacy is-

* Seven anthropologists were involved in this research, including two staff members at the Center for Survey Methods Research and five who participated under contract. I would like to acknowledge the contributions of Alisu Shoua-Glousberg, Betsy Strick, Jessica Skolnikff, Susan Trencher, Bhavani Arabandi, and Melinda Crowley to this research.

sues, and to show how these concerns are interconnected[1]. Our aim is not numerical assessment of different points of view, although we were interested in the diversity that arose within our data. Following the model of research on undercounted populations commonly adopted at the Census Bureau, our original research plan focused on members of ethnic groups previously underrepresented in the census and surveys. But the overall picture with respect to the basic decision model is one of similarity rather than diversity between these groups. Fully delineating the cultural, historical, and socioeconomic influences present in this overall picture would require further research.

2. Methods

The research on which this study is based occurred in two phases. In the preliminary phase, we concentrated on privacy issues surrounding demographic surveys. The second phase addressed issues of interest to the decennial census.

Respondents

In Phase I, interviews were carried out with 39 respondents, all of whom had previously participated (at least for one month) in the Current Population Survey. (We chose this recruiting strategy in order to assess respondents' experiences and preferences in a large panel survey.) Interviews were carried out in northern Virginia, Los Angeles, the Boston area, and Chicago. Interviews included 27 non-Hispanic white, 6 African American, 3 Pacific Islander, and 3 Hispanic respondents.

[1] It should be noted that our aim was not to account for all reasons why respondents refuse surveys or survey questions. Our respondents told us about non-privacy-related reasons for refusing to participate in surveys, including time constraints or questionnaire difficulty and the like, but these reasons are not discussed in this chapter.

Table 1. Respondents by Race and Ethnicity

	Non-Hispanic White	African American	American Indian	Asian	Pacific Islander	Hispanic	Multi-racial	Total
Phase I	27	6	0	0	3	3	0	39
Phase II	10	15	17	14	0	20	5	81
Total	37	21	17	14	3	23	5	120

In Phase II of the research, we recruited through a wide variety of citizen or social service groups, some of which had served as partnership groups in the decennial census. In addition, an ethnographer with ties to an American Indian group in Oakland, California, arranged interviews with us in that community. In Phase II, 81 interviews were carried out in a variety of locations, including Washington, D.C., Chicago, San Diego, Los Angeles, Oakland, Miami, and northern Virginia. We interviewed 15 African American, 20 Hispanic, 17 American Indian, 14 Asian, and 10 non-Hispanic white respondents, and 5 respondents who reported more than one race. We also used personal contacts to identify respondents who could be considered technologically sophisticated: 2 African American, 4 white, and 7 Asian individuals. By occupation, the latter group comprised 7 software engineers, analysts, or consultants; 2 employees of data mining companies; 1 structural engineer; 1 accountant; 1 sociologist; and 1 Internet services worker.

In recruiting by ethnic group, we followed established Census Bureau practice in studying issues that are considered to be 'barriers to enumeration' in the decennial census and surveys. Our institutional sponsors had specifically asked us to focus on such ethnic groups. One of the findings of this research was that diversity does not pattern neatly into these ethnic categories.

Other diverse elements were built into the recruiting, or discovered fortuitously, as we proceeded. For example, we specifically recruited a few additional respondents with jobs in information technology after that appeared to be a fruitful avenue to pursue. Finding generational differences among urban American Indians was the result of working through two different sources for scheduling interviews in two cities with different histories of migration. Our aim in recruiting through partnership groups and other social agencies was to reach respondents who might have been reluctant to cooperate with the census, or who would have been reluctant to speak directly with our ethnographers. Recruiting through such groups has often been useful to us in the past. However, in requesting this help from local social agencies it is not possible to precisely control other elements of the recruiting, such as age, gender, or education. Thus, we can observe some of the diversity that such factors may cause, although we recognize that our recruiting strategy does not optimize analysis of them.

The Ethnographic Interview

For Phase I, a semistructured research protocol was drawn up for use by the anthropologists connected with this research. Because the aims of the interview were generally exploratory, interviewers used flexible probes to follow up interesting lines of discussion. The Phase I protocol included debriefing about participation in the Current Population Survey, questions about other experiences with requests for information and how respondents decide whether to reveal information, and questions about means of controlling information. Next, a set of eight vignettes was administered, each of which described circumstances in which the central character has to decide whether or not to divulge information. These vignettes served to expand the set of topics under discussion to include subjects of particular interest to the research. The circumstances of these vignettes suggested revealing information over the Internet, revealing information over the telephone, revealing information through an agent, risks associated with giving information, issues of information sharing, and respondents' belief in assurances of confidentiality. The main aim of these vignettes was to elicit the reasoning processes that respondents applied to the decisions faced by the central character in the vignette. This protocol was pretested with five interviews.

The Phase II research protocol focused on somewhat different issues. We debriefed respondents about their experiences with the decennial census, which had taken place only two months before the start of ethnographic interviewing. We added questions eliciting respondents' understanding of specific privacy terms and concepts, their sense of whether privacy has increased or decreased, and their reactions to various modes of questionnaire administration. The vignettes were revised somewhat and new ones were created to elicit responses on several new topics of interest. In particular, we were interested in assessing respondents' knowledge of and reactions to issues of data sharing[2]. This new research protocol was pretested with 10 interviews.

3. A Decision Model for Revealing Information in Surveys and Censuses

This section describes the general schema that respondents use for deciding which information to reveal about themselves and their families in particular circumstances.

Situational Decision Making

In planning this research, we began with the naive concept that information on certain identifiable topics would be rarely revealed, and that information on other topics would be readily revealed in almost all circumstances. For the most part, however, this is not an adequate conceptualization for the way our respondents dealt with privacy. Instead, they made a complex assessment of who was asking and what the consequences of answering might be, given their own particular circumstances. Thus, information is not private or public in itself, but is revealed or withheld as a result of a situational judgment.

[2] The Phase I protocol also used several card sorting tasks. The Phase II protocol included a section assessing respondents' understanding of specific confidentiality language used in demographic surveys and decennial census contexts. These data will be included in the final report of the project.

Even simple demographic information can be treated as highly private. For example, an actress told us that she never reveals her age because it may affect her ability to find work. Conversely, even highly sensitive material may be revealed if the situational judgment indicates a need for it. A number of respondents easily imagined answering survey questions about the number of their sexual partners for a survey with a medical purpose.

Sponsorship and Authenticity

In deciding whether to answer, respondents are very concerned with knowing to whom they are giving information. This judgment resolves into two related questions. First, respondents must determine whether they approve of the individual or organization collecting the data. We refer to this as the *sponsorship* of the question. An additional assessment must be made. Questions are often answered through agents authorized by the sponsor, such as interviewers, or through remote collection devices, such as questions on a website or mailed questionnaires. Respondents are aware that these 'agents' may be misrepresented. Thus, for respondents, the *authenticity* of the agent or collection device presents a second question.

Establishing Bona Fides: Authenticity. According to some of our respondents, it is impossible to know whether people who ask for information are really who they say they are. This is an example:

> With what people can do with the computer any more and the way people have found ways to skirt laws concerning impersonating various agencies, it's entirely too easy for somebody to put together a form that implies a connection with a legitimate business that isn't really that...and they can name themselves the FBI, which stands for...Fred's Business Institute, and just put FBI at the top.

Being certain of the questioner can be even harder over the telephone, because it is easy to misrepresent an identity there. (This is why some respondents said they will not answer any telephone survey questions.) Some were also aware that an identifying logo on a questionnaire or an ID card can be faked. Respondents reassured themselves about this in a variety of ways. In the case of government surveys, respondents looked for something that marked the data collection as 'official'. Badges were mentioned, and advance letters were viewed as a mark of the serious intent of the sponsor. One respondent said he looked for some kind of notary seal or watermark on the letter or the survey form itself.

Beyond this, respondents' search for authenticity in sponsorship became more personalized. For some respondents, deciding that an interview was legitimate required an additional personal assessment of the interviewer, based on the interviewer's behavior and bearing. One respondent described this to us as a 'leap of trust'. In fact, we were struck by how many of our respondents described their interviewers in very positive terms. Interviewers were described as 'nice', 'agreeable', and 'bubbly'. Respondents recalled jokes that were made, mutual interests in dogs or travel, flattering questions about home schooling, and the like. These transactions tend to transform an anonymous relationship into a personal one. This may have had the function of preventing boredom and burnout in a set of repetitive questions. It may also have had an effect on respondents' acceptance of the legitimacy of the interview. Anonymous relations are subject to mistrust, but once they were transformed into personal ones, benefit of the doubt could be given and the 'leap of trust' made.

Sponsorship. Respondents wanted to know who was collecting the information and whether they approved of the agency in question. If they do not approve, they will not agree to cooperate. For example, many of our respondents did not like marketing research questions, and said they would not answer any questions at all for such a sponsor. One respondent said she would not answer questions for the Centers for Disease Control, because she disliked the research they do on 'certain diseases'. Respondents did not require specific or accurate information to judge a sponsor, and often used what they were able to deduce from the agency's name. (An example was a respondent who had a positive reaction to a 'health department' because 'health sounds better than disease'.) Thus, respondents seem to be forming their judgments of sponsoring organizations, and what they are entitled to ask, on somewhat vague and inferential grounds.

In general, collecting personal information is widely considered a legitimate function, and our respondents concede wide rights to a variety of superordinate authorities to collect it. Our respondents often mentally take on the role of authorities in these agencies when considering whether to divulge information. People reason that insurance companies have a right to information about prior health conditions, mortgage lenders should have access to information about credit histories, and so forth. Respondents even took this attitude with information that they might regard as inappropriate to discuss with acquaintances, if they could see a benign use for the information. One of our vignettes described a preschool that asked parents about how they discipline their children and how often they quarrel. Although this could be easily marked as information that should 'stay in the family', many respondents thought the school might be better able to educate the children if it had this sensitive information.

Government Sponsors. We asked respondents how comfortable they would be in revealing information to particular government agencies, including a variety of agencies on a state or local level. Government organizations were generally perceived as having helpful or benign goals, and their rights to collect information tended therefore to be accepted. The Census Bureau is widely seen as having positive intentions. Respondents tended to search for reasonable purposes to collect specific data, if they were not immediately apparent, and to conclude 'they must have a good reason' to ask. However, a few government agencies were not granted this credit. Among our respondents, these included the Internal Revenue Service and the Immigration and Naturalization Service (INS), police agencies (such as the Federal Bureau of Investigation and local police), and the Bureau of Indian Affairs.

Recruiting in Phase II attempted to locate respondents who might have a more negative attitude toward sharing information with government organizations. The mistrust we encountered varied considerably among groups. Mistrust of government generally was highest among immigrants, African Americans, and American Indians, but we found these attitudes to some extent in all groups.

This mistrust is generally connected with the consequences that particular agencies control, such as being deported by the INS if one is an undocumented immigrant. However, the mistrust is also connected with a belief that the government is 'monitoring' or 'tracking' individuals. (Respondents were often familiar with the idea from media sources, but only a minority believed in it. Others thought that the technical potential was there, but were not sure if it really occurred.) The following is an example of a respondent who believes that the federal government is tracking individuals. The respondent is a white middle-class homemaker. Here, the issuance of Social Security numbers to infants is taken as evidence of government tracking of individuals.

> The federal government keeps track of everybody...You didn't have to have a Social Security card until you started work. Now it's required...as soon as a child is born. And you don't think that's a way the government is tracking individuals? You'd better believe they are!

It is interesting to note that belief in government 'monitoring' of individuals does not mean automatic refusal to divulge information to government agencies. If respondents believe there is a good purpose to be achieved by giving the information, they will cooperate despite their suspicions. Thus, in agreeing to cooperate with the census, respondents see the benefits to the community as more salient then the vague and rather distant risk of adding to government files on themselves. Other ideas support cooperation in the face of significant suspicion about data files on individuals. One idea was that, even if tracking occurs, there would be no reason to single out the respondents' personal data. That is, their data will not call attention to them because their lives are unnoteworthy, average, or even rather boring. Related to that was the notion that surveillance could turn up nothing that could harm them because they 'aren't doing anything anyway'. This implies a belief that the government is using stored data primarily to find lawbreakers.

Commercial Sponsors. Although respondents see legitimate sponsors in commercial organizations such as bank and insurance companies, these organizations were not given nearly the same latitude to ask questions as were government agencies. Information considered 'necessary' to the specific transaction is understood to be the commercial organization's business, even if that information is considered highly sensitive (like income or credit history). Beyond this, respondents are not likely to give the benefit of the doubt, as they are willing to do for agencies such as the Census Bureau. For example, many respondents said they answered questions directly related to a product that they have purchased or used, but would not answer ancillary questions, such as those about their lifestyle, preferences, or other purchasing habits. Respondents could see why a company might have a legitimate stake in knowing how satisfied a customer is with a recent purchase, but they could not understand why that entitles the company to information about their levels of education.

Respondents resented commercial enterprises that collect information to sell it to others. They often complained about not receiving any profit from information that they regard themselves as owning[3]. In addition, they dislike attempts to collect information to market things to them at a later time, although this tends to be associated with the annoyance they feel at junk mail and junk telephone calls. Many respondents said they refuse to cooperate with any marketing questions at all.

Relevance of Questions to the Sponsor's Purpose or to Self

The decision schema for divulging information required a judgment about the relevance of the specific questions to a legitimate purpose of the questioner. If the requested information was not viewed as relevant to a legitimate purpose, respondents regarded it as 'none of their [the agency's] business', or decided that 'they don't need to know that'. Thus, respondents mobilized a set of assumptions about what questions should be asked to serve the survey's intended purpose. Respondents formed these impressions from general knowledge about the sponsoring agency (however vague and inferential), explanations given to them at the start of the survey, and prior experience with similar data collections.

Once the subject matter of the data collection went beyond the respondents' assumptions about what they should be asked, they told us they often refuse to answer. Such questions are considered 'unnecessary', 'nosy', or 'a fishing expedition'. The requested information might not be considered sensitive, but the question broke the boundaries to which the respondents believed they had agreed. This is one reason respondents object to questionnaire supplements that are included in some panel surveys. Most of our Phase I respondents had been exposed to a supplement in the Current Population Survey asking them about tobacco use. Some respondents were uncomfortable with these questions because the official topic of the interview was employment, and tobacco questions were unrelated to that end. This is also why respondents refused to answer certain questions in market research surveys when the subject was expanded beyond a product or service they had actually used.

It is worth noting that some of the recent complaints about the 'intrusiveness' of the long form in Census 2000 may have had a similar basis. The advertising

---

[3] In fact, information collected by a seller in the course of a transaction legally belongs to the seller, although we never encountered a respondent who was aware of this.

campaign that accompanied the census and general discussion of the event were effective in informing our respondents about one legitimate purpose of the census: that of counting everyone in the United States. Questions about commuting or housing (which might have caused no difficulty in a survey with a different primary purpose) appeared unrelated to the count. Thus, some long form questions failed respondents' test of 'relevance' and were perceived as breaking a privacy boundary.

Respondents were also concerned with the relevance of questions to the particular circumstances of their lives and interests. One respondent told us that she would not participate in a survey by a school board because she does not have any children. Other respondents refused political polls because they were not interested in politics and did not know much about it. In both these instances, the refusals were based on the notion that their answers could not be 'helpful', and would therefore be irrelevant to the purposes of the survey.

Assessing the Consequences of Giving Information

Another important way respondents assessed questions was by examining the consequences that might flow from providing certain information to certain sponsors. These consequences can be generally described as benefits and risks. The negative consequences of giving information are described here as risk rather than cost, because respondents seemed more concerned with harm than with effort or expense.

Benefits. Possible benefits are an important reason for respondents to provide information. These respondents were familiar with trading information for particular benefits in many venues, including insurance and job applications, applications for loans and mortgages, and paperwork for social service agencies. In these circumstances, respondents told us over and over, 'you have no choice' but to give the information, even if the questions seem nosy or insensitive. It was our impression that poor people in our society are very used to trading information for benefits, but respondents in all social classes are familiar with the experience.

Respondents trade off risks and benefits in a wide variety of less critical situations. One of our vignettes described supermarket 'club' cards, which collect marketing information in return for discounted merchandise. The trade-off was apparent in how respondents reasoned about this situation. Even if they were highly protective of information and conscious of privacy, they did not think that the risks (primarily getting more junk mail) outweighed the benefits (store discounts).

Altruistic Benefits. Benefits are not always seen as personal gain. Our respondents also took into account certain altruistic benefits of providing information. They were motivated by a sense of doing a good for their community, however they defined it, or for society at large. When we asked why people participated in the recent census, probably the most frequent answer was in terms of 'being a good citizen' or because the census was 'important' for their local area or ethnic group.

Another value is described as 'having one's voice heard' or one's life circumstances represented in the data. For example, members of ethnic minorities, or those who identified with a social condition such as being a single mother, sometimes said they participated in surveys to make sure their group was represented in the data. These abstract concepts appeared highly salient to some respondents, and were sometimes enough to counteract potently resented risks. For example, one respondent who was a firm believer in the evils of the federal government collecting information had not only answered the census herself, but talked to her neighbors about its importance. She made an exception for the census because of her understanding of its importance to her town[4].

However, understanding these benefits in the abstract does not necessarily mean that respondents will agree to provide information. If respondents perceived the benefits as too marginal, or themselves as unable to share in them, they might refuse to participate even if the information requested was not particularly private. For example, one homeless man in Oakland told us that he had not filled out a census form in 2000, although he had learned that the census brings money into the community, because he was unable to find a shelter program that would accept him. Others told us they did not believe that money would come into their disadvantaged communities as a result of the census, regardless of what was promised, on the basis of past experience. They were skeptical of the advertising campaigns that had stressed such benefits. As one respondent put it, they had cooperated with the census in 1990, and they had yet to see any schools built in their neighborhood.

Risks. Negative consequences seemed relatively more salient to our respondents than did positive ones. To a great extent, the possibility of negative consequences controlled what data respondents felt comfortable in revealing. Four important kinds of consequences stood out: physical danger, loss of control of data, fraud, and getting in trouble with government authorities.

Some respondents who mentioned the possibility of physical danger as a consequence of providing information considered work and home address highly private because of the possibility of a stalker finding the respondent. We also heard that some elderly respondents prefer not to answer in-person surveys because they are afraid of having anyone come to their door.

The consequence about which our respondents worried most was fraud. Information that could be used to access or defraud financial accounts was almost universally highly protected. It included Social Security number, credit card numbers, bank account numbers, and the like. Collectively, this information was often termed 'your numbers'. This worry focused on the actions of criminals. Some respondents had heard of cases of identity theft, in which a fraudulent individual creates debt in someone else's name. Many had also personally experienced

---

4 This dichotomy between dislike for the federal government and cooperation with the census as a value to one's local community has also been described in other Census Bureau ethnographic research centering on minority members of Generation X. See Crowley (2001).

difficulties in which financial information was misused. Anecdotes about problems with Internet purchases, credit accessed by strangers, and hucksters trying to elicit financial information or Social Security numbers over the telephone were not uncommon. Most people believed that eventually they would be able to 'clear their names' but realized that it might be a long and costly process.

Loss of control of data also concerned our respondents. They were concerned that data given willingly to one source might be transferred to third parties without their knowledge or consent. Upswings in junk mail and phone calls are the most frequently mentioned evidence that this has occurred. Respondents note, for instance, that house purchasers are flooded with advertising for gardening tools, and new parents with calls from diaper services. Overstuffed mailboxes and interrupted dinners may seem relatively minor consequences, but they represented real irritants to respondents because, as they say, 'now you know your data is out there'.

In fact, it was the very uncertainty of having data 'floating around' that could be distressing. Respondents expressed this anxiety in such terms as 'you don't know when it will come back to haunt you'. In the words of another respondent, 'it's just the unknowing. You don't know what could become of it'. Concerns about loss of control of data are at the root of many of our respondents' attitudes toward providing information over the Internet. Despite wide differences in technological knowledge and experience, almost all of our respondents perceived some risk in providing data across the Internet. Concerns about data being 'out there' appear to be intensified by this technology.

4. Data Sharing as a Perceived Risk

In our second research protocol, we attempted to address ideas about data sharing more specifically by building in questions about information technology, using vignettes that addressed the possible use of administrative records to replace a survey and the sale of data by commercial enterprises, and by probing for ideas about government keeping and sharing data.

The reaction to having personal information in a large database was not universally negative, even if the respondents understood that the data could be sold or shared. We have previously described a vignette about grocery store 'clubs' that collect data on purchases. Even highly privacy-conscious individuals did not see these databases as problematic. Respondents reasoned that 1) they did not care who knew what groceries they purchased, 2) it was worth trading the information for the discounts, 3) 'embarrassing' purchases could be paid for in cash, and 4) nothing bad had happened yet. Thus, it appears that the kind of data involved and the manageability of risks control attitudes toward data sharing.

The transfer of more sensitive data, especially without permission, can elicit a very different response. Our second protocol included a vignette that described a pharmacy that was selling customer prescription information to drug companies and researchers. Prescription drugs are more sensitive than groceries, and may in fact be thought of as confidential information. Transfer of these data was often rejected by respondents, if they could see an alternative way for the vignette character to get necessary prescriptions filled. They said they would particularly resent this practice occurring without prior notification and permission. Most people believed that this information belonged to them and that they had rights to control its disposal.

The 'Big Computer'

The vignettes described above elicited reactions to specific instances of data sharing. Our discussions also revealed a sense in which data sharing is thought to occur on a larger scale. Respondents' picture of this data sharing was somewhat vague, but it was experienced as threatening. We came to refer to this as the 'big computer' theory. Essentially, it is the idea that data available on a computer in one location will (sooner or later) be available in all computers everywhere. The imagery associated with this cultural representation is supported by the media. Respondents alluded to news stories involving hackers accessing data from private industry and from secure government agencies. They also mentioned movie images picturing police or federal agencies tapping into huge databases to reveal a person's history, face, and whereabouts.

Respondents' awareness of the actual public accessibility of data varied somewhat. Some respondents were quite knowledgeable about actual public sources of information. They pointed out that in most of the states where we interviewed, Social Security numbers are used as driver's license numbers and are copied down everywhere. Others mentioned that names and addresses are available on voter registration lists and that it is easy to get credit information from credit bureaus. They knew or had heard that it was possible to look up individuals on the Internet. Thus, to some extent, respondents' belief in the public accessibility of data about them is based in fact.

Although 'big computer' ideas were not exclusive to views of government, they arose most commonly in that association. The exchange of data between agencies was often seen as happening easily and quickly through centralized files. (One respondent described for us the computers sitting on our desks in the Census Bureau where we could pull up data about anyone in whom we were interested.) In general, our respondents believed that 'government' shares data among different agencies and levels, regardless of any promises of confidentiality that are given. They assumed that information they give will make its way back to interested authorities. Thus, it is assumed that the Immigration and Naturalization Service will find out about illegal aliens and the Internal Revenue Service about tax evaders, regardless of who is collecting the data or what promises they give. One of our vignettes described an undocumented immigrant who is asked about his immigration status in a survey and is promised confidentiality. Most of our respondents thought that answering truthfully would be very risky. (Two of them thought he should answer the questions truthfully because they thought he should be deported, and this information would facilitate it.) A similar vignette described a man who fixes cars off the books and is asked about his income in a survey. Again, most respondents did not think he should reveal his cash payments in any survey, because the information would then be available to the IRS.

In particular, law enforcement and the courts were assumed to be able to get whatever data they want. Respondents recounted anecdotes of local police being able to find individuals by using data they had given to other agencies (such as credit bureaus, housing authorities, or social service agencies), and this is taken as proof of widespread information sharing. While some respondents see this as a major risk factor, others think it may be proper, because it allows malefactors to be caught (or, in one case, a respondent's runaway son to be located). Belief in police powers to get information may not entirely invalidate a respondent's belief in claims of confidentiality. Assurances of confidentiality may be assumed to apply to normal circumstances and to the behavior of average agency personnel. However, if a high level employee of a law enforcement agency demands data normally considered confidential, some respondents had no doubt it would be made available.

The belief that all government information is available to all government agencies had some interesting ramifications. Although many of our respondents try to be careful with their Social Security numbers, they often thought there was no additional risk in supplying them on a government survey. As they pointed out, 'they [the government] gave it to me in the first place'. Many respondents were puzzled by why a government survey would ask for a Social Security number in the first place. One respondent thought that asking for a Social Security number might make him suspicious. If they have to ask your Social Security number, he reasoned, they may not actually be a legitimate government agency.

Government data sharing is implicit in ideas about government 'tracking' of individuals, alluded to previously. The purposes of this tracking were not clear to our respondents, however. When respondents were probed about the specific purposes of this government tracking, two themes emerged. One was an interest in keeping track of the location of individuals (perhaps rendering them more accessible to the police). The other was government interest in monitoring wages, assets, and other financial transactions, presumably with the purpose of uncovering lies on income tax forms.

Administrative Records Use

Concerns about data sharing have a strong effect on respondents' attitudes about the use of administrative records. One vignette described a government survey in which a character could either fill out 40 questions on each family member or give permission for the same data to be acquired from other agencies. The reaction to this vignette was interesting. Forty questions seemed burdensome to most respondents. But the idea of allowing an agency *carte blanche* in one's data files was disturbing, and many respondents who were initially tempted by the saving of time changed their minds. Others were angered by the suggestion of using other records, and they were inclined to refuse both the survey and the permission to look elsewhere. The risks associated with the administrative records use were the following: (1) Respondents had reservations about the accuracy of data already in the files. They are afraid the data might contain 'mistakes', or that outdated information about them might be perpetuated, causing problems at a later time. Some worried about being somehow held responsible for the incorrect data. (2) Some respondents were concerned about discovery of contradictions between various sources. Respondents frequently modify the data they give for specific purposes, so they are aware that they may not have reported exactly the same information in every venue. These respondents regarded the suggestion to use data from other records as an opportunity to 'check' their answers. (3) Some respondents think it is more risky to have all data in one central location, and deliberately follow a strategy of revealing only part of the data in any one venue. Because of these concerns, most respondents felt that they had more control and less risk by refusing permission to get data from administrative records sources and filling out even a burdensome questionnaire themselves. A few respondents, however, felt that the choice as presented was unrealistic. Because they believed in the wide availability of data among government agencies, they assumed the administrative records data would be checked even if they refused permission.

Respondents are not just afraid of intentional data sharing by agencies that collect it. They are also concerned about data being 'shared' because of the bad behavior of individuals. Hackers are a prime example of this concern. Respondents believed that even if an agency has an official policy of confidentiality, people with bad intentions can access agency computers from outside and steal data that everyone thinks are protected. This is one reason that assurances of confidentiality were not completely convincing to respondents; they did not believe that reputable organizations are effective at protecting themselves from intruders.

5. Managing Information

Respondents' concerns about information led them to attempt to control what others can find out about them. Because respondents are highly concerned with fraud, most salient are their efforts to protect 'their numbers'. When we asked if there was anything that people did or avoided doing to protect information about themselves, respondents told us about cutting horizontally through the numbers on out-of-date credit cards, carefully destroying 'preapproved' credit card applications, using shredders for bank or credit card statements, whispering their driver's license number (hence, Social Security number) in stores, hiding check deposit slips, and always hanging up on anyone who seemed to want this information over the telephone.

When the requested information in a survey or application was beyond respondents' comfort level, some of them recommended asking the sponsor 'if you really need that information', or if you 'have to give those answers'. This suggestion often arose if the questions provided access to a benefit; respondents explained that they were trying to establish whether they could leave the answer blank and still be eligible. For example, if prior medical conditions might affect access to insurance, these respondents will try to negotiate how much information they are required to reveal. But other respondents believed in the wide accessibility of the information and reasoned that the authority would inevitably uncover any omitted information, and they would then be subject to additional penalties.

Respondents also told us they use 'don't know' and 'not applicable' options to manage information, especially if they think the questions are irrelevant to the legitimate purpose of a questionnaire. Thus, these responses can represent privacy-based refusals of information. Another commonly mentioned technique for protecting information was lying. We were struck in these interviews by how often respondents reported that they lied in response to requests for information. This is particularly true if the questions are beyond the boundaries of what the respondent considers to be legitimate. Respondents felt little or no compunction about telling these lies. Sometimes these lies presented what respondents regard as partial truth, such as reporting only some of off-the-books income. In other cases, respondents presented outright untruths, such as giving an incorrect ZIP Code or phone number to prevent marketers from finding them. One respondent enjoyed making up bizarre answers to marketing surveys in shopping malls, as a kind of game.

6. Diversity in Privacy Beliefs/Behaviors

Technological Awareness

Because of the importance of computers and data sharing to respondents' anxieties about information, we wanted to interview persons with a variety of levels of technological sophistication. We had many respondents who had never used computers or the Internet (although most had friends and relations who were users). We also recruited some respondents who worked in the computer industry or in data mining. On the whole, the basic ideas governing the decision to reveal information did not differ greatly between the two ends of the spectrum. That is, there is considerable commonality about assessing sponsorship, relevance, and attending to trade-offs between risk and benefit. But some differences did emerge.

The more technologically sophisticated respondents had a somewhat different attitude toward the Internet. They were more familiar with it, and one respondent told us that she felt comfortable answering questions on the Internet because she 'thinks better on a computer'. Yet technologically knowledgeable respondents, like other respondents, believed that information supplied over the Internet was at risk. For example, one respondent employed as an analyst in a commercial data mining firm refused to make purchases over the Internet and did not want to give sensitive survey information in that mode either:

> Maybe my answer would be different five years from now, but the security is still not very good. When I type messages, when I type e-mail, I assume that everybody in the world is reading what I'm writing...so if I transmit something I just assume everybody's seeing it.

Other such respondents were more certain of their ability to determine a secure site and were therefore more willing to use the Internet for important transactions. They looked for a privacy policy on a website. (Although they might not read it completely, they liked the fact that it was there.) They looked for specific icons indicating secure sites, although this was not entirely reassuring because icons are easily made. A few mentioned checking the encryption programs that were in use before they supplied information. But even these assurances did not convince them that the Internet was entirely without risk, and they too were primarily worried about fraud.

Because they did not consider the Internet entirely secure, their attention focused on the ways in which it is possible to mitigate negative consequences if they occurred. The strategies they had developed were primarily what distinguished the more technologically sophisticated respondents in our data. First, they were very aware of the policies protecting consumers over the Internet, which led them to choose certain credit cards or retailers that eliminate all financial liability in cases of misuse of finances. Like less sophisticated consumers, they preferred dealing with established organizations with good reputations. Second, they appeared much more willing than our other respondents to divest themselves of compromised identifiers. One respondent called this strategy 'ditching'. These more sophisticated respondents consider it relatively easy to 'ditch' information such as a credit card number, a bank account, an e-mail address, or a post office box. The process of getting new identifiers did not seem like a 'hassle' to those who embraced this strategy, although it was anxiety-provoking for other respondents.

Another difference worth noting is that these respondents had a greater awareness of the existence of extensive private, as opposed to government, databases on individuals. Unlike other respondents, they mentioned information maintained by large corporations on their customers, industrywide databases such as a 'pooled cooperative database of catalogue merchandisers', and a shared industrywide database with information on about 60 to 70 million households. The extent of this private data sharing sometimes made respondents who were aware of it give up on attempting to control information at all. This is one response:

> Yeah, well, so what? I mean, what are they going to do about it?...I would imagine they must rent that information out or something...because they must, they just must. Everybody's doing it to everybody else, they must....Well, here is my personal belief....Everybody, anybody who has any information about me will sell it if they can find somebody to buy it. So asking for consent is kind of meaningless....I mean the sense in which they're disclosing that they're doing it, I mean that element of this seems unnecessary to me because I'm assuming everybody is doing it. And, knock yourself out, is my attitude....I mean before I worked here I never gave it an ounce of thought. And I certainly wasn't aware of the extent to which people are selling information to each other.

Group Differences

One original aim of the research was to discover differences in cultural beliefs and attitudes about privacy that might affect the response patterns of particular groups. We had expected to discover differences in the definition of privacy or different patterns of protected information among these groups. For this reason, recruiting in Phase II stressed American Indian, African American, Asian, and Hispanic respondents. On the whole, however, the emerging picture does not indicate significant differences in approach to privacy that can be ascribed to ethnicity. Faced with the same data collections and similar risks and benefits, there is considerable commonality in the way these groups approach revealing information. Respondents in all groups process decisions about revealing information by thinking about sponsorship, relevance, risks, and benefits. Fraud and loss of control over data are important concerns to all groups. It should be noted that our research protocols tended to stress decision making in practical situations where data were being collected by a commercial or governmental organization. This focus was appropriate to the research, but it was not designed to elicit differences in the social construction of privacy in interpersonal interaction within the home or in community contexts. A complete account of ethnic differences in the interpersonal aspects of privacy awaits further research.

Nevertheless, we are able to suggest some patterns in which these groups may be said to diverge from the common account already rendered. These include (1) the effects of the particular groups' experiences with government on their attitudes toward data collection, (2) the contrast between privacy in communally based cultures and individually based cultures, (3) the effects of social class on decisions about privacy, and (4) different privacy sensitivities. The descriptions below should be considered suggested hypotheses for further research.

Experiences With Government. The historical relations between groups and the government appear to have a strong effect on attitudes toward data collections. For example, among American Indian respondents, beliefs in government tracking of individuals seemed to be somewhat more common. One view was that 'keeping track' of or 'keeping tabs' on the population was the purpose of the decennial census. The census seemed to be a redundant effort to one such respondent, because the federal government 'knows where I am, they can find me'. One respondent mentioned the Certificate of Degree of Indian Blood, which the federal government gives to enrolled members of federally registered tribes, and which controls access to such benefits as the Indian Health Service. Another mentioned a belief that the FBI maintains files on Indians, especially if they had connections with certain Indian political movements.

The 'tracking' idea seemed particularly troubling to a few American Indian respondents. Here is an example of a respondent who found the census intimidating:

> I really feel intimidated that I have to let the government know where I live, what I do, how many in my family. Almost, basically, running my life now. And I don't like it. I hate it. Somehow I think that with all the computer technology they have now, that they could track you... That's actually how I feel, like in the wild how they tag the animals and then they could tell where you're at, how far you've traveled in a given time, and...with the census thing, I don't know why they do that. I still don't know. Why they have to track everybody like that, why do they have to know who lives in your house, what do you guys do every day....

It should be noted that these negative views of the census were more common among one group of American Indian respondents than the other we interviewed. The group where these ideas were more salient lived in the Oakland area, and were somewhat older than the group interviewed in Los Angeles. Further, they were part of a community formed initially in the 1950s and 1960s as a result of a federal relocation project. Some of the respondents were children of American Indians who had been removed from their homes and sent to school in the San Francisco Bay area. They still remember and resent this experience. By contrast, the group in Los Angeles was younger and generally better educated. The attitudes of the second group do not show as great a sensitivity to the relationship between the federal government and American Indians. Most were quite positive about the census.

Other attitudes toward the government also reflect particular relationships between the government and a local ethnic community. In particular, Hispanic and Asian respondents frequently mentioned problems with the INS in connection with concerns about the census. Some of the Hispanic interviewees were in San Diego, and respondents there reported negative interactions with agents of the federal government because of the proximity of the border. One respondent had been detained crossing back into the United States after a visit to Mexico while officials demanded documentation, including birth certificates, pay stubs, and rent receipts. This made the respondent feel powerless, but, as she remarked, 'they're federal agents, they can do it'.

Perceptions of our government may be mitigated by contrast to more repressive governments in countries from which our respondents emigrated. For example, one

Cuban who had recently come to Florida found this country more private than her homeland:

> In this country I think there's quite a bit of privacy in people's lives. That's one of the things I like best. Because in the country I come from, they look even into the toenail of your big toe. So you want to get away from there because there's so much they want to know about your life. They have you under surveillance. You feel completely asphyxiated, like you have no privacy even in your own home.

Communally Based Cultures. There is some suggestion in these data that concepts of privacy may be influenced by the sense of community that exists within a particular group. This was evident in a group of respondents who were immigrants from India. Most respondents in this research believe that privacy has decreased in recent years, but these Indian respondents sense more privacy in America than in India. There, family and acquaintances expect to know all sorts of personal details about one's life, such as income or plans to have children, and apparently have wide rights to inquire about such matters. For example:

> In the U.S. everything is pretty private. There is more privacy in this country than back home in India. There is a lot of socializing that goes on in our lives back home. So information passes around pretty fast. And it's quite common...like, for example, the kind of money you make or the sources of our income or the relations we have with other people, general things. These things are open to a certain extent in the community.

In general, they said they liked the new sense of privacy. They also noted, however, that protecting information was more of a concern in the United States, and that they had learned to worry about it after they came here:

> I never paid so much attention to privacy before coming to America, and I never actually thought of privacy as such a show-stopping, life-critical thing. But after watching so many Hollywood films, I probably think that privacy could be protected closely...It's a big deal, is what I'm feeling right now....

In dealing with decisions to reveal information, these respondents appeared to be primarily influenced by their new concerns, and in fact sounded much like the rest of our respondents in describing the decision to divulge information.

A different sense of communally held information also emerged from our interviews with a group of poor African Americans. Their assessments of whether particular information is in the public or private realm tended to take account of the ease with which information spreads within tightly knit communities. Thus, information that might be considered damaging, such as having more than the allowed number of residents in an apartment, was not regarded as private, because everyone in the neighborhood knew about it. It should be noted that just because this information was considered in some sense public did not mean that they were willing to reveal it in a government survey.

Social Class. Social class appears to have a considerable impact on the responses to matters of privacy. First, as we have already mentioned, the poorest respondents may not be influenced by messages couched in terms of the benefits, because they may have ample evidence that resources tend not to be funneled to them or to their communities. Second, being poor may restrict one's options in attempting to protect privacy. This became clear in the reactions of most of the group of poor African Americans to a vignette in which the central character has to decide what to do about a pharmacy that is selling information about drug purchases. The typical response of the more affluent respondents was to suggest changing pharmacies. However, these poorer respondents sensed that this was not possible if no other pharmacy existed or gave credit in the neighborhood. In a sense, this is an example of trading information for benefits, described earlier. It should be kept in mind that when options are restricted, protecting privacy may take a lower priority.

Another difference that social class may create occurs in reactions to various modes of question administration. In general, more affluent respondents preferred modes of administration that allow them to stay in control of their time and living space; thus, mail and, for some, the Internet, are preferred modes. Among our less affluent respondents, face-to-face interviews tended to be more highly valued. Respondents said they like to be able to assess an interviewer in person, to decide if the interviewer is trustworthy. Perhaps they had more confidence in their ability to read individuals than to determine if written promises of confidentiality are dependable. In addition, some respondents see the interviewer as a source of explanations of difficult material and a possible helper if giving the information proves damaging.

Different Privacy Sensitivities. Matters of sexuality may be a salient privacy concern with some Hispanic respondents. There is evidence that the Spanish term *privado* (private) tends to elicit associations with sexuality that the English term 'private' does not. Thus *privado* was defined as 'things about a married couple that no one should intrude into', or 'intimate things. In couples'. This should be taken into account in Spanish translations of privacy statements.

Our American Indian respondents indicated certain sensitivities that other respondents did not mention. Some mentioned being asked questions about their children, particularly their names, as problematic. This may have been a result of the history, mentioned earlier, of children being separated from their families by the government. In addition, these respondents mentioned matters of religion and spirituality as issues they wanted to keep very private. They expressed a concern that the wider society might find their religious practices odd or different, and thus they were very concerned with keeping them within the family or the community. The sensitivity to revealing names may also have religious connotations for some American Indian groups, where names may have a sacred connotation.

7. Conclusions

Diversity and Commonality in Privacy Beliefs

Overall, this research indicates that there is a wide common area of agreement among respondents in the way they make decisions about requests for information. Our major findings, such as the concern with sponsorship, relevance, and risks and benefits, can be found in every group we interviewed. All are concerned with the possibility of fraud and give high priority to the protection of financial resources. Suspicion of the security of the Internet occurs in all groups, regardless of the degree of experience with the mode or technical expertise with computers. Thus, diversity does not stand as clearly as the commonality as a result of this research.

However, we did find it useful to think about the cultural, ethnic, and social class diversity. First, we need to look for attitudinal and behavioral variation within the very broadly defined ethnic categories that tend to structure our research. Defining the research units as 'American Indians' or 'Asians' is less appropriate than defining them as 'urban Indians in the Bay area' or 'middle-class immigrants from India'. A complete cultural account of reactions to privacy in surveys, structured around variation within ethnic communities, would require a much larger research project than we were able to do here.

Second, the differences we see between groups may reflect factors such as social class or contingent aspects of a specific historical relationship with government rather than ethnically specific beliefs and definitions. Future research is needed to examine these dimensions of difference directly, rather than assuming differences to be coextensive with ethnic community. This strategy highlights similarity of social situation and relationship to power as explanations for privacy behaviors, rather than emphasizing ethnic differences from a 'mainstream culture'.

The Construction of Privacy

The control of information is central to our respondents' understanding of privacy. This stress on control is consistent with general American values. Loss of control over information is resented or provokes anxiety. Respondents' sensitivities to the control of information determine their attitudes toward issues such as data sharing between agencies and affect attitudes toward different modes of questionnaire administration.

Because privacy is evaluated situationally, it is not possible to define in advance a set of topics that are perceived as breaches of privacy in all situations. As we have indicated here, perception of a legitimate need for the information is a critical factor in determining whether the request for information will trigger privacy concerns. Respondents require a sense of the legitimate, beneficial uses of information before they release it. Benefits to a particular community or to society as a whole do serve as motivations, and as such they are critical to communicate—although, as we have seen, this strategy can backfire with the most socially marginal individuals. This observation leads to the conclusion that explaining the uses of the data to respondents merits considerable attention on the part of questionnaire designers. We suggest that these explanations must be on the level of the specific information that is requested, not on the level of the entire data collection. Thus, text could be inserted in appropriate sections in a questionnaire, explaining the need for, say, housing or income information. This would be preferable to including a more general privacy statement in a cover or advance letter in a form that does not specifically address question content. The strategy of inserting specific privacy language where it is needed has been shown to work well with topic-based interviewer-administered questionnaires. (Moore and coauthors (2001) found that inserting an explanation of the need for income data in wording that was sensitive to privacy concerns reduced the level of nonresponse[5].)

Another conclusion is that including topical material that is only distantly related to the publicly known purpose of the data collection may be a risky strategy. The relationship of the content of topical modules to the respondent's perception of the overall purpose of the study should be considered in deciding where to include it.

Respondents also assess risks. It is important to note that assurances of confidentiality are generally not enough to counter a defined risk, especially when the stakes are considered to be high. The public perception is that data are widely shared among government agencies. We found little that would serve to counteract this belief among people who had a great deal to lose. However, it might be useful to directly address concerns about data sharing in explanations of confidentiality. These explanations should include more than descriptions of official policy; they should also explain how data are protected from intrusion by outsiders.

These conclusions suggest another limitation of assurances of confidentiality. If the privacy reaction results from a failure to see the legitimate purpose of the data collection, assurances of confidentiality are actually irrelevant to encouraging participation. That is, if a question is assessed as 'none of their business', it does not help to assure the respondent that the information will not be given to a third party. Reactions to privacy cannot always be managed with assurances of confidentiality. Other avenues, primarily explaining better the legitimate need for the information, must be pursued.

[5] See Moore and Loomis (2001). The wording of the text inserted before the topic-based income questions read as follows: "The next questions are about income. We know that people aren't used to talking about their income but we ask these questions to get an OVERALL statistical picture of your community and the nation, NOT to find out about you personally."

References

Crowley, Melinda (2001) *Generation X Speaks Out on Civic Engagement and the Census: An Ethnographic Approach,* Report for the Center for Survey Methods Research, Statistical Research Division, Washington, D.C.: U.S. Bureau of the Census.

Moore, Jeffrey C., and Laura S Loomis (2001) "Reducing Income Nonresponse in a Topic-Based Interview." Paper prepared for the 2001 Meetings of the American Association for Public Opinion Research, Montreal, May 17-20, 2001.

Confidentiality, Disclosure, and Data Access: Theory and Practical Application for Statistical Agencies
Pat Doyle, Julia I. Lane, Jules J.M. Theeuwes and Laura M. Zayatz (Eds)

Chapter 16

Business Perceptions of Confidentiality*

Nick Greenia
Internal Revenue Service

J. Bradford Jensen
U.S. Bureau of the Census

Julia Lane
American University,
The Urban Institute,
and
U.S. Bureau of the Census

1. Introduction

The core mandate of statistical agencies is to collect data—on a multiplicity of entities, including businesses and households. In so doing, each agency enters into an implicit pact with its respondents that the data not only will be used to fulfill the agency's mission, but also will be protected from unauthorized access and use. This promise of confidentiality is not only a legal and ethical mandate, but also an important contributor to optimal data quality and response rates. However, little is known about the substance of this pact—particularly with respect to businesses. This ignorance can have serious potential consequences. If government misconstrues its role or the nature of the pact, and consequently businesses do not trust government to protect their data because they either mistrust or misunderstand the pact, it will be difficult for government not only to maintain high quality and timely response rates but also to frame new ideas on data collection, protection, and access.

* This study would not have been possible without the extraordinary efforts put into the development of the survey by Diane Willimack and Kristin Stettler, U.S. Bureau of the Census. The authors would also like to thank Dr. Frederick Knickerbocker for both reviewing and facilitating this work and Faye Schwartz for excellent research assistance. All views expressed represent those of the authors, and not necessarily the institutions they represent.

This study reviews the current state of knowledge of business perceptions and presents the results of a survey of businesses on the topic. It is one of the first quantitative analyses of businesses' perceptions of the sensitivity of different types of data and their assessment of the protection provided by different agencies—as well as their assessment of the quality of statistical work performed by the agencies. It also examines the business community's knowledge of the financial and criminal penalties associated with breaches of confidentiality and its willingness to permit data to be shared among different federal and private agencies.

This information can be used not only for the maintenance and improvement of current collection systems but also for framing ideas on new data collection, such as data sharing initiatives, and access systems, such as secure data analysis sites. In each of these cases, the concerns of businesses are likely to differ from those of households. Businesses, unlike households, are more likely to have multiple requests from different government agencies for information, so data sharing across agencies may be an attractive way of reducing response burden. Similarly, businesses may be more aware of the importance of data quality for their own research and planning needs, and may favor access by outside researchers in controlled sites to further this aim.

The study is particularly timely given the heightened interest in confidentiality and privacy issues. Media attention has increasingly focused on the creeping loss of confidentiality protections—for example, the private marketing of personal dossiers compiled from consumers' electronic sales records as well as the dissemination of medical records. The agencies that constitute the federal statistical community should do all that is possible to convey to the respondent community that they have addressed both real and perceived concerns.

This chapter is organized into four core sections. The first describes and discusses the role of business confidentiality in statistical data collection. This is followed by a description of the questionnaire design and the sample frame. The next section provides the quantitative results. The conclusion offers some preliminary suggestions for extensions of this research that can ultimately be used to affect decisions on data reporting, collection, protection, and access.

2. Business Confidentiality Issues

In general we need less government intrusion into business as well as personal life. Still sensitive about releasing financial data. (Survey Respondent)

Because I don't understand the 'system' I am generally distrusting—especially of any government agency. (Survey Respondent)

The lack of quantitative research does not mean that no attention has been paid to the confidentiality[1] of business data provided for federal statistical purposes. In 1992, the U.S. Office of Management and Budget (OMB) established a working group that noted differences between household and business perceptions of confidentiality[2] and also identified several factors that were likely to affect businesses' trust in the protection afforded their data. These factors are more fully developed in a series of papers by Willimack and colleagues (1999). This chapter examines a subset of these factors—the sensitivity of individual items queried; the perceived benefits of the data collection (*e.g.*, survey objectives); the costs of data collection (*e.g.*, survey completion time); and the protection provided respondents.

Sensitivity

Statistical agencies might use knowledge about what types of data businesses consider to be most sensitive to accord different levels of protection and permit broader access to, and analysis of, subsets of data. It is well known that different types of household data have different levels of sensitivity—item response levels on income measures vary substantially from those on age and number of children. Although no hard evidence exists, the sensitivity of business data is likely to be different for a number of reasons: the existence of publicly available information (including commercial advertisements), the structure of business entities, and the existence of competitors.

Businesses, unlike households, are often routinely required for administrative or regulatory purposes to provide information, some of which quickly becomes available to the public (Willimack *et al.* 1999). For example, publicly traded companies are required annually to provide extensive financial information to the U.S. Securities and Exchange Commission, and much of that information also ends up on commercially available datasets such as Compustat. Further, all employers sponsoring an employee benefit plan (retirement, health, *etc.*) are required to file an annual Form 5500 series information return so that the Internal Revenue Service (IRS), the Department of Labor, the Pension Benefits Guaranty Corporation, and the Social Security Administration can administer their respective provisions of the 1974 Employee Retirement Income Security Act. Private corporations, such as ABI/INFORM, use information in the yellow pages to create business lists with name, address, employment, payroll, and industry information on the universe of

[1] The protection of confidentiality in this chapter is defined as the restriction of access to information about the individual party/entity once it has been provided—for statistical or administrative purposes—to a second party charged with the collection responsibility. The confidentiality protection responsibility is traditionally viewed as residing with the collecting party, even when the law permits third and fourth parties to access the data. Indeed, the consequences of any breach of confidentiality would almost always be borne by the collecting party in the form of reduced response rates and less precise responses.

[2] These are similar to the differences between collecting household and business data generally (see, *e.g.*, Cox and Chinnappa 1995).

businesses. In sum, virtually all the data provided for these various purposes are publicly available, raising the question of whether confidential datasets containing subsets of the provided information really need to—or even can—protect all their information equally. It may well be that the overriding issue for businesses in such cases is how to avoid further reporting burden, rather than how to obtain maximal 'confidentiality' protections for data already in the public domain.

Other items, not in the public domain, may not be sensitive simply because of the structure of business entities. For example, the taxpayer identification number (the SSN or Social Security number), which is quite sensitive for individuals, may not be sensitive for firms, because the Employer Identification Number or EIN (assigned by the IRS) often appears on publicly available datasets and also may be changed several times over the business's lifetime. This difference makes businesses more elusive to track and monitor over time than individuals and hence does not enable instant access to complete lifetime data.

The inherently competitive nature of business is also something to be considered in analyzing the sensitivity of data. Many businesses may consider some information, such as name and mailing address, less sensitive than individuals and households do—and, in fact, may promote its dissemination—but may consider other items that are necessary for profitable strategic planning, such as sales at the establishment level or trade secrets, very sensitive indeed.

Another dimension along which household respondents and business respondents may differ is the time sensitivity of records. While individuals may feel that their personal information, such as medical records or earnings histories, remains sensitive throughout their life, the nature of business respondents might mean that after some period the information is no longer sensitive. If particular types of data provide a competitive edge in a rapidly changing business world, then data that are more than five years old may need much less protection than current data. Other types of data could provide information in legal cases regardless of the time frame and thus may need more protection. The survey questionnaire examines business sensitivities to both the types of information collected and the period for which the data should be protected.

Perceived Benefits to Respondents

> *Have no clue what the government collects and what they do with it— this information should be for private industry to use, collect, buy, and sell— not a role for federal monies.* (Survey Respondent)

> *I would be willing to provide information to any agency if they could show how it would benefit the public or my company. I am a health care provider; it costs me money to provide extra reports, but if they can use what we already have, it works!* (Survey Respondent)

The work by Gerber (Chapter 15, this volume) supports the notion that respondents are more willing to provide data if they understand the benefits that are derived from the data collection. One concern raised by the OMB working group, and supported by the research of Willimack *et al.* (forthcoming), is that the direct benefits to businesses of the data that are collected from them by government statistical agencies may not always be readily apparent to them. In fact, the major producers of statistical data on businesses, such as the Census Bureau, the Bureau of Labor Statistics, and the Bureau of Economic Analysis, have as their primary mandate to produce data on the economy for government policy makers (*e.g.,* Congress and the administration), not for businesses. Although the Census Bureau notes that businesses can use economic census data to study their industry, gauge competition, calculate market share, and study business markets, among other things, these data are not intended to be a substitute for data used in market research and competitive analyses carried out by private sector firms. A brief perusal of private company websites such as Dun & Bradstreet, American Business Information, and the Donnelly Information Files provides convincing evidence that businesses can access and analyze detailed and quite current firm-level information on competitors[3] as an alternative to using aggregate, federal government data, albeit at considerable expense. The survey explores businesses' perceptions about the usefulness of government data products and the correlation between business perceptions about usefulness and confidentiality.

Respondent Burden and Data Sharing

The burden imposed on businesses by the federal government's requirements that they fill out surveys and censuses is a clear concern of both OMB and Congress (per the Paperwork Reduction Act of 1995), yet the full cost of such an imposition is neither easy to measure nor well known. Examples merely suggest the magnitude of the burden. To illustrate, the Census Bureau's Economics Directorate has estimated that the cost to business of filling out surveys in non-economic census years is about 2 million hours; economic census years add an extra 5 million hour burden.

The burden associated with filling out surveys is evident from business response rates. Mail response rates of 1 percent to 2 percent can mean a highly successful mailing for some credit card offers. The response rates to market research surveys are usually much higher, but still range from 10 percent to 15 percent[4]. Our own anecdotal experience suggests a general reluctance to provide any information. We had originally intended to explore the topic of confidentiality with the chief information officers of the 200 largest U.S. corporations, but the initial set of 15 phone calls revealed that not one company was willing to participate in a voluntary sur-

---

3 See, for example, http://www.mscnet.com/prodserv/nationaldatabases/index.htm.
4 See, for example, DSS Research, http://www.dssresearch.com/.

vey. In fact, almost none of these calls made it past the initial company screener. Federal statistical agencies, which typically are able to invoke mandatory compliance rules, enjoy much higher response rates, but there is evidence that U.S. businesses are particularly unhappy about providing the same or similar information to different agencies (Nichols *et al.* 1999).

One means of reducing respondent burden is to share more data among administrative and statistical agencies. While data sharing does, to some degree, compromise the confidentiality of respondent data, businesses might be willing to trade less protection for lower respondent burden. The survey explores business perceptions on this issue.

Knowledge of Protection Provided Respondents

Work by Gerber (Chapter 15, this volume) demonstrates that individual respondents are more likely to provide answers to 'bona fide' government agencies, and that they are less willing to provide data if they believe that data are 'floating around'. Although federal statistical agencies make much of the legal protections that are afforded respondents by statute and practice, it is not known whether businesses are aware of or value these protections. This lack of knowledge may be due to the decentralized nature of the U.S. statistical system, since data collection instruments from different agencies carry different privacy and confidentiality protection statements or pledges[5]. Alternatively, it may reflect trust in the federal statistical system, which has a long history of data protection and no history of breaches of confidentiality.

This situation raises several questions. Do business respondents understand what is meant by confidentiality protection? Do they realize and understand the variation that exists across different collection agencies? An additional implicit question is how and even whether the federal statistical system should address—or even acknowledge—the authorized secondary disclosures that occur among the collection agencies[6]. Should the agencies spell out what data are re-disclosed, to whom, and for what purpose? Should they reaffirm that these secondary disclosures are not to the public, and are only for statistical, not regulatory purposes? How much 'comprehensive' information should be provided to respondents to bolster their informed consent status, without adversely affecting response rates? Should the same data collected from different sources (administrative or survey) be treated differently from a confidentiality standpoint?

[5] See Appendix A to this chapter for the privacy and confidentiality pledges published by the federal statistical agencies, the Census Bureau, and the federal income tax agency, the Internal Revenue Service (IRS).

[6] For example, every year the IRS provides to Census the universe extracts of records with limited item content, primarily to reduce both respondent burden and collection costs. While such redisclosures are authorized by statute (U.S.Code, Title 26, Section 6103), such arrangements are often not mentioned at all or else only briefly in the actual collection instruments used by these agencies.

Suggesting answers to any of the foregoing questions is complicated by the absence of data about perceptions of the problem within the respondent community itself—a reasonable point of departure in any discussion of the subject. Confounding the issue are findings such as those by Singer *et al.* (1999) suggesting that the volume of information supplied in survey questions can itself adversely influence response rates. Obviously, this is not in the ultimate interest of either respondents or policy makers, so the questionnaire instructions tried to strike a balance between information and brevity.

3. Questionnaire Design and Data Collection

Although this brief discussion raises many questions, it is clear that there is no quantitative base on which to answer them. We developed a survey questionnaire to begin to answer a subset of these questions, while recognizing that this process represents only a first step toward the development of a much broader research agenda. We administered the survey through the mail using a commercially available business database (Dun & Bradstreet) to designate the respondent population.

Questionnaire Design

The questionnaire itself was designed to inform two discrete components of the confidentiality knowledge base[7]:

(1) What kinds of data/information do businesses consider sensitive—and for how long are they perceived to be sensitive?
(2) What are businesses' perceptions of collection agencies' ability to collect and protect data? In particular, do businesses believe there are differences in the quality of data collection and protection across both private and public agencies, and are they aware of (and do they have confidence in the efficacy of) penalties for disclosure violations?

These are complex concepts—particularly given that most businesses may not think much about confidentiality issues (Nichols *et al.* 1999; Willamack *et al.* 1999). Because information in itself is known to influence perceptions, we decided to define confidentiality only minimally in this first pass at data collection[8]. This strategy (minimally defining 'confidential') also parallels research conducted on

[7] A copy of the questionnaire is provided in Appendix B to this chapter.
[8] Nevertheless, the questionnaire does use the item responses themselves to begin to establish differences in perceived definitions. For example, item 7 seeks answers to core questions that frame business belief systems regarding confidentiality, including whether statistical collection agencies release identifiable data to anyone, keep collected business data confidential, release any collected data outside government, and share collected data with other agencies.

household perceptions of confidentiality to identify the dimensions of confidentiality as understood by respondents.

The first set of questions deals directly with whether businesses believe that different types of data have different levels of sensitivity—despite the fact that statistical agencies treat all these data with virtually the same level of protection. For example, we ask whether the business considers its primary identifiers—name, address, and phone number—to be sensitive. We then ask its views on the sensitivity of amount items such as employment, payroll, sales, profits, and tax liability. We expect, *a priori*, the former items (name, *etc.*) to be less sensitive, because they are typically available in the phone book and are even advertised to promote the firm or its activities, and the latter set of items to be more sensitive. We also ask whether similar data are more sensitive at the establishment or company level.

Building on this foundation, we then probe whether businesses feel that there is a time dimension to the sensitivity of their data (again, statistical agencies typically treat business data as sensitive regardless of the age of the data). That is, are some items considered sensitive longer than other items? We differentiate again between types of data and age of the data (1, 5, 10, and 30 years).

The next set of questions asks whether businesses are more or less concerned about the data collecting agent/recipient—whether they distinguish among federal regulatory or statistical agencies, not-for-profit researchers, for-profit researchers, other businesses, or the general public.

We then attempt to capture how businesses feel about the performance of federal statistical agencies, and later correlate this with their other responses. In particular, we ask whether the federal government is better than the private sector at collecting, providing, and protecting information. This is followed by a more detailed set of questions to find out how businesses perceive the protection provided by federal statistical agencies—whether their data are kept confidential, or whether their data are disclosed to or shared with other agencies. We also try to test our idea of respondent cynicism regarding government trust by asking whether the respondent believes that any federal agency, including the IRS, can access data provided by businesses any time it wants. This question is particularly relevant given the work by Gerber (Chapter 15), which finds that individual respondents fear the release of data to the IRS.

An important—and hitherto unexplored—issue is whether businesses are aware of the legal and financial penalties imposed by federal agencies on employees who divulge confidential information without authorization. We address not only this contingency but also whether these penalties should be increased or more stringently enforced, and, if either, whether the business would be more inclined to provide confidential data as a result.

The last set of questions generally investigates whether respondents might view responsible data sharing as a solution to some of their concerns about burden and invites respondents to identify any other concerns they may have about the collection or use of their data by federal statistical agencies.

Data Collection and Response Rates

The sample frame for the survey was obtained from Dun & Bradstreet's (D&B's) commercial database, which represents 11.3 million U.S. businesses[9]. This database has several advantages. It provided us with information on the name and title of up to four levels of management (*e.g.,* the owner, chief executive officer, chief information officer, and chief financial officer), information on the industry, number of employees, and whether the business is single- or multi-unit. The information also is quite current—updated daily. Dun & Bradstreet implements a number of checks to improve the quality of the data. In particular, it conducts either an on-site visit or a telephone investigation to each business in the D&B database at least once a year. The average age of a record in D&B's U.S. business database is 7.5 months. The most obvious disadvantages are that the sample frame is not necessarily representative of all businesses in the United States, and the quality of this sample frame relative to the major federal sampling frames in the country (the Business Register of the Bureau of Labor Statistics and the Business Register at the Census Bureau) is unknown.

From this frame, we selected a stratified random sample of 5,000 cases; 1,250 in each of four strata defined by the number of employees—0–49 employees 50–249, 250–499, and 500 plus. Of the 5,000 total businesses, 2,530 were multi-unit businesses with headquarters locations[10], and 2,470 were single-unit companies with only one business. While the geographic coverage was national, and the sample was representative of Dun & Bradstreet's database, we do not attempt to weight the sample to assign any kind of representativeness to it, and we view our results simply as preliminary evidence.

We conducted two types of pilots. The first was a mailout of 25 questionnaires to a random subset of the survey sample. The second was a set of cognitive interviews conducted by survey methodologists from the Census Bureau on a subset of 8 out of 25 respondents who were interviewed by telephone about their understanding of the questions. These results were used to clarify and reformat some of the survey questions.

We employed the Standard Total Design Method methodology (Dillman 1978). We sent out the first wave of questionnaires by first class mail on November 28, 2000. A stamped, first class envelope was provided to the respondents. A follow-up reminder postcard was sent to nonrespondents on December 8, and a second mailing of the questionnaire and cover letter was transmitted on December 28.

Of the 5,000 questionnaires, 213 were returned as undeliverable, so that the response total of 509 questionnaires resulted in a response rate of just over 10

[9] This total is derived from the Dun & Bradstreet report and differs from the size of the Census Bureau's Business Register, most likely because of differences in the definition of a 'business'. See Appendix C to this chapter for more summary information on Dun & Bradstreet's database.

[10] We did not ask Dun & Bradstreet to strip out subsidiaries from the file.

percent[11]. While this response rate is quite reasonable for a private sector survey of businesses, the response rates for mandatory, government surveys often exceed 80 percent. In hindsight, we erred in not excluding subsidiaries or the nonheadquarters units of multi-unit enterprises from our sample. The response rate for this group (929 survey units) was only 6 percent; that of other survey units was almost 12 percent.

We provide a more detailed analysis of the response rates by different types of business in Table 1. Response rates decline as business size increases—possibly because the survey did not reach the appropriate person, because large businesses are already burdened by numerous surveys, or because small businesses are more interested in voicing their confidentiality issues than are large ones. The effect of firm size is quite substantial: Even controlling for industry characteristics, the likelihood of response from firms with employment of 250-499 drops 4 percentage points; that of firms in the largest size class drops 5 percentage points.

There are some substantive differences in response rates across industries as well—businesses in the service sector were far more likely to respond than those in most other sectors, notably manufacturing. Again, even controlling for size of firm, the likelihood of firms in the manufacturing sector to respond to the survey was more than 4 percentage points lower than in the service sector, and the same held true when manufacturing firms were compared with firms in transportation, communications and public utilities, and retail trade.

Clearly, one of the early lessons learned from this relatively low response rate is that it is very difficult to conduct business surveys privately, even with government support, on such a sensitive topic as confidentiality issues. In addition, the low response rate means that the discussion of the survey results that follows should be taken as a preliminary to further research, rather than a conclusive analysis.

4. Survey Results

Sensitivity of Data Items

Many of our *a priori* notions were upheld with regard to types of data considered sensitive[12], as Table 2 shows. Entity information—such as name and address—was not considered very or extremely sensitive, except when the data pertained to the company's employees. This is not surprising, given that it is obvious that the information exists in the public domain (through Dun & Bradstreet). Interestingly, of all the data types queried in the survey, respondents considered the employee identity data to be the most sensitive, and the larger the company, the more sensitive it con-

[11] We did not separate out subsidiaries or non-headquarters responses.

[12] There is little variation across industry classifications, so we do not continue to report these results. They are available from the authors on request.

Table 1. Response Rates and Sample Counts by Industry and Employment Size Class

Major Industry	0-49	50-249	250-499	500+	Total
Agriculture, Mining and Contract Construction	10.10%	12.80%	7.60%	3.80%	9.60%
	199	109	66	52	426
Manufacturing	8.20%	9.90%	8.00%	7.00%	8.10%
	85	232	289	355	961
Transportation, Communication and Public Utilities	7.32%	16.95%	4.84%	13.40%	11.20%
	41	59	62	100	262
Wholesale Trade	13.50%	13.70%	3.60%	11.50%	10.50%
	74	95	84	52	305
Retail Trade	10.90%	6.50%	9.50%	5.80%	8.40%
	221	215	137	104	677
Finance, Insurance and Real Estate	19.20%	9.20%	8.70%	8.10%	10.40%
	73	109	126	123	431
Services	17.20%	14.70%	11.90%	8.90%	13.20%
	459	382	444	440	1,725
Total	13.63%	11.66%	9.19%	8.26%	10.64%
	1,152	1,201	1,208	1,226	4,787

Table 2. Percentage Indicating That These Data Items Are 'Very or Extremely Sensitive' (Question 1)

Employment Size Class	0-49	50-249	250-499	500+	Total
Name, address and phone number of your company	9.20	4.70	1.00	2.10	4.60
Type of industry/business operated	6.90	3.20	2.00	5.10	4.40
Name, home address and phone number of employees	78.50	86.40	89.00	87.80	85.00
Number of establishments	13.10	7.10	3.00	4.10	7.30
Number of managers and executives	15.60	14.30	13.00	19.40	15.50
Number of non-managerial employees	23.40	14.30	12.00	14.30	16.40
Your company's total payroll	69.00	66.70	55.00	53.60	61.90
Your company's total sales	59.20	52.40	40.00	35.10	47.90
Your company's total operating costs	66.20	67.50	57.00	52.00	61.50
Your company's total profits	73.80	77.00	67.00	58.80	70.00
Your company's total tax liability	70.00	77.80	77.00	64.30	72.50

sidered these data. One reason sensitivity increased with company size might be that such companies have legal departments that are more aware of the serious problems that can result from disclosing information on the identity of employees—such as workplace or domestic violence from estranged spouses.

Employment size data also were not considered sensitive, even when distinguished by categories of management and nonmanagement. This result is not surprising, given the public availability of this information. Indeed, Dun & Bradstreet—from which we obtained the sample frame data—provided employment data on the records we obtained for the survey sample.

Not too surprisingly, financial data, such as company payroll, operating costs, profits, and tax liability, were considered quite sensitive, with tax data lagging only employee entity information in perceived sensitivity. For multi-unit respondents, the types of data surveyed were generally considered slightly more sensitive at the company level than at the establishment level. Both payroll and profits data, which we already knew most businesses considered very sensitive, were considered much more sensitive at the company level—by a factor of almost two—which, at least for publicly traded companies, is surprising given that a lot of company-level information is reported to other, publicly available, sources.

The variation in response across firms' employment size classes is worth noting because of the quite distinct views exhibited. Broadly speaking, smaller firms seemed to believe that financial information is much more sensitive than did larger firms, perhaps because their data are less likely to be publicly available, or because they are less likely to have been sampled in a government survey.

Time Sensitivity of Data

Not all data need to be current in order to be useful. While some data users, such as policy makers, analysts, and researchers, need access to entity-level data to enhance their understanding of the working of the economy or society, these data can be historical. However, agencies impose lengthy time limitations on the confidentiality of some important data or do not release data at all. For example, the Census Bureau maintains the confidentiality protections on business data it collects for 30 years, and it never releases federal tax data, which constitute a significant portion of its business register.

We examined businesses' perceptions of the time sensitivity of their data in question 4, and report the results in Table 3. Two general impressions stand out. First, the sensitivity of data is clearly affected by their age. Our respondents had a wide variety of views spanning the spectrum of available time periods—and the views were clearly different depending on the data items concerned. Second, there is a great deal of either indifference to or misinformation about the issue. Unlike the earlier questions, a substantial proportion of responding businesses skipped this question. However, because this phenomenon applied particularly to the company-

Table 3. Time Period After Which Data Are No Longer Sensitive (Question 4)

	After 1 year	After 5 years	After 10 years	After 30 years	Don't know, not applicable, no response
Number of employees at each establishment	27.03%	28.76%	10.42%	4.05%	29.73%
Establishment-level payroll	13.83%	39.46%	19.50%	12.47%	14.74%
Establishment-level sales	16.36%	37.05%	19.09%	11.59%	15.91%
Establishment-level profits	8.30%	27.80%	20.27%	14.86%	28.76%
Number of employees at the company	36.10%	23.17%	7.53%	5.79%	27.41%
Company-level payroll	12.77%	30.06%	17.88%	12.57%	26.72%
Company-level sales	18.66%	26.52%	16.11%	10.61%	28.09%
Company-level profits	11.20%	24.17%	21.22%	15.72%	27.70%

level questions, it is possible that single-unit businesses felt they had already answered the question.

Turning to the details, the least sensitive data item was again the number of employees at both the establishment and the company levels. Well over half the respondents, and almost 80 percent of those with an opinion, believed that this item was no longer sensitive after five years. As we would expect, the most sensitive data element in terms of timing was profits—only 8 percent thought establishment-level profits were not sensitive after one year; 11 percent had the same view of company-level profits. Only a little over one-third of respondents, and more than half of those having an opinion, thought that company and establishment-level profits were no longer sensitive after five years.

This time qualification of data sensitivity could be a useful avenue for future research, particularly if certain companies would permit expanded researcher access to some of their financial data after a suitable period.

Trust and Respect for Federal Agencies

> *I have very little faith that the government can accurately gather such data and less that it can maintain it securely.* (Survey Respondent)

> *As a highly regulated industry, we provide REAMS of data regularly to a variety of federal agencies—both operational and statistical. We have never had a problem.* (Survey Respondent)

An important component of confidentiality is the trust that respondents have in the agency that collects the data. Because the U.S. statistical system is somewhat fragmented in nature, we asked respondents about their degree of trust in the statistical system in general, and specific types of agencies in particular. We also asked how respondents felt about providing data to other entities, to create a benchmark for comparison. Once again, the responses provided useful information about not only the level of trust but also the differences in trust across agencies and the differences by size of firm.

In general, respondents seemed least concerned about providing data to the core federal statistical agencies—such as the Census Bureau and the Bureau of Labor Statistics. Respondents seemed most concerned about providing data to other businesses (presumably, their competitors), followed closely by commercial or for-profit researchers, which they perhaps viewed as funded by their competitors or at least possibly supporting competitors' interests (especially if their competitors provided no data at all).

The level of trust is quite interesting. Even though the federal statistical agencies score the highest in relative trust, almost one-third of respondents were very or extremely concerned about providing data to core statistical agencies such as the Census Bureau. (At the same time, almost the same proportion—37 percent—

reported feeling not at all concerned about this issue.) The analysis by size class reveals that, while the concern is widespread, by and large small businesses seemed more concerned than large businesses.

While the responses to Question 5 indicated that it matters who collects the data, the responses to Question 6 were intended to help us understand why. Instead, we got apparently conflicting answers. After having just reported that the core statistical agencies are the most trusted as data collectors, respondents seem to change their minds and say the federal government is worse than the private sector at everything: collecting data, protecting the data's confidentiality, and converting the data into useful information. One interpretation of such apparently contradictory information is that the term 'federal government' has attained demon status with many businesses, which understand 'federal government' to be virtually synonymous with 'anti-business'. Future surveys might well substitute 'federal statistical agencies' (such as the Census Bureau and the Bureau of Labor Statistics) for the term 'federal government' to see if the same results hold.

We expected to see a link between businesses' trust in protection of data and their concern about providing data. To verify this notion, we correlated the responses to Question 5—particularly the response to federal statistical agencies—with the responses to Question 6[13]. Those businesses that seemed least concerned about providing data were in all cases significantly more likely to agree to the third statement in Question 6—that the federal government was better at protecting business data. This finding was independent of size of business and of industry classification.

Respondent Burden and Data Sharing Concerns

The fragmented nature of the U.S. statistical system has generated a fundamental respondent burden problem in that multiple statistical agencies (*e.g.,* the Bureau of Labor Statistics and the Bureau of the Census) request similar or identical data from businesses but often are not permitted to share these data. In Questions 7 and 10, we directly asked businesses whether they knew about the extent and legality of data sharing across agencies, and the responses provided two very interesting results. First, respondents seemed evenly split on whether they believe that federal statistical agencies keep the data confidential (the first and second rows of Table 6). Second, they believe that data sharing already occurs (the third row of Table 6), and in general, there is no problem with the core federal statistical agencies sharing data among themselves. The clear implication is not only that respondents did not think such data sharing is a problem, they think it is currently occurring and probably authorized! Obviously, this interpretation has to be treated with considerable caution because of the small response rate for this survey.

[13] These detailed correlations are not reported in the text, but are available from the authors on request.

Table 4. Percentage of Respondents Who Were Very or Extremely Concerned About Providing Business Data to the Following Entities (Question 5)

Type of recipient	0-49	50-249	250-499	500+	Total
Federal Regulatory Authority (EPA, SEC, FTC)	48.40	43.20	25.00	32.00	38.20
Federal Statistical Agency (BLS, Census, BEA)	34.10	33.10	20.00	27.80	29.30
Not-for-profit researchers (universities, think tanks, research organizations)	42.20	44.80	41.00	37.50	41.60
For-profit researchers	60.90	62.40	68.70	56.70	62.10
Other businesses	73.60	76.00	77.80	67.70	73.90
The general public	70.50	73.60	66.00	53.60	66.70

Table 5. Percentage of Respondents Who Somewhat or Strongly Agreed With the Following Statements (Question 6)

	0-49	50-249	250-499	500+	Total
The federal government is better than the private sector at collecting business data for statistical purposes.	32.60	23.80	38.00	33.30	31.50
The federal government is better than the private sector at providing useful information to government policy makers.	38.80	26.20	31.30	29.20	31.60
The federal government is better than the private sector at protecting business data from being released to those without authority to have it.	35.90	39.20	42.00	24.00	35.60

Table 6. Percentage of Businesses Somewhat or Strongly Agreeing With the Following Statements (Question 7)

	0-49	50-249	250-499	500+	Total
I believe that federal statistical agencies keep data provided by businesses confidential.	48.10	69.80	69.70	60.80	61.60
I believe that federal statistical agencies do not release information by which a company or its data can be identified.	41.10	66.40	65.70	57.70	57.10
I believe that federal statistical agencies do not share data provided by businesses with other government agencies.	21.10	21.60	27.60	21.60	22.80
I believe that federal statistical agencies do not release data provided by businesses to people outside the government.	28.30	41.70	43.30	38.50	37.60
I believe that any federal agency, such as the Internal Revenue Service, Small Business Administration and Federal Trade Commission, can access business data my company has provided to other federal agencies whenever it wants.	71.10	68.50	67.30	73.20	70.00

What is not in doubt is that most respondents were cynical in their views of the IRS—indeed any federal agency—when it comes to whether they can and do access any government data whenever they wish. This view of the world is particularly discouraging in light of the fact that federal statistical agencies are extremely careful to protect their data from even the perception that the IRS has access, and the fact that the IRS is circumspect not only about accessing any other agency's data (only through authorized channels, if they exist) but also about its own employees accessing tax data when they have no need to know[14]. It is safe to say that the education and public relations tasks before the IRS, and to a lesser extent before all federal statistical agencies, suggested by such survey responses are considerable to the point of being daunting.

Knowledge of Penalties

> *Training, discipline, and top-quality people management skills go farther than threats of fines and prison. Hiring and retaining personnel with high morals is better than jails. The government is generally weak in effective supervision and developing objectives and holding personnel accountable for goals and accomplishment. Many large bureaucracies are weak in this area.* (Survey Respondent)

> *I firmly believe all information should be highly confidential and sensitive at all times. Large penalties should continue to be enforced.* (Survey Respondent)

One way to address concerns about disclosure is to impose stiff fines and penalties for breaches of confidentiality. Indeed, Title 13 of the U.S. Code, which governs the Census Bureau, provides for fines of up to $250,000 and prison time of up to five years if such a violation occurs. While federal administrators believe these penalties are both an important deterrent and an important reassurance to the public, in Question 8 we directly ask businesses whether they know that such penalties exist. To our surprise, only 25 percent of respondents knew they existed; 55 percent did not know; the balance were unsure.

Given the existence of penalties, however, very few—fewer than one in four—respondents believed that the penalties were enforced (Question 9; see Table 7). It is of interest that small firms generally seemed more trusting than large firms in the efficacy of penalties. Although the link between enforcement and response rates has often been touted, our preliminary results suggest that this link is complex. While penalties and convictions would reassure only about half the respondents, we established that almost 80 percent of the respondents do not believe current penalties are enforced. Because the second question deals with higher penalties (not necessarily their enforcement) and the third question deals with conviction,

[14] Annual mandatory briefings at the IRS remind employees that unauthorized browsing or access of tax information is reason for dismissal and even criminal prosecution.

Table 7. Percentage of Respondents Somewhat or Strongly Agreeing With the Following Statements (Question 9)

	0-49	50-249	250-499	500+	Total
Federal penalties for the release of business data by federal statistical agencies or employees to those without authority to have it are adequately enforced.	28.10	23.80	20.20	16.50	22.70
Higher penalties would make my company more willing to provide business data to federal statistical agencies.	51.90	50.80	43.40	43.30	47.90
A higher likelihood of conviction would make my company more willing to provide business data to federal statistical agencies.	51.90	58.70	47.50	53.60	53.20

which is part of enforcement, it is possible that people who are skeptical about current enforcement would continue to be skeptical about it even with more stringent penalties.

5. Summary

I feel comfortable with how things currently work. (Survey Respondent)

We hope this chapter will serve as a starting point for the development of a continuing dialogue among business respondents, data collectors, researchers/analysts, and policy/decision makers—to inform respondents about data protection, to make decisions about data releases and data protection, and to monitor trends over time.

We began by noting that statistical agencies have an implicit pact with their respondents. Both the quantitative results and the written comments that we included in this analysis suggest that the current state of the pact is an uneasy one in many respects—but not all. While it would be ideal to have all respondents write in comments like the one that begins this section, unfortunately this comment was not representative of the general tenor of responses to the survey. Agencies have some work on their hands to convey the extent to which data are already protected. It is not clear what work needs to be done, however. The relatively low survey response rate, combined with some inconsistency in the responses, demonstrates that while we now know more, we still know very little about business perceptions of confidentiality. A full-scale survey mounted by a federal statistical agency, with both its imprimatur and the advantage of its extensive sampling frame, is probably necessary to fill the knowledge gap. Existing surveys and censuses would also make reasonable vehicles for these purposes, often to their own immediate benefit.

We also noted that agencies had a mission to disseminate data. This mission might meet with even more success than it does now, without compromising confidentiality, if businesses were asked whether their less sensitive data might be released to academics, policy makers, and other researchers. Some business data items, such as company name, address, and employment size, might be candidates for release, as might other data items old enough to be nonsensitive. If access were granted to more researchers and analysts, the entire statistical knowledge base could benefit from exposure to a richer institutional skill set. Additional benefits would include, of course, the reduction of both respondent burden and data collection costs.

Respondents also indicated that they distinguish among collection entities in the statistical data process—they do not regard the federal statistical community, commercial survey takers, and nonprofit researchers as homogeneous. Thus, statistical agencies might want to consider the possibility of different degrees of confidentiality protection for different types of analysts. Moreover, based on our survey responses, it seems that much work needs to occur in the federal statistical community to inform the respondent community not only of current practice, but of the relationship of current practice to legal authorization and the degree of enforcement.

Where might this kind of research lead? Perception studies could be used to monitor response climate, routinized so that surveys/censuses and other data gathering constructs (including administrative data uses and time series linkages) are not undertaken without an understanding of how these factors affect firms' perceptions. The results from such studies could be used to demonstrate good-faith efforts on the part of the statistical community. They could also facilitate new collection systems, particularly the collection and storage of, and access to, administrative data. Finally, such surveys could be used to evaluate the value and believability of accountability standards of statistical agencies.

References

Cox, Brenda G., and B. Nanjamma Chinnappa (1995) 'Unique Features of Business Surveys', Chapter 1 in B. Cox *et al.* (eds.) *Business Survey Methods*. New York: John Wiley & Sons.

Dillman, Donald A. (1978) *Mail and Telephone Surveys: The Total Design Method,* New York: John Wiley & Sons.

Gerber, Eleanor (2002) Chapter 15, this volume.

Nichols, E., Diane K. Willimack, and S. Sudman (1999) 'Balancing Confidentiality and Burden Concerns in Censuses and Surveys of Large Businesses', paper presented to the Washington Statistical Society, Washington, D.C.: U.S. Bureau of the Census, September.

Singer, Eleanor, Stanley Presser, and J. Vanhoewyk (1999) 'Public Attitudes Toward Data Sharing By Federal Agencies', in *Record Linkage Techniques—1997 Proceedings of an International Workshop and Exposition,* Washington, D.C.: National Academy Press.

Wallman, Katherine, and Jerry Coffey (1999) 'Sharing Statistical Information for Statistical Purposes', in *Record Linkage Techniques—1997 Proceedings of an International Workshop and Exposition,* Washington, D.C.: National Academy Press.

Willimack, Diane K., Elizabeth Nichols, and Seymour Sudman (forthcoming) 'Understanding Unit and Item Nonresponse In Business Surveys', in R.M. Groves *et al.* (eds) *Survey Nonresponse*, New York: John Wiley & Sons.

Willimack, Diane K., Seymour Sudman, Elizabeth Nichols, and Thomas Mesenbourg (1999) 'Cognitive Research on Large Company Reporting Practices: Preliminary Findings and Implication for Data Collectors and Users', mimeo, Washington, D.C.: U.S. Bureau of the Census.

Appendix A: Privacy and Confidentiality Pledges by Federal Agencies

1. Standard Confidentiality Pledges

Internal Revenue Service

The following statement provides the Service's policy on confidentiality for tax data provided on business income tax returns; *e.g.,* the corporate Form 1120 series.

Generally, tax returns and return information are confidential, as required by section 6103 [of Title 26].

Census Bureau

The following confidentiality pledges are representative of the pledges used in business/employer-based censuses and surveys.

Mandatory—Census: Current for 2002 Economic Census
YOUR RESPONSE IS REQUIRED BY LAW. Title 13, United States Code, requires businesses and other organizations that receive this questionnaire to answer the questions and return the report to the U.S. Census Bureau. By the same law, YOUR CENSUS REPORT IS CONFIDENTIAL. It may be seen only by U.S. Census Bureau employees and may be used only for statistical purposes. Further, copies retained in respondents' files are immune from legal process.

Mandatory—Survey: SA-42A Annual Trade Survey 1999
NOTICE—Response to this inquiry is required by law (Title 13, U.S. Code). By the same law, your report to the Census Bureau is confidential. It may be seen only by sworn Census employees and may be used only for statistical purposes. The law also provides that copies retained in your files are immune from legal process.

Voluntary—Survey: Advance Monthly Retail Trade Report
NOTICE—Your report to the Census Bureau is confidential by law (Title 13, U.S. Code). It may be seen only by sworn Census employees and may be used only for statistical purposes from which no firm or establishment may be identified. The law also provides that copies retained in your files are immune from legal process.

Voluntary—Survey: Business and Professional Classification Report
NOTICE—Your report to the Census Bureau is confidential by law (Title 13, U.S. Code). It may be seen only by sworn Census employees and may be used only for statistical purposes from which no firm or establishment may be identified. The law also provides that copies retained in your files are immune from legal process.

2. Expiration Period

Internal Revenue Service

There is no statute of limitations for federal return information; it is considered confidential in perpetuity.

Census Bureau

Title 44, United States Code (U.S.C.), which includes the National Archives Act (the 'Archives Act'), governs the ultimate disposition of personal and business information. Under the Archives Act, the National Archives and Records Administration (NARA) determines what government records have permanent value. The NARA is obliged to preserve records that have such value. It has the authority to take legal possession of such records after a specified period of time. A 1952 interagency agreement between the Census Bureau and the NARA established the 72-year rule, which covers population census and survey records. In 1978, Congress recognized this agreement through an amendment to Title 44, which incorporates the 1952 agreement by reference, thus providing for the 72-year protection unless the two agencies mutually agree on some other arrangement.

Specifically, the retention periods for Title 13 permanent records are reflected in Title 44 U.S.C., Section 2108(a) and (b).

Title 44, Section 2108(b) states

> With regard to the census and survey records of the Bureau of the Census containing data identifying individuals enumerated in population censuses, any release pursuant to this section of such identifying information contained in such records shall be made by the Archivist pursuant to the specifications and agreements set forth in the exchange of correspondence on or about the date of October 10, 1952, between the Director of the Bureau of the Census and the Archivist of the United States, together with all amendments thereto, now or hereafter entered into between the Director of the Bureau of the Census and the Archivist of the United States. Such amendments, if any, shall be published in the [Federal] Register.

Section 2108(a) of Title 44 permits the Archivist to withhold other government records from examination or use for 30 years if they are subject to statutory restrictions. Data collected by the Census Bureau in the various economic censuses and surveys are also classified as confidential by Title 13.

Appendix B: Text of Questionnaire on Business Perceptions of Confidentiality

Form CQM-46 OMB No. 0670-0760

Businesses' Perceptions of Confidentiality

1. Some types of business data may be considered more sensitive than other types of data. The sensitivity of data may be related to its strategic, legal or security importance to the business or to whether it is released to those without authority to have it.

For your company, please indicate the sensitivity for each of the following types of data. *Circle one response for each line. DK = Don't Know. NA = Not Applicable.*

	Not at all Sensitive	Somewhat Sensitive	Very Sensitive	Extremely Sensitive	
a. Name, address and phone number of your company	1	2	3	4	DK/NA
b. Type of industry/business operated	1	2	3	4	DK/NA
c. Name, home address and phone number of employees	1	2	3	4	DK/NA
d. Number of establishments (locations/ stores/plants) operated by your company	1	2	3	4	DK/NA
e. Number of managers and executives	1	2	3	4	DK/NA
f. Number of non-managerial employees	1	2	3	4	DK/NA
g. Your company's total payroll	1	2	3	4	DK/NA
h. Your company's total sales	1	2	3	4	DK/NA
i. Your company's total operating costs	1	2	3	4	DK/NA
j. Your company's total profits	1	2	3	4	DK/NA
k. Your company's total tax liability	1	2	3	4	DK/NA

2. Does your company have more than one location/store/plant? *Mark one.*

   ☐ Yes, multiple locations/stores/plants - continue with Item 3.

   ☐ No, only one location/store/plant - GO TO Item 4 on next page.

3. For each of the following types of data, please indicate which level of data is more sensitive: establishment/location-level data or company-level data. *Circle one response for each line. If your company has only 1 location/store/plant, please go to Item 4 on next page.*

	Establishment/Location-Level Data	Company-Level Data	
a. Number of employees	1	2	DK/NA
b. Payroll	1	2	DK/NA
c. Sales	1	2	DK/NA
d. Profits	1	2	DK/NA

4. For each of the following types of data, please indicate the time period after which they are no longer sensitive to the operation of your business. *Circle one response for each line.*

		After 1 year	After 5 years	After 10 years	After 30 years	
a.	Number of employees at each establishment	1	2	3	4	DK/NA
b.	Establishment-level payroll	1	2	3	4	DK/NA
c.	Establishment-level sales	1	2	3	4	DK/NA
d.	Establishment-level profits	1	2	3	4	DK/NA
e.	Number of employees at the company	1	2	3	4	DK/NA
f.	Company-level payroll	1	2	3	4	DK/NA
g.	Company-level sales	1	2	3	4	DK/NA
h.	Company-level profits	1	2	3	4	DK/NA

5. In your opinion, how concerned is your company about providing business data (e.g., number of employees, payroll, sales, profits, etc.) to each of the following? *Circle one response for each line.*

	Not at all Concerned	Somewhat Concerned	Very Concerned	Extremely Concerned	
a. Federal regulatory authorities (e.g., Environmental Protection Agency, Securities and Exchange Commission and Federal Trade Commission)	1	2	3	4	DK/NA
b. Federal statistical agencies (e.g., Census Bureau, Bureau of Labor Statistics and Bureau of Economic Analysis)	1	2	3	4	DK/NA
c. Not-for-profit researchers (e.g., universities, think tanks and research organizations)	1	2	3	4	DK/NA
d. For-profit researchers (e.g., market researchers and consulting firms)	1	2	3	4	DK/NA
e. Other businesses	1	2	3	4	DK/NA
f. The general public	1	2	3	4	DK/NA

6. Business data are collected for statistical purposes by both the federal government and the private sector (e.g., universities, think tanks, research organizations, market researchers, consulting firms, etc.). For each statement below, please indicate if you strongly agree, somewhat agree, somewhat disagree or strongly disagree. *Circle one response for each line.*

	Strongly Agree	Somewhat Agree	Somewhat Disagree	Strongly Disagree	
a. The federal government is better than the private sector at collecting business data for statistical purposes.	1	2	3	4	DK/NA
b. The federal government is better than the private sector at providing useful information to government policy makers.	1	2	3	4	DK/NA
c. The federal government is better than the private sector at protecting business data from being released to those without authority to have it.	1	2	3	4	DK/NA

7. For each statement below, please indicate if you strongly agree, somewhat agree, somewhat disagree or strongly disagree. *Circle one for each line.*

	Strongly Agree	Somewhat Agree	Somewhat Disagree	Strongly Disagree	
a. I believe that federal statistical agencies (e.g., Census Bureau, Bureau of Labor Statistics and Bureau of Economic Analysis) keep data provided by businesses confidential.	1	2	3	4	DK/NA
b. I believe that federal statistical agencies do not release information by which a company or its data can be identified.	1	2	3	4	DK/NA
c. I believe that federal statistical agencies do not share data provided by businesses with other government agencies.	1	2	3	4	DK/NA
d. I believe that federal statistical agencies do not release data provided by businesses to people outside the government.	1	2	3	4	DK/NA
e. I believe that any federal agency, such as the Internal Revenue Service, Small Business Administration and Federal Trade Commission, can access business data my company has provided to other federal agencies whenever it wants.	1	2	3	4	DK/NA

8. Currently, federal regulations do provide penalties, such as fines and prison time, for the release of business data by federal statistical agencies or employees to those without authority to have it.

   Were you aware that these penalties existed? *Mark one.*

   ☐ Yes

   ☐ No

   ☐ DK

9. For each statement below, please indicate if you strongly agree, somewhat agree, somewhat disagree or strongly disagree. *Circle one for each line.*

	Strongly Agree	Somewhat Agree	Somewhat Disagree	Strongly Disagree	
a. Federal penalties for the release of business data by federal statistical agencies or employees to those without authority to have it are adequately enforced.	1	2	3	4	DK/NA
b. Higher penalties would make my company more willing to provide business data to federal statistical agencies.	1	2	3	4	DK/NA
c. A higher likelihood of conviction would make my company more willing to provide business data to federal statistical agencies.	1	2	3	4	DK/NA

10. Currently, except as authorized by law, federal statistical agencies, such as the Census Bureau, Bureau of Labor Statistics and Bureau of Economic Analysis, cannot share the business data they collect with any other government agencies. If data sharing were legal, among which federal statistical agencies would your company be willing to allow its data to be shared, for statistical purposes only? *Please circle one response option for each line.*

a.	Census Bureau	Yes	No	DK/NA
b.	Bureau of Labor Statistics	Yes	No	DK/NA
c.	Bureau of Economic Analysis	Yes	No	DK/NA
d.	Statistics of Income Division of Internal Revenue Service	Yes	No	DK/NA
e.	Other federal agencies			
	Please specify: ____________________	Yes	No	DK/NA
	____________________	Yes	No	DK/NA
	____________________	Yes	No	DK/NA

11. Please use this space to provide any additional comments you may have about the sensitivity of company data, confidentiality pledges, federal statistical agency use and protection of business data, or related topics.

12. Please identify the job position/title of the person completing this questionnaire. *Mark one.*

    ☐ Chief Financial Officer

    ☐ Controller/Accounting Manager

    ☐ Chief Information Officer

    ☐ Manager of Financial/External Reporting

    ☐ Chief Executive Officer

    ☐ Owner

    ☐ Other, Please specify:

    ____

13. Optional

    Name: ____

    Phone Number: ____

    Would you like a copy of the results of this study? *Mark one.*

    ☐ Yes

    ☐ No

    Would you be willing to discuss your answers? *Mark one.*

    ☐ Yes

    ☐ No

Thank you for your time.

7184930

Appendix C: Information on the Dun & Bradstreet Database

Geographic Coverage

Dun & Bradstreet's Worldbase database is a global marketing database that contains more than 53 million business records, including linkage, in 200 countries. The U.S. marketing database includes 11.3 million business establishments in the 50 states and the territories of Puerto Rico and the Virgin Islands.

Frequency of Master Database Update

Under D&B's Full File Maintenance Strategy, all businesses in the U.S. marketing database are investigated via either a site visit or a telephone investigation at least once a year. Many records are touched much more frequently—as often as several times a month—through various triggered maintenance programs. The average age of a record in D&B's U.S. business database is 7.5 months.

Source of Data Elements

Dun & Bradstreet leverages a variety of sources for new record identification, including the following:

- Directly from the business principal during investigation.
- Response to inquiries from D&B credit customers.
- Customer files—D&B partners with many of its customers to identify new businesses and add them to the marketing file after investigating and enhancing the information.
- Personalized investigations—These new businesses enter our database as a result of customer requests for specific investigations.
- Public record sources (local, state, and federal), including new business registrations, corporate charter details, and public bulk source files.
- D&B bounty program that uses D&B associates in more than 50 field offices throughout the country by motivating them to proactively identify new business start-ups in their area.
- Private third party sources; many of these files are acquired via niche-specific compilers of information.
- Business directories.

Source: Dun & Bradstreet marketing information.

Name Index

Subject Index